Entomology and Nematology

Entomology and Nematology

Edited by **Carlos Wyatt**

SYRAWOOD
PUBLISHING HOUSE

New York

Published by Syrawood Publishing House,
750 Third Avenue, 9th Floor,
New York, NY 10017, USA
www.syrawoodpublishinghouse.com

Entomology and Nematology
Edited by Carlos Wyatt

International Standard Book Number: 978-1-68286-131-8 (Hardback)

The publisher's policy is to use permanent paper from mills that operate a sustainable forestry policy. Furthermore, the publisher ensures that the text paper and cover boards used have met acceptable environmental accreditation standards.

Trademark Notice: Registered trademark of products or corporate names are used only for explanation and identification without intent to infringe.

Printed in the United States of America.

Contents

Preface

Over the recent decade, advancements and applications have progressed exponentially. This has led to the increased interest in this field and projects are being conducted to enhance knowledge. The main objective of this book is to present some of the critical challenges and provide insights into possible solutions. This book will answer the varied questions that arise in the field and also provide an increased scope for furthering studies.

This book aims to provide a cohesive knowledge on the interdisciplinary fields of entomology and nematology. It discerns the current progress of these fields and highlights some of their key concepts and applications for further research and observations. It includes some of the vital pieces of work being conducted across the world, on various topics related to morphology and physiology of insects and nematodes, taxonomy and forensic entomology, etc. The extensive content of this book provides the readers with a thorough understanding of the subject for all the graduate and post graduate students, researchers, etc.

I hope that this book, with its visionary approach, will be a valuable addition and will promote interest among readers. Each of the authors has provided their extraordinary competence in their specific fields by providing different perspectives as they come from diverse nations and regions. I thank them for their contributions.

Editor

Annotated catalogue of whiteflies (Hemiptera: Sternorrhyncha: Aleyrodidae) from Arasbaran, Northwestern Iran

Hassan Ghahari[1]*, Shaaban Abd-Rabou[2], Jiri Zahradnik[3] and Hadi Ostovan[4]

[1]Department of Agriculture, Islamic Azad University, Shahre Rey Branch, Tehran, Iran.
[2]Plant Protection Research Institute, Ministry of Agriculture, Dokki-Giza, Egypt.
[3]Podebradova 498, 512 51, Lomnice nad Popelkou, Czech Republic.
[4]Department of Entomology, Islamic Azad University, Fars Science and Research Branch, Iran.

The fauna of whiteflies (Hemiptera: Sternorrhyncha: Aleyrodidae) was studied in Arasbaran region, Northwestern Iran. A total of 25 species of 15 genera were collected which of these, two species including, *Tetralicia ericae* Harrison and *Trialeurodes ericae* Bink-Moenen are new records for Iran.

Key words: Aleyrodidae, fauna, new record, arasbaran, Iran.

INTRODUCTION

Whiteflies belonging to the order Hemiptera and comprise a single super family, Aleyrodoidea, within the suborder Sternorrhyncha. They are all placed in a single family, Aleyrodidae, and are small sap-sucking insects whose adults bear a remarkable superficial resemblence to tiny moths. Adult whiteflies are very small insects, most measuring 1 - 3 mm in body length. A structure known as a 'vasiform orifice' is unique to aleyrodids, and comprises the anus, a 'lingula' which ejects excreta, and an 'operculum' which partially or wholly covers the orifice itself. The vasiform orifice is present in all larval stages, as well in the adults (Mound and Halsey, 1978; Gerling, 1990).

Amongst the Sternorrhyncha, whiteflies appear to be a recently evolved group, with the oldest known fossil remains (not recognizably belonging to one of the two modern subfamilies) being from Lebanese amber from the lower Cretaceous, 135 million years ago (Schlee, 1970). Whiteflies with modern affinities are thus known from a period during which angiosperm plants underwent great diversification (Campbell et al., 1994). Many the rare species such as *Selaginella* (Mound et al., 1994) that habitually feed on ferns, and on 'fern allies' are exceptions to the rule. The great majority of whiteflies in existence today colonize only dicotyledonous angiosperms and a smaller, but significant, feed on monocots,

particularly grasses and palms. The list of cultivated plants colonized by whiteflies is extensive, but a great many records concern the relatively few highly polyphagous whitefly species (Mound and Halsey, 1978; Carver and Reid, 1996). In the geographical area covered by this study, whiteflies are primarily pests of vegetable crops (especially in greenhouses), citrus and ornamental plants (Martin, 1999).

The systematic of both subfamilies is currently based almost entirely on the puparial stage, and adults in isolation can be identified only rarely. This situation has arisen, in part, because puparia are often discovered in the absence of adult insects. However, adult characters have been used with most success in the least speciose subfamily, Aleurodicinae, but a fundamental appraisal is much needed before adults are likely to be used more widely in whitefly systematics. The use of modern molecular techniques also promises to assist our understanding of the systematics of this insect group (Martin et al., 2000).

Arasbaran is an important region in East Azarbayjan province. This biosphere reserve situated in the north of Iran at the border to Armenia and Azerbaijan belongs to the Caucasus Iranian Highlands. In-between the Caspian, Caucasus and Mediterranean region, the area covers mountains up to 2,200 meters, high alpine meadows, semi-arid steppes, rangelands and forests, rivers and springs.

*Corresponding author. E-mail: h_ghahhari@yahoo.com.

Arasbaran is the territory of about 23,500 nomads who are mainly living in the buffer and transition zones (2000). Economic activities in the biosphere reserve are mainly agriculture, animal husbandry, horticulture, apiculture, handicrafts and tourism, but business activities can also be found in urbanized areas. The location of Arasbaran is 38°40' to 39°08'N; 46°39' to 47°02'E and its Altitude (metres above sea level) is +250 to +2,887.

The fauna of Iranian whiteflies is very diverse but rather unknown. Although, there have been a number of publiccations of whiteflies in different regions, but there has been no account of the group across the whole region. In the present research, the fauna of these important pests I studied in Arasbaran region.

MATERIALS AND METHODS

In connection with this study, the puparia of whiteflies were collected on host plants' leaves from different regions of Arasbaran region (East Azarbayjan province, Northwestern Iran). Slide-mounted material was prepared as the method of Martin (1985). Detailed information about synonymies and distribution of quoted species are available in this work (Evans, 2005; Martin and Mound, 2007).

RESULTS AND DISCUSSIONS

In a total of 25 species of 15 genera including, *Paraleyrodes, Acaudaleyrodes, Aleuroclava, Aleurolobus, Aleurothrixus, Aleuroviggianus, Aleyrodes, Asterobemisia, Bemisia, Bulgarialeurodes, Dialeurodes, Parabemisia, Siphoninus, Tetralicia* and *Trialeurodes* were collected from Arasbaran region. Of these, two species including, *Tetralicia ericae* Harrison and *Trialeurodes ericae* Bink-Moenen are new records for Iran. The list of species is presented as follow with the synonymies and distributional data.

Family aleyrodidae Westwood 1840

The family Aleyrodidae (Hemiptera) includes 161 genera and 1556 species in 3 extant (living) subfamilies (Aleurodicinae, Aleyrodinae and Udamosellinae), and one fossil (non-living) subfamily (Bernaeinae). The identification of genera and species is largely based upon characteristics present in the fourth stage nymph, known as the puparium. The family Udamosellinae, also known as "giant whiteflies" includes only 2 Neotropical species. Most whitefly species can be classified into their respective subfamilies (Aleyrodinae or Aleurodicinae) with the following key:

Key to subfamilies of aleyrodidae (Puparium) [Adapted from Evans, 2007]

1- Puparium usually with compound or agglomerate pores present, a claw present at the apex of each thoracic leg; lingula usually very long, extending past the vasiform orifice with two or more pairs of setae at its apex...Aleurodicinae
1b- Pupa without compound pores present (simple pores rarely present), thoracic legs with adhesive or circular disc at the apices of the legs; lingula usually not long and extending past the vasiform orifice and with 1 pair of setae ....Aleyrodinae

Key to the subfamilies of aleyrodidae (Adults) [Adapted from Evans, 2007]

1- Forewing usually with a forked, central vein (Rs present, R1 and media veins strongly developed), forewing of *Paraleyrodes* with a single vein, males with 3 and females with four antennal segments; tarsal paronychium thin and spine-like; females with 4 and males with 3 ventral abdominal plates, respectively...Aleurodicinae.
1b- Forewing with a single non-forked central vein (Rs present, R1 usually short or absent), tarsal paronychium thick and blade-like; females with 2 and males with 4 ventral abdominal plates, respectively ....Aleyrodinae.

Species list of aleyrodidae from arasbaran

In a total of 25 whitefly species from 15 genera and two Subfamilies Aleurodicinae and Aleyrodinae were collected and identified from different regions of Arasbaran. The list of species is below:

I. Subfamily Aleurodicinae Quaintance and baker 1913

Aleurodicinae Quaintance and Baker 1913: 25.

Genus *Paraleyrodes* Quaintance 1909

Paraleyrodes Quaintance 1909: 169 - 170. Type species: *Aleurodes perseae* Quaintance 1900, by monotypy.

Paraleyrodes minei Iaccarino 1990

Paraleyrodes minei Iaccarino 1990: 132. Holotype male. Syria: Tartous, 17.viii.1988, on *Citrus aurantium* (Rutaceae), Martin 1996: 1856.

Material: Khodafarin, July 2007 on *Piper* sp. (Piperaceae).

Distribution: Belize, Benin, Bermuda, Guatemala, Hawaii, Honduras, Hong Kong, Iran, Israel, Lebanon, Mexico, Morocco, Puerto Rico, Spain, Syria, Turkey, USA.

II. Subfamily Aleyrodinae Westwood 1840

Aleyrodinae Westwood 1840: 442.

Genus *Acaudaleyrodes* Takahashi 1951

Acaudaleyrodes Takahashi 1951a: 382. Type-species: *Acaudaleyrodes pauliani* Takahashi 1951, by monotypye.

Acaudaleyrodes rachipora (Singh) 1931

Aleurotrachelus rachipora Singh 1931: Syntypes. India: Pusa and Dholi (Bihar), Navsari (Baroda), Miani (Punjab, on *Cassia fistula* (Fabaceae), *Euphorbia pilulifera* (Euphorbiaceae), *Bauhinia* sp. (Fabaceae) and *Dalbergia sissoo* (Fabaceae).

Acaudaleyrodes rachipora (Singh); Russell 1962: 64

Aleurotrachelus citri Priesner and Hosny 1934. Syntypes. Egypt: Behera, on *Citrus* spp. (Rutaceae), *Punica granatum* (Punicaceae) and other plants; synonymy according to Jesudasan and David 1991: 242.
Acaudaleyrodes citri (Priesner and Hosny); Russell 1962: 64.
Aleurotrachelus alhagi Priesner and Hosny 1934. Syntype. Egypt, Minya, Luxor-Karnak, on *Alhagi* sp. (Fabaceae); Mound 1965: 119.

Material: Kalibar, August 2006 on *Punica granatum* (Punicaceae).

Distribution: Cameroon, Canary Islands, Chad, Cyperus, Egypt, India, Iran, Iraq, Israel, Jordan, Kenya, Liberia, Madagascar, Niger, Nigeria, Saudi Arabia, Sierra Leon, South Africa, Sudan.

Genus *Aleuroclava* Singh 1931

Aleuroclava Singh 1931: 90. Type species: *Aleuroclava complex* Singh 1931, by monotypy.
Aleuromigda Singh 1931. Nomen nudum, no type species designated.
Aleurotuberculatus Takahashi 1932: 20. Type species. *Aleurotuberculatus gordoniae* Takahashi 1932, by original designation; synonymy according to Martin 1999: 31.
Japaneyrodes Zahradnik 1962: 13. Type species. *Aleurotuberculatus trachelospermi* Takahashi 1938, by original designation; synonymy according to Mound and Halsey 1978: 78.
Hindaleyrodes Meganathan and David 1994: 37. Type species. *Hinaleyrodes hindustanicus*, by monotypy; synonymy according to Martin and Mound 2007: 9.
Martiniella Jesudasan and David 1990: 7. Type species. *Aleurotuberculatus canagae* Corbett 1935, by original

designation; synonymy according to Martin 1999: 31; Manzari and Quicke 2006: 2470.
Taiwanaleyrodes Takahashi 1932: 28. Type species. *Taiwanaleyrodes meliosmae* Takahashi 1932, by monotypy; synonymy according to Manzari and Quicke 2006: 2470.
Note: Martin 1999: 31 synonymized *Martiniella* Jesudasan and David 1990 with *Aleuroclava*, stating that the characters used to separate it from *Aleuroclava* - the very much enlarged, jointed cephalic and first abdominal setae were also present in species of *Taiwanaleyrodes* and *Dialeurodes*, and that this character has been seen to vary among samples. Sundararaj and Dubey 2004: 358 considered *Martiniella* to be a valid genus based upon its differentiated type of setae.

Aleuroclava neolitseae (Takahashi) 1934

Aleurotuberculatus neolitseae Takahashi 1934: 55. Syntypes. Taiwan: on *Neolitsea acuminatissima* (Lauraceae), TARI.
Aleuroclava neolitseae (Takahashi); Martin 1999: 31.

Material: Ahar, July 2007 on *Szygium* sp. (Myrtaceae).

Distribution: Iran, Malaysia, New Guinea, Sarawak, Sulawesi, Taiwan.

Genus *Aleurolobus* Quaintance and Baker 1914

Aleurolobus Quaintance and Baker 1914: 108. Type species: *Aleurodes marlatti* Quaintance 1903, by original designation.
Neoaleurolobus Takahashi 1951b: 5. Type species. *Aleurolobus musae* Corbett 1935, by monotypy; synonymy according to Regu and David 1993: 32; Martin and Mound 2007: 13.
Rositaleyrodes Meganathan and David 1994: 48. Type species. *Aleurolobus opilismeni* Takahashi 1931, by monotypy; synonymy according to Manzari and Quicke 2006: 2471; Martin and Mound 2007: 13.

Aleurolobus marlatti (Quaintance) 1903

Aleurodes marlatti Quaintance 1903: 61. Lectotype (designated Martin 1999:43). Japan: on orange [*Citrus* sp. (Rutaceae)], USNM.
Aleurolobus marlatti (Quaintance); Quaintance & Baker 1914: 109.
Aleurolobus niloticus Priesner and Hosny 1934: 1. Syntypes. Egypt: on *Zizyphus spina-christi* (Rhamnaceae), USNM; Bink-Moenen 1983: 50; synonymy according to Martin 1999: 43.
Aleurolobus ravisei Cohic 1968: 95. Syntypes. Congo: on *Hymenocardia acida* (Euphorbiaceae), CORSTOM; synonymized with *A. niloticus* by Bink-Moenen 1983: 50.

Material: Khomarloo, September 2007 on *Hedera* p. (Araliaceae).
Distribution: Chad, China, Egypt, India, Iran, Israel, Japan, Java, Jordan, Malaysia, Philippines, Saudi Arabia, Taiwan.

Aleurolobus moundi **David and Subramaniam 1976**

Aleurolobus moundi David and Subramaniam 1976: 161. Holotype. India: on *Bassia* sp. (Chenopodiaceae), ZSI.

Material: Aras boundary, September 2006 on *Euphorbia* sp. (Euphorbiaceae).

Distribution: Iran, India.

Aleurolobus olivinus **(Silvestri) 1911**

Aleurodes olivinus Silvestri 1911: 214. Syntypes. Italy: on *Olea* sp. (Oleaceae), IESP.
Aleurolobus olivinus (Silvestri); Quaintance and Baker 1915.

Material: Kalibar, August 2006 on *Olea ferruginea* (Oleaceae).

Distribution: China, Cyperus, Egypt, France, Israel, Italy, Morocco, Spain.

Aleurolobus selangorensis **Corbett 1935**

Aleurolobus selangorensis Corbett 1935b: 819. Syntypes. Malaya: on undetermined plant; Martin 1985: 317.

Material: Ahar, September 2007 on *Vitex pseudo-negundo* (Verbenaceae).

Distribution: Iran, Malaya, Papau New Guinea.

Genus *Aleurothrixus* **Quaintance and Baker 1914**

Aleurothrixus Quaintance and Baker 1914: 103. Type species. *Aleyrodes howardi* Quaintance 1907: 91, junior synonym of *Aleurodes floccosa* Maskell 1896: 432.
Aleurothrixus (*Philodamus*) Quaintance and Baker 1917: 404. Type species. *Aleyrodes interrogationis* Bemis 1904, by monotypy.
Hempelia Sampson and Drews 1941: 166. Type species. *Hempelia chivelensis* Sampson and Drews 1941, by monotypy; synonymy according to Martin 2005: 20.

Aleurothrixus floccosus **(Maskell) 1895**

Aleurodes floccosa Maskell 1895: 432. Syntypes. Cuba: on *Citrus* sp. (Rutaceae), NZAC.
Aleurothrixus floccosus (Maskell); Quaintance and Baker

1914: 91.
Aleyrodes horridus Hempel 1899: 394. Syntypes. Brazil: on *Psidium guajava* (Myrtaceae).
Aleurothrixus horridus (Hempel); Quaintance and Baker 1914: 103.
Aleyrodes howardi Quaintance 1907: 91. Syntypes. Cuba: on *Citrus* sp., USNM; synonymized by Costa Lima 1942: 425.

Material: Khodaafrin, July 2007 on: *Cordia* sp. (Boraginaceae).

Distribution: Angola, Antigua, Argentina, Bahamas, Barbados, Bermuda, Bolivia, Brazil, Canary Islands, Chile, Colombia, Congo, Costa Rica, Colombia, Cuba, Dominica, Dominican Republic, Ecuador, El Salvador, France, Gabon, Guadeloupe, Guam, Guatemala, Guinea, Guyana, Haiti, Honduras, India, Iran, Italy, Jamaica, Japan, Liberia, Madeira, Martinique, Mexico, Montserrat, Morocco, Nicaragua, Nigeria, Panama, Paraguay, Portugal, Puerto Rico, Reunion, Sicily, Spain, Suriname, Tahiti, Taiwan, Thailand, Trinidad & Tobago, Uruguay, USA, Venezuela, Virgin Islands.

Genus *Aleuroviggianus* **Iaccarino 1982**

Aleuroviggianus Iaccarino 1982: 37. Type species. *Aleuroviggianus adrianae* Iaccarino 1982, by original designation.

Aleuroviggianus adrianae **Iaccarino 1982**

Aleuroviggianus adrianae Iaccarino 1982: 38. Holotype. Italy: Portici, on *Quercus ilex* (Fagaceae), UNP.

Material: Ahar, September 2007 on *Quercus macranthera* (Fagaceae).

Distribution: France, Greece, Iran, Italy, Spain.

Aleuroviggianus halperini **Bink-Moenen 1992**

Aleuroviggianus halperini Bink-Moenen 1992, in Bink-Moenen and Gerling 1991: 14. Holotype. Israel: Mt. Meron, ix.1976, R. Neeman, on *Quercus calliprinos* (Fagaceae), BMCol.

Material: Kalibar, August 2006 on *Quercus petraea* (Fagaceae).

Distribution: Crete, Greece, Iran, Rhodes, Turkey.

Genus *Aleyrodes* **Latreille 1796**

Aleyrodes Latreielle 1796: 93. Type species. *Phalaena* (*Tinea*) *proletella* Linnaeus 1758: 537, by monotypy.
Conantulus Goux 1987: 65. Type species. *Conantulus*

lacombiensis Goux 1988, by monotypy; synonymy according to Martin 1999: 53.

Aleyrodes lonicerae Walker 1852

Aleyrodes lonicerae Walker 1852: 1092. Syntypes. England: on *Lonicera periclymenum* (Caprifoliaceae).
Aleyrodes borchsenii Danzig 1966: 371. Holotype. USSR: southern Maritime Territory, on *Urtica* sp. (Urticaceae), ZIN; Danzig 2004.
Aleyrodes fragariae Walker 1852: 1092. Syntypes. England: on *Fragaria* sp. (Rosaceae); Ossiannilsson 1955: 193.
Conantulus lacombiensis Goux 1987: 65; Martin 1999: 53.
Aleyrodes menthae Haupt 1934: 139. Syntypes. Germany: on *Mentha piperita* (Labiatae); Ossiannilsson 1955: 193.
Aleyrodes spiraeae Douglas 1894b: 73. Syntypes. Germany: on *Lonicera xylosteum* (Caprifoliaceae); Mound 1966: 406.
Aleyrodes rubi Signoret 1868: 382. Syntypes. France: on *Rubus fruticosus* (Rosaceae); Trehan 1940: 608.

Material: Aynalo, June 2007 on *Solanum melongena* (Solanaceae).

Distribution: Austria, England, Finland, France, Germany, Hungary, Iran, Israel, Italy, Korea, Poland, Portugal, Russia, Sweden, Switzerland, Turkey, USSR, Yugoslavia.

Aleyrodes proletella (Linnaeus) 1758

Phalaena (Tinea) proletella Linnaeus 1758: 537. Syntypes. Europe: on *Brassica* sp. (Brassicaceae).
Aleyrodes brassicae Walker 1852: 1092. Syntypes. England: on cabbage (*Brassica* sp.); Haupt 1935: 256.
Aleyrodes chelidonii Latreille 1807; Walker 1852: 1092.
Aleyrodes euphorbiae Low 1867: 746. Syntypes. Austria: on *Euphorbia peplus* (Euphorbiaceae); Zahradnik 1991: 113.
Coccus preanthis Schrank 1801: 147.
Aleyrodes prenanthis (Schrank) 1801; Cockerell 1902: 281.
Aleyrodes youngi Hempel 1901: 385. Syntypes. Brazil: on cabbage [*Brassica* sp. (Brassicaceae)].

Material: Khodafarin, July 2006 on *Brassica oleracea* (Brassicaceae).

Distribution: Angola, Austria, Azores, Belgium, Bermuda, Brazil, Canary Islands, Czech Republic, Egypt, England, Finland, France, Germany, Hong Kong, Hungary, Iran, Italy, Kenya, Mexico, Mozambique, New Zealand, Puerto Rico, Poland, Portugal, Russia, Sierra Leon, Spain, Sweden, Switzerland, USA (intercepted in USA but not known to be established), USSR, Virgin Islands, Yugoslavia, Zimbabwe.

Genus Asterobemisia Trehan 1940

Asterobemisia Trehan 1940: 591. Type species. *Aleurodes carpini* Koch 1857, by monotypy. [*Aleyrodes* [*sic*] *carpini* Koch].
Bemisia (Neobemisia) Visnya 1941: 8. Type species: *Bemisia yanagicola* Takahashi 1934, by original designation; Mound and Halsey 1978: 104.
Neobemisia Visnya; synonymy according to Zahradnik 1961: 61.

Asterobemisia carpini (Koch) 1857

Aleurodes carpini Koch 1857: 327. Syntypes. Germany (west): on *Carpinus betulus* (Betulaceae), BMNH.
Asterobemisia carpini (Koch) 1857; Trehan 1940: 593.
Aleurodes avellanae Signoret 1868: 385. Lectotype (designated by Zahradnik 1961: 437). France: on *Corylus avellana* (Betulaceae), IESP; Mound and Halsey 1978: 105.
Aleurochiton avellanae (Signoret); Harrison 1920: 59; Zahradnik 1956: 44.
Bemisia (Neobemisia) avellanae (Signoret); Visnya 1941: 8.
Aleurodes ribium Douglas 1888: 265. Lectotype. unknown host and locality, BMNH.
Bemisia (Neobemisia) ribium (Douglas); Visnya 1941: 9.
Aleurodes rubicola Douglas 1891: 200. Lectotype. England: on bramble [*Rubus* sp. (Rosaceae)] leaves, BMNH; Trehan 1939: 266.
Aleurochiton vaccinii Kunow 1880: 46. Syntypes. Konisberg Prov, on *Vaccinium uliginosum* (Ericaceae), BMNH.

Material: Horand, October 2006 on *Corylus avellana* (Betulaceae).

Distribution: Austria, Costa Rica, Czech Republic, Denmark, England, Finland, France, Germany, Hungary, Iran, Italy, Korea, Moldavia, Netherlands, Spain, Sweden, Taiwan, USSR, Former Yugoslavia.

Genus Bemisia Quaintance and Baker 1914

Bemisia Quaintance & Baker 1914: 99. Type species. *Aleurodes inconspicua* Quaintance 1900: 28 (junior synonym of *Aleurodes tabaci* Gennadius 1889).
Cortesiana Goux 1988: 63. Type species. *Cortesiana restonicae* Goux 1988, by monotypy; synonymy according to Martin 1999: 54.
Roucasia Goux 1940: 45. Type species. *Roucasia ovata* Goux 1940, by monotypy; synonymy according to Danzig 1964: 326.

Bemisia afer (Priesner and Hosny) 1934

Dialeurodoides afer Priesner and Hosny 1934: 6. Syntypes. Egypt, Kom Ombo, 4.vii.1931, on *Lawsonia alba* (Lythraceae), USNM.
Bemisia afer (Priesner and Hosny); Habib and Farag 1970: 8.
Bemisia (Neobemisia) afra [sic] (Priesner and Hosny); Visnya 1941: 8.
Bemisia citricola Gomez-Menor 1945: 293. Syntypes. Spain, Orihuela, Alicante, on *Citrus limonium*, *Citrus aurantium* (Rutaceae), *Eucalyptus* sp. (Myrtaceae), *Morus* sp. (Moraceae), *Cynanchum acutum*, *Laurus nobilis* (Lauraceae).
Bemisia hancocki Corbett 1936. Syntypes. Uganda: 1934, G. Hancock, on cotton [*Gossypium* sp. (Malvaceae)]; BMNH; synonymy according to Bink-Moenen 1983: 95.
Bemisia (Neobemisia) hancocki Corbett; Visnya 1941: 8.

Material: Khodafarin, July 2006 on *Gossypium hirsutum* (Malvaceae).

Distribution: Australia, Brazil, Cameroon, Chad, China, Congo, Egypt, Guinea, India, Iran, Israel, Italy, Ivory Coast, Kenya, Korea, Madagascar, Mulawi, New Guinea, Niger, Nigeria, Pakistan, Sicily, Sierre Leon, Spain, South Africa, Sudan, Uganda, Zaire.

Bemisia tabaci (Gennadius) 1889

Aleurodes tabaci Gennadius 1889: 1-3. Syntypes. Greece: on tobacco [*Nicotiana* sp. (Solanaceae)], USNM.
Bemisia tabaci (Gennadius); Takahashi 1936: 110.
Bemisia argentifolii Bellows and Perring 1994, in Bellows et al., 1994; synonymy according to De Barro et al., 2005: 201.
Bemisia achyranthes Singh 1931: 82. Syntypes. India: on *Achyranthes aspera* [synonmized with *B. gossypiperda* by Corbett 1935b: 783].
Bemisia argentifolii Bellows and Perring 1994, in Bellows *et al.* 1994. Holotype pupal case. USA: California, xii.1992, stock culture, on *Phaseolus limensis* (Fabaceae).
Bemisia bahiana Bondar 1928: 30. Syntypes. Brazil: on *Nicotiana tabacum*.
Bemisia costa-limai Bondar 1928: 27. Syntypes. Brazil: on *Euphorbia hirtella* (Euphorbiaceae).
Bemisia emiliae Corbett 1926: 273. Syntypes. Sri Lanka: on *Emilia sonchifolia* (Asteraceae).
Bemisia goldingi Corbett 1935c: 249. Syntypes. Nigeria: on cotton [*Gossypium* sp. (Malvaceae)].
Bemisia gossypiperda Misra and Singh 1929: 1. Syntypes. India: on many plants.
Bemisia gossypiperda var *mosaicivectura* Ghesquiere 1934: 30. Syntypes. Zaire: on *Jatropha multifida* (Euphorbiaceae) and *Manihot* sp. (Euphorbiaceae).

Bemisia hibisci Takahashi 1933: 17. Syntypes. Taiwan: on *Hibiscus rosa-sinensis* (Malvaceae).
Aleurodes inconspicua Quaintance 1900: 28. Syntypes. USA; Florida, Barlow, on *Physalis* sp. (Solanaceae), USNM; Russell 1957: 122.
Bemisia longispina Priesner & Hosny 1934: 6. Syntypes. Egypt: on *Psidium guajava* (Myrtaceae).
Bemisia lonicerae Takahashi 1957: 16. Syntypes. Japan: on *Lonicera japonica* (Caprifoliaceae).
Bemisia manihotis Frappa 1938: 30. Syntypes. Madagascar: on *Manihot* sp. (Euphorbiaceae).
Bemisia minima Danzig 1964: 638. Holotype. USSR: Caucasian Black Sea coast, on *Elsholtzia patrini*.
Bemisia miniscula Danzig 1964: 640. Holotype. USSR: Adzharia, on *Cissus salvifolius* (Vitaceae).
Bemisia nigeriensis Corbett 1935c: 250. Syntypes. Nigeria: on cassava [*Manihot* sp. (Euphorbiaceae)].
Bemisia rhodesiansis Corbett 1936: 22. Syntypes. Rhodesia: on tobacco [*Nicotiana* sp.].
Bemisia signata Bondar 1928: 29. Syntypes. Brazil: on *Nicotiana glauca*.
Bemisia vayssierei Frappa 1939: 255. Syntypes. Madagascar: on tobacco [*Nicotiana* sp.].
Cortesiana restonicae Goux 1988. Holotype. Corsica.

Material: Abshahmad, June 2006 on *Nerium oleander* (Apocynaceae); Khomarloo, Augut 2006 on *Nerium oleander* (Apocynaceae); Aras boundary, September 2006 on *Plantago* sp. (Plantaginaceae); Ahar, July 2007 on *Beta vulgaris* (Chenopodiaceae); Kalibar, July 2008 on *Brassica campestris* (Brassicaceae).

Distribution: Virtually worldwide; Afghanistan, Algeria, Andaman and Nicobar Islands, Argentina, Australia, Barbados, Brazil, Cameroon, Canary Island, Chile, Caroline Islands, Central African Republic, Chad, China, Colombia, Congo, Cuba, Cyprus, Dominican Republic, Ecuador, Egypt, El Salvador, England, Ethiopia, Fiji, France, French Guiana, Guadeloupe, Haiti, Honduras, Gabon, Gambia, Ghana, Greece, Grenada, Guatemala, Guam, Guyana, Hawaii, Hong Kong, India, Iran, Iraq, Israel, Italy, Ivory Coast, Jamaica, Japan, Jordan, Kenya, Korea, Lebanon, Liberia, Libya, Madagascar, Malaysia, Mariana Islands, Mauritius, Mexico, Mozambique, Netherlands, New Guinea, Nicaragua, Nigeria, Pakistan, Panama, Peru, Philippines, Portugal, Puerto Rico, Romania, Saipan, Saudi Arabia, Senegal, Seychelles, Singapore, Sierra Leone, South Africa, Spain, Sri Lanka, Sudan, Sumatra, Syria, Tahiti, Taiwan, Thailand, Trinidad and Tobago, Turkey, Uganda, United Kingdom, USA, Venezuela, Virgin Islands, Yemen, Zaire, Zimbabwe.

Genus *Bulgarialeurodes* Corbett 1936

Bulgarialeurodes Corbett 1936: 18. Type species: *Bulgarialeurodes rosae* Corbett 1936 (syn. *Aleurodes cotesii* Maskell 1896), by monotypy.

Bulgarialeurodes cotesii (Maskell) 1895
Aleurodes cotesii Maskell 1895: 427. Syntypes. Pakistan: on *Rosa* sp. (Rosaceae), ADSIR.
Bulgarialeurodes cotesii (Maskell); Russell 1960: 30.
Aleurodes rosae Kiriukhin 1947: 10. Syntypes. Iran: on *Rosa* spp.; synonymy according to Russell 1960: 30.
Bulgarialeurodes rosae Corbett 1936: 18. Syntypes. Bulgaria: on *Rosa damascena*; synonymy according to Russell 1960a: 30.

Material: Kalibar, August 2008 on *Rosa canina* (Rosaceae).

Distribution: Afghanistan, Bulgaria, Iran, Pakistan, Romania, Turkmenistan, USSR, Yugoslavia.

Genus *Dialeurodes* cockerel 1902

Aleyrodes (*Dialeurodes*) Cockerell 1902: 283. Type species. *Aleyrodes citri* Riley & Howard 1893, by original designation, a synomym of *A. citri* Ashmead 1885: 704.
Dialeurodes Cockerell; full genus, Quaintance and Baker 1914: 97.
Kanakarajiella David and Sundararaj 1993. Type species. *Dialeurodes vulgaris* Singh 1931, by original designation; synonymy according to Martin and Mound 2007: 28.
Lankaleurodes David 1993: 23. Type species. *Dialeurodes radiipuncta* Quaintance & Baker 1917, by original designation; synonymy according to Martin and Mound 2007: 28.
Shanthiniae David 2000, in P.M.M. David 2000: 125. Type species - *Shanthiniae sheryli* David 2000, by monotypy and original designation; synonymy according to Martin and Mound 2007: 28.
Comment: Martin and Mound (2007) tentatively listed species in the following subgenera of *Dialeurodes* as being in the genus *Dialeurodes*: *Dialeurodes* (*Dialeuronomada*) Quaintance and Baker 1917: 51; *Dialeurodes* (*Rabdostigma*) Quaintance and Baker 1917: 426, *Dialeurodes* (*Gigaleurodes*) Quaintance and Baker 1917: 426, and *Dialeurodes* (*Dialeuroplata*) Quaintance and Baker 1917: 435.
Dialeuronomada Quaintance and Baker; full genus by Sundararaj and David 1991.

Dialeurodes kirkaldyi (Kotinsky) 1907

Aleyrodes kirkaldyi Kotinsky 1907: 95-96. Syntypes. USA: Hawaii, on undetermined trailing shrub, *Beaumontia grandifolia*, *Morinda citrifolia* (Rubiaceae) and *Jasminum grandiflorum* (Oleaceae), USNM.
Dialeurodes kirkaldyi (Kotinsky); Quaintance and Baker 1914: 98.
Dialeurodes yercaudensis Jesudasan and David 1991: 307. Holotype pupal case. India: on *Ligustrum walkeri* (Oleaceae), IDAV; synonymy according to Sundararaj and Dubey 2006.

Material: Khodafarin, July 2007 on *Malva sylvestris* (Malvaceae).

Distribution: Andaman and Nicobar Islands, Australia, Azores, Bahamas, Barbados, Burma, Caroline Islands, China, Cook Islands, Costa Rica, Cuba, Egypt, Fiji, Ghana, Greece, Guam, Guyana, Hawaii, Hong Kong, India, Iran, Israel, Jamaica, Japan, Lebanon, Malaysia, Mexico, Pakistan, Philippines, Puerto Rico, Samoa, Sri Lanka, Syria, Tahiti, Taiwan, Thailand, Trinidad, Turkey, UK, USA, Virgin Islands.

Genus *Parabemisia* Takahashi 1952

Parabemisia Takahashi and Mamet 1952: 21. Type species. *Parabemisia maculata* Takahashi 1952, by original designation.

Parabemisia myricae (Kuwana) 1927

Bemisia myricae Kuwana 1927: 249. Syntypes. Japan: on *Myrica rubra* (Myricaceae), *Morus alba* (Moraceae), *Citrus* spp. (Rutaceae) and other plants, TARI.
Parabemisia myricae (Kuwana); Takahashi 1952: 24.

Material: Khomarloo, September 2007 on *Lantana camara* (Verbenaceae).

Distribution: China, Egypt, Hawaii, India, Iran, Israel, Italy, Japan, Morocco, Spain, Taiwan, Turkey, USA, Venezuela.

Genus *Siphoninus* Silvestri 1915

Siphoninus Silvestri 1915: 245. Type species: *Siphoninus finitimus* Silvestri 1915, regarded by Mound and Halsey 1978: 191 as synonym of *S. phillyreae*, by original designation.

Siphoninus immaculatus (Heeger) 1856

Aleurodes immaculatus Heeger 1856: 33. Syntypes. Germany? On *Hedera helix* (Araliaceae).
Aleurochiton immaculatus (Heeger); Quaintance and Baker 1914: 105.
Trialeurodes immaculatus (Heeger); Quaintance and Baker 1915.
Siphoninus immaculatus (Heeger); Trehan 1940: 601.
Aleurodes immaculatus Heeger 1856: 33. Syntypes. Germany? On *Hedera helix*.
Siphoninus heegeri Haupt 1935: 259; synonymy according to Zahradnik 1963: 9.

Material: Khodafarin, August 2008 on *Fraxinus* sp. (Oleaceae).

Distribution: Austria, Czech Republic, England, Germany, Hungary, Iran, Italy, Sweden, USSR.

Siphoninus phillyreae (Haliday) 1835

Aleurodes phillyreae Haliday 1835: 119. Syntypes. Ireland: on *Phillyrea latifolia* (Oleaceae), HAL.
Trialeurodes phillyreae (Haliday); Quaintance and Baker 1915.
Siphoninus phillyreae (Haliday); Silvestri 1915: 247.
Siphoninus phillyreae inequalis Goux 1949: 11. Syntypes. France: on pear [*Pyrus* sp. (Rosaceae)].
Siphoninus phillyreae mulititubulatus Goux 1949: 11. Syntypes. Corsica: on *Olea europea* (Oleaceae).
Siphoninus phillyreae mulititubulatus Goux; Mound and Halsey 1978: 192.
Aleurodes dubia Heeger 1859: 223. Syntypes, Germany?, on *Fraxinus* sp. (Oleaceae); Frauenfeld 1867: 796.
Aleurochiton dubius (Heeger); Quaintance and Baker 1914: 105.
Siphoninus dubiosa Haupt 1935: 259; synonymy according to Zahradnik 1963: 9.
Aleurodes phylliceae Bouche 1851: 110. Syntypes. Southern Europe: on *Phillyrea latifolia*? (Oleaceae).
Aleurodes phylliceae Bouche 1851; Frauenfeld 1867: 786.
Asterochiton phillyreae (Haliday); Quaintance and Baker 1914: 105.
Siphoninus finitimus Silvestri 1915: 245. Syntypes. Eritrea: on *Olea chrysophylla* (Oleaceae), IESP.
Siphoninus finitimus Silvestri; Mound and Halsey 1978: 192.
Siphoninus granati Priesner and Hosny 1932: 1. Syntypes. Egypt: Meadi, 16.viii.1931, Priesner and Hosny, on *Punica granatum* (Punicaceae), EDAC.
Siphoninus granati Priesner and Hosny; Mound and Halsey 1978: 192.
Trialeurodes inaequalis Gautier 1923: 339. Syntypes. France: on *Pyrus* sp. (Rosaceae); synonymy according to Mound and Halsey 1978: 192.

Material: Khodafarin, September 2006, on *Malus communis* (Rosaceae).

Distribution: Australia, Bulgaria, Cameroon, Corsica, Cyprus, England, Egypt, Eritrea, Ethiopia, Finland, France, Germany, Greece, Hungary, Iran, Ireland, Israel, Italy, Java, Jordan, Mexico, Peru, Spain, Sudan, Syria, Taiwan, USA, USSR, Venezuela, Yugoslavia, Zaire.

Genus Tetralicia Harrison 1917

Tetralicia Harrison 1917: 60. Type species. *Tetralicia ericae* Harrison 1917, by monotypy.

Tetralicia ericae Harrison 1917

Tetralicia ericae Harrison 1917: 61. Syntypes. England: on *Erica tetralix* (Ericaceae), BMNH; Bink-Moenen 1976.

Material: Khodafarin, September 2007 on *Erica* sp. (Ericaceae). New record for Iran.

Distribution: Austria, Czech Republic, England, Italy, Netherlands, Sweden, Scotland, USSR, Wales.

Genus Trialeurodes Cockerell 1902

Aleyrodes (*Trialeurodes*) Cockerell 1902: 283. Type species. *Aleurodes pergandei* Quaintance 1900, by original designation.
Trialeurodes Cockerell; full genus by Quaintance and Baker 1915.
Aleyrodes (*Asterochiton*) Maskell; misidentification; Kirkaldy 1907: 43; Quaintance and Baker 1914: 104.
Aleurodes (*Ogivaleurodes*) Goux 1948: 31. Types species. *Aleurodes lauri*, by monotypy; synonymy according to Mound and Halsey 1978: 205.
Gymnaleurodes Sampson and Drews 1940: 29. Types species. *Gymnaleurodes bellissima*, by monotypy; synonymy according to Sampson 1943: 209.

Trialeurodes ericae Bink-Moenen 1976

Trialeurodes ericae Bink-Moenen 1976: 17. Holotype. The Netherlands: on *Erica tetralix* (Ericaceae), NHM (Rapisarda 1986: 497); Bink-Moenen 1989: 176.
Trialeurodes (*Ericaleyrodes*) *ericae* Bink-Moenen; Rapisarda 1986: 490.

Material: Ahar, August 2008 on *Erica* sp. (Ericaceae). New record for Iran.

Distribution: Crete, Corsica, France, Majorca, Netherlands, Spain.

Trialeurodes lauri (Signoret) 1882

Aleurodes lauri Signoret 1882: CLVIII. Syntypes. Greece: Athens (Grenadius), on *Laurus nobilis* (Lauraceae).
Aleuroparadoxus lauri (Signoret); Silvestri 1934: 399.
Trialeurodes lauri (Signoret); Russell 1947: 6.
Aleyrodes (*Ogivaleurodes*) *lauri* (Signoret); Goux 1948: 31.
Ogivaleurodes lauri (Signoret); Goux 1951: 12.
Trialeurodes lauri (Signoret); Zahradnick 1963: 232.
Trialeurodes klemmi Takahashi 1940: 148. Syntypes. Yugoslavia: Rab, on *Laurus nobilis*; synonymy according to Russell 1947: 6.

Material: Kalibar, October 2007 on *Laurus nobilis* (Lauraceae).

Distribution: Australia, Belgium, Cyprus, France, Greece, Iran, Israel, Italy, Lebanon, Luxembourg, Switzerland, Turkey, USSR, Yugoslavia.

Trialeurodes ricini (Misra) 1924

Aleyrodes ricini Misra 1924: 131. Syntypes. India: on *Ricinus communis* (Euphorbiaceae), USNM.
Trialeurodes ricini (Misra); Singh 1931: 46.
Trialeurodes rara Singh 1931: 47. Syntypes. India, on *Breynia* sp. (Euphorbiaceae); synonymy according to Bink-Moenen 1983: 185.
Trialeurodes desmodii Corbett 1935c: 243. Syntypes. Sierra Leone, on *Desmodium lasiocarpum* (Fabaceae), BMNH; synonymy [with *T. rara*] according to Mound & Halsey 1978: 217.
Trialeurodes lubia El Khidir & Khalifa 1962: 47. Holotype. Sudan: on *Dolichos lablab* (Fabaceae); synonymy with *T. rara* according to Mound 1965: 157.

Material: Kalibar, July 2006 on *Ipomoea* sp. (Convolvulaceae).

Distribution: Andaman and Nicobar Islands, Cambodia, Cameroon, Central African Republic, Chad, India, Iran, Israel, Ivory Coast, Gabon, Madagascar, Malaya, Malaysia, Nigeria, Pakistan, Sierra Leon, Saudi Arabia, Sri Lanka, Sudan, Thailand, Turkey, Uganda, Zaire, Zimbabwe.

Trialeurodes vaporariorum (Westwood) 1856

Aleurodes vaporariorum Westwood 1856: 852. Syntypes. England: on *Gonolobus* sp. (Asclepiadaceae), *Tecoma velutina* (Bignoniaceae), other plants, OUMNH.
Asterochiton vaporariorum (Westwood); Quaintance and Baker 1914: 105.
Trialeurodes vaporariorum (Westwood); Quaintance and Baker 1915.
Aleurodes glacialis Bemis (misidentification in part); Bemis 1904: 518.
Asterochiton lecanioides Maskell 1879 (in part); Maskell 1879: 215. Syntypes. New Zealand: on *Pittosporum eugenioides* and *Polypodium billardieri*, ADSIR; synonymy according to Quaintance and Baker 1914: 105.
Trialeurodes mossopi Corbett 1935a: 9. Syntypes. Rhodesia: on *Phaseolus vulgaris* (Fabaceae), BMNH; synonymy according to Russell 1948: 44.
Aleurodes nicotiana Maskell 1895: 436. Syntypes. Mexico: on *Nicotiana tabacum* (Solanaceae), USNM; synonymy according to Quaintance and Baker 1914: 105.
Asterochiton papillifer Maskell 1890b: 173. Syntypes. New Zealand: on *Pittosporum eugenioides* (Pittosporaceae), ADSIR; synonymy according to Quaintance and Baker 1914: 105.
Trialeurodes sesbania Corbett 1936: 19. Syntypes. Australia: on *Sesbania tripeti* (Fabaceae), BMNH; synonymy according to Russell 1948: 44.
Aleurodes sonchi Kotinsky 1907: 97. Syntypes. USA: Hawaii, Honolulu, on *Sonchus oleraceus* (Asteraceae), USNM; synonymy according to Baker and Moles 1923: 645.

Asterochiton sonchi (Kotinsky); Quaintance and Baker 1914: 105.
Trialeurodes sonchi (Kotinsky); Quaintance and Baker 1914: 105.

Material: Aynalo, June 2007 on *Solanum melongena* (Solanaceae); Khomarloo, September 2007 on *Cucumis sativus* (Cucurbitaceae); Aras boundary, September 2006 on *Mentha* sp. (Labiatae); Horand, October 2006 on *Cucurbita pepo* (Cucurbitaceae).

Distribution: Worldwide; Argentina, Australia, Austria, Azores, Bangladesh, Belgium, Bermuda, Brazil, Bulgaria, Canada, Canary Islands, Chile, Colombia, Costa Rica, Cuba, Denmark, Dominican Republic, Ecuador, El Salvador, Ethiopia, France, Germany, Greece, Guadeloupe, Guatemala, Hawaii, Honduras, Hungary, Hong Kong, Korea, India, Indonesia, Iran, Ireland, Italy, Israel, Jamaica, Japan, Jordan, Kenya, Mexico, Netherlands, New Guinea, New Zealand, Norway, Peru, Philippines, Poland, Portugal, Puerto Rico, Reunion, South Africa, Spain, Sri Lanka, Turkey, United Kingdom, USA, Venezuela, Zimbabwe.

Arasbaran is rather large region of northwestern Iran with various climates. Although totally 25 species were collected from this region, but surely the same surveys must be continued in this region and neighboring areas. We expect that continuing the faunistic survey in Arasbaran will be resulted to many other species and new country records.

ACKNOWLEDGEMENTS

The authors are indebted to Dr. F.M. Iaccarino (Universita degli studi di Napoli, Italy) and R. Bink-Moenen (Zuider-eng 6, 6721 HH Bennekom, The Netherlands) for invaluable helps in progress the project. We are thanks to Mr. M. Havaskary (Tehran Islamic Azad University) and Mr. Azad Ahmadi (Shahre Rey Islamic Azad University) for loaning many specimens and Mr. H. Mohebbi for determining the host plants. The research was supported by Shahre Rey Islamic Azad University and Fars Science and Research Branch.

REFERENCES

Ashmead WH (1885). The orange *Aleurodes* (*Aleurodes citri* n. sp.). Florida Dispatch 2: 704.
Bellows TS Jr, Perring TM, Gill RJ, Headrick DH (1994). Description of a species of *Bemisia* (Homoptera: Aleyrodidae). Annals of the Entomological Society of America 87: 195-206.
Bemis FE (1904). The aleyrodids or mealy-winged flies of California with reference to other American species. Proceedings of the United States National Museum 27: 471-537.
Bink-Moenen RM (1976). A new whitefly of *Erica tetralix*: *Trialeurodes ericae* sp. n. (Homoptera: Aleyrodidae). Entomol. Ber. Ams. 36: 17-19.
Bink-Moenen RM (1983). Revision of the African whiteflies (Aleyrodidae). Monograficen van de Nederlandse Entomologisclie Vercniging, Amsterdam 10: 1-211.

Bink-Moenen RM (1989). A new species and new records of European whiteflies (Homoptera: Aleyrodidae) from heathers (*Erica* spp.). Entomologist's Gazette 40(2): 173-181.

Bink-Moenen RM (1992). Whitefly from mediterranean evergreen oaks (Homoptera: Aleyrodidae). Syst. Entomol. 17(1): 21-40.

Bink-Moenen RM, Gerling D (1991). Aleyrodidae of Israel. Bollettino del Laboratorio di Entomologia Agraria 'Fillipo Silvestri' 47: 3-49.

Bondar G (1928). Aleyrodideos do Brasil (2a contribuicao). Bolm. Labpath. Veg. Est. Bahia. 5: 1-37.

Bouche J Fr (1851). Beschreibung zwei neuer Arten der Gattung

Bouche J Fr (1851). Beschreibung zwei neuer Arten der Gattung *Aleurodes*. Stettin Entomol. Ztg. 12: 108-110.

Campbell BC, Steffen-Campbell JD, Gill RJ (1994). Evolutionary origin of whiteflies (Hemiptera: Sternorrhyncha: Aleyrodidae) inferred from 18S rDNA sequences. Insect Mol. Biol. 3(2):73-88.

Carver M, Reid IA (1996). Aleyrodidae (Hemiptera: Sternorrhyncha) of Australia. Systematic catalogue, host-plant spectra, distribution, natural enemies and biological control. Technical Paper, Division of Entomology, Commonwealth Scientific and Industrial Research Organization, Canberra 37: 55

Cockerell TDA (1902). The classification of the Aleyrodidae. Proceedings on the Academy of Natural Sciences of Philadelphia 54: 279-283.

Cohic F (1968). Contribution a' l'etude des aleurodes africains (4' Note). Cah. Off. Rech. Sci. Tech. Outre-Mer (Biologie) 6: 63-143.

Corbett GH (1926). Contribution towards our knowledge of the Aleyrodidae of Ceylon. Bull. Entomol. Res. 16: 267-284.

Corbett GH (1935a). Three new aleurodids (Hem.). Stylops 4: 8-10.

Corbett GH (1935b). Malayan Aleurodidae. Journal of Federated Malay States Museums 17: 722-852.

Corbett GH (1935c). On new Aleurodidae (Hem.). Annals and Magazine of Natural History (10) 16: 240-252.

Corbett GH (1936). New Aleurodidae (Hem.). Proceedings of the Royal Entomological Society of London (B) 5: 18-22.

Costa Lima AD (1942). Sobre aleirodideos do genero '*Aleurothrixus*' (Homoptera). Revista Brasileira de Biologia. 2:419-426.

Danzig EM (1964). The whiteflies (Homoptera: Alevrodoidea) of the Caucasus. Entomologicheskoe Obozrenie 43: 633-646. [English translation in Entomological Review. Washington 43: 325-330.]

Danzig EM (1966). The whiteflies (Homoptera: Aleyrodoidea) of the southern Primor'ye (Soviet Far East). Entomologicheskoe Obozrenie 45: 364-386. (English translation in Entomological Review. Washington 45: 197-209.]

Danzig EM (2004). *Aleyrodes borchsenii* Danzig, 1966 is a junior synonym of *A. lonicerae* Walker, 1852 (Homoptera, Aleyrodidae). Zoosystematica Rossica 13: 114.

David BV (1993). The whitefly of Sri Lanka (Homoptera: Aleyrodidae). Frederick Institute of Plant Protection and Toxicology, Entomology Series 3: 1-32.

David PMM (2000). Three new genera of whiteflies *Mohanasundaramiella*, *Shanthiniae* and *Vasantharajiella* (Aleyrodidae: Homoptera) from India. J. Bombay Nat. Hist. Soc. 97(1): 123-130.

David BV, Subramaniam TR (1976). Studies on some Indian Aleyrodidae. Record of Zoological Survey of India 70: 133-233.

David BV, Sundararaj R (1993). Studies on Dialeurodini (Aleyrodidae: Homoptera) of India: *Kanakarajiella* gen. nov. J. Entomol. Res. 17(4): 289-295.

De Barro PJ, Trueman JWH, Frohlich DR (2005). *Bemisia argentifolii* is a race of *B. tabaci* (Hemiptera: Aleyrodidae): the molecular genetic differentiation of *B. tabaci* populations around the world. Bull. Entomol. Res. 95: 193-203.

Douglas JW (1888). Description of new species of Aleurodes, *Aleurodes ribium*. Entomologist's Monthly Magazine 24: 265-267.

Douglas JW (1891). A new species of *Aleurodes*? Entomologist's Monthly Magazine 27: 200.

Douglas JW (1894). A new species of *Aleurodes*. *Aleurodes spireae*. Entomologist's Monthly Magazine 30: 73-74.

El Khidir E, Khalifa A (1962). A new aleyrodid from the Sudan. Proceedings of the Royal Entomological Society of London (B) 31: 47-51.

Evans G (2007). Last modified November 28, 2007 Online: The whiteflies of the world and their host plants and natural enemies. Version 2007-11-28, (http://www.sel.barc.usda.gov:591/1WF/whitefly _catalog.htm)

Frappa C (1938). Description de *Bemisia manihotis*, n. sp. (Hem. Hom. Aleyrodidae) nuisible au manioc à Madagascar. Bulletin de la Société Entomologique de France 43: 30-32.

Frappa C (1939). Note sur une nouvelle espèce d'aleurode nuisible aux plantations de tabac de la Tsiribihina. Bulletin Économique Trimest. de Madagascar 16: 254-259.

Frauenfeld GR (1867). Zoologische Miscellen XIII. Ueber Aleurodes und Thrips, vorzüglich im Warmhause. Verhandlungen der Zoologische-Botanischen Gesellschaft in Wien 17: 793-800.

Gautier C (1923). Un aleurode parasite du poirier et du trêne *Trialeurodes inaequalis* n. sp. (Hém. Aleurodidae). Annales de la Société Entomologique de France 91: 337-350

Gennadius P (1889). Disease of tobacco plantations in Trikonia. The aleurodi of tobacco. (In Greek). Ellenike Georgia 5: 1-3.

Gerling D (1990). Whiteflies: their bionomics, pest status and management. Wimborne, U.K. Intercept, 348 pp.

Gomez-Menor J (1945). Aleurodidos de interes agricola. Boln. Patol. Veg. Entomol. Agric. 13: 161-198.

Goux L (1940). Contribution a 1'etude des aleurodes (Hem.: Aleyrodidae) de la France II. Description de deux especes nouvelles de Marseille. Bulletin de la Societe Entomologiilue de France 45: 45-48.

Goux L (1948). Contribution à l'étude des aleurodes (Hem. Aleyrodidae) de la France V. L'aleurode du laurier sauce (*Laurus nobilis* L.). Annales de la Société d'Histoire Naturelle de Toulon 2: 30-34.

Goux L (1949). Contribution a l'etude des aleurodes (Hem.: Aleyrodidae) de la France. VI. Le genre *Siphoninus* Silvestri. Bull. Mens. Soc. Linn. Lyon N.S. 18: 7-12.

Goux L (1988). Aleurodes de France - VII. Description de deux especes nouvelles constituant des genres nouveaux. Bulletin de la Societe Linnccnne de Provence 39: 63-66.

Habib A, Farag FA (1970). Studies on nine common aleurodids of Egypt. Bulletin de la Societe Entomologique d'Egypte 54: I-41.

Haliday AH (1835). *Aleyrodes phillyreae*. Entomol. Mag. 2: 119-120.

Harrison JWH (1917). New and rare British Aleurodidae. Entomologist 53: 255-257.

Harrison JWH (1920). Notes and records. Aleyrodidae. Vasculum 6: 59.

Haupt H (1934). Neues uber die Homoptera - Aleurodina. Dentsche Entomologisclie Zeitschrift. Berlin 1934, pp. 127-141.

Haupt H (1935). Schmetterlings - od. Mottenlause, Aleurodina. in Die Tienuelt Mitteleuropas 4: 253-260.

Heeger E (1856). Beitrage zur Naturgeschichte der Insecten. Naturgeschichte der *Aleurodes immaculata* Steph. Sitziingberichte der Kaiserlichen Akademie der Wissenschaften. Mathematisclie-natuncissensc Jiaftliche klasse. Wien 18: 33-36.

Heeger E (1859). Beitrage zur Naturgeschichte der Insecten. Naturgeschichte der *Aleurodes dubia* Stephens. Sitzungberichte der Kaiserlichen Akademie der Wissenschaften. Mathematische-naturwissenschaftliche klasse. Wien 24: 223-226.

Hempel A (1899). Descriptions of three new species of Aleurodidae from Brazil. Psyche. Cambridge (Massachusetts) 8: 394-395.

Hempel A (1901). A preliminary report on some new Brazilian Hemiptera. Annals and Magazine of Natural History 8: 383–391.

Iaccarino FM (1982). Descrizione di *Aleitroviggvinus adrianae*, n. gen, n. sp. (Homoptera: Alevrodidae). Bollettino del laboratorio di Entoir.ologia Agraria 'Filippo Silvestri' 39: 37-45.

Iaccarino FM (1990). Description of *Paraleyrodes minei* n. sp. (Homoptera: Aleyrodidae), a new aleyrodid of citrus, in Syria. Bollettino del Laboratorio di Entomologia Agraria 'Filippo Silvestri' 46: 131-149.

Jesudasan RWA, David BV (1990). Revision of two whitefly genera, *Aleuroclava* Singh and *Aleurotuberculatus* Takahashi (Homoptera: Aleyrodidae). Frederick Institute of Plant Protection and Toxicology, Entomological Series 2: 1-13.

Jesudasan RWA, David BV (1991). Taxonomic studies on Indian Aleyrodidae (Insecta: Homoptera). Oriental Insects 25: 231-434.

Kiriukhin G (1947). Quelques Aleurododea de l'Iran. Entomologie Phytopath. Appl. 5: 8-10 (In Persian, 5: 22-28; French summary).

Kirkaldy GW (1907). A catalogue of the hemipterous family Aleyrodidae. Bull. Bd Commnrs Agric. For. Hawaii Div. Entomol. 2: 1-92.

Koch CL (1857). Die Pflanzenlaŭse Aphiden. Nürnberg, p.330

Kotinsky J (1907). Aleyrodidae of Hawaii and Fiji with descriptions of new species. Bulletin. Board of Commisioners of Agriculture and Forestry, Hawaii, Division of Entomology 2: 93-102.

Künow G (1880). Zweineue Schildlaŭse. Entomologische Nachrichten. Berlin 6: 46-47.

Kuwana I (1927). On the genus *Bemisia* (Family Aleyrodidae) found in Japan, with description of a new species. Annotationes zoologicae japonensis 11: 245-253.

Latreille PA (1796). Précis des charactères génériques des insectes, disposés dans un ordre naturel. Brive p.201

Latreille PA (1807). Genera Crustaceorum et Insectorum. Paris, 4 volumes.

Linnaeus C (1758). Systema Naturae. Uppsala, p. 824.

Löw F (1867). Zoologische Notizen Zweite Serie. Verhandlungen der Zoologische-Botanischen Gesellschaft in Wien 17: 745–752.

Manzari Sh and Quicke DLJ (2006). A cladistic analysis of whiteflies, subfamily Aleyrodinae (Hemiptera: Sternorrhyncha: Aleyrodidae). J. Nat. Hist. 40: 2423-2554.

Martin JH (1985). The Whitefly of the Guinea (Homoptera: Aleyrodidae). Bull. Br. Mus. Nat. Hist. 50 (3): 303-351.

Martin JH (1996). Neotropical whiteflies of the subfamily Aleurodicinae established in the western Palaearctic (Homoptera: Aleyrodidae). Journal of Natural History 30(12): 1849-1859.

Martin JH (1999). The whitefly fauna of Australia (Sternorrhyncha: Aleyrodidae), a taxonomic account and identification guide. Technical Paper, CSIRO Entomology 38: 1-197.

Martin JH (2005). Whiteflies of Belize (Hemiptera, Aleyrodidae). Part 2-introduction and account of the subfamily Aleyroinae Westwood. Zootaxa 1098: 1-116.

Martin JH, Mifsud D, Rapisarda C (2000). The whiteflies (Hemiptera: Aleyrodidae) of Europe and the Mediterranean Basin. Bulletin of Entomological Research 90(5): 407-448.

Martin JH, Mound LA (2007). An annotated check list of whiteflies (Insecta: Hemiptera: Aleyrodidae). Zootaxa 1492: 1-84.

Maskell, WM (1879). On some Coccidae in New Zealand. Transactions and Proceedings of the New Zealand Institute, 11:187-228.

Maskell MW (1890). On some Aleyrodidae from New Zealand and Fiji. Transactions and Proceedings of the New Zealand Institute (1889) 22: 170-176.

Maskell WM (1895). Contributions towards a monograf of Aleurodidae, a family of Hemiptera-Homoptera. Trans. Proc. N. Z. Inst. 28: 411-449.

Maskell WM (1896). *Aleurodes eugeniae*, a new species of bug. Indian Museum Notes 4: 52-53.

Meganathan P, David BV (1994). Aleyrodidae fauna (Aleyrodidae: Homoptera) of Silent valley, A tropical evergreen rain-forest, in Kerala, India. FIPPAT Entomological Series 5: 1-66.

Misra CS (1924). The citrus whitefly, *Dialeurodes citri* in India and its parasite, together with the life history of *Aleurodes ricini*, n. sp. Report of Proceedings of Entomological Meetings at Pusa 1923, 129-135.

Misra CS, Singh KL (1929). The cotton whitefly (*Bemisia gossypiperda* n. sp.). Bulletin of the Agricultural Research Institute Pusa 196: 1-7.

Mound LA (1965). An introduction to the Aleyrodidae of Western Africa (Homoptera). Bull. Br. Mus. Nat. Hist. 17: 113-160.

Mound LA (1966). A revision of the British Aleyrodidae (Hemiptera: Homoptera) Bull. Br. Mus. Nat. Hist. 17: 397-428.

Mound LA, Halsey SH (1978). Whitefly of the world. 340 pp. British Museum (Natural History)/John Wiley & Sons, Chichester.

Mound LA, Martin JH, Polaszek A (1994). The insect fauna of *Selaginella* (Pteridophyta: Lycopsida), with descriptions of three new species. J. Nat. Hist. 28: 1403-1415.

Ossiannilsson F (1955). Till Kannedomom om de svenska mjollossen (Hemiptera: Homoptera: Aleyrodina). Opusc. Entomol. 20: 192-199.

Priesner H, Hosny M (1932). Contributions to a knowledge of the whiteflies (Aleurodidae) of Egypt (I). Bulletin. Ministry of Agriculture, Egypt. Technical and Scientific Service 121: 1-8.

Priesner H, Hosny M (1934). Contributions to a knowledge of the whiteflies (Aleurodidae) of Egypt (II). Bulletin. Ministry of Agriculture, Egypt. Technical and Scientific Service 139: 1-21.

Quaintance AL (1900). Contribution towards a monograph of the American Aleurodidae. Technical Series, US Department of Agriculture Bureau of Entomology 8: 9-64.

Quaintance AL (1903). New oriental Aleurodidae. Canadian Entomologist 35: 61-64.

Quaintance AL (1907). The more important Aleyrodidae infesting economics plants, with description of a new species infesting the orange. U.S.D.A. Bur. Entomol. Tech. Ser. 12: 89-94.

Quaintance AL (1909). A new genus of Aleyrodidae, with remarks on *Aleyrodes nubifera* Berger and *Aleyrodes citri* Riley & Howard. Technical Series, US Department of Agriculture Bureau of Entomology 12: 169-174.

Quaintance AL, Baker AC (1914). Classification of the Aleyrodidae Part II. Technical Series, US Department of Agriculture Bureau of Entomology 27: 95-109.

Quaintance AL, Baker AC (1915). Classification of the Aleyrodidae - contents and index. Technical Series, US Department of Agriculture Bureau of Entomology 27: i-xi, 111-114.

Quaintance AL, Baker AC (1917). Contribution to our knowledge of the whiteflies of the subfamily Aleyrodinae (Aleyrodidae). Proceedings of the United States National Museum 51: 335-445.

Rapisarda C (1986). *Trialeurodes (Ericaleyrodes) sardiniae*, subgen. n., sp. n.: a new heather-feeding whitefly (Homoptera, Aleyrodidae). Frustula entomologica, nuova serie 7-8 (1984-85): 487-499.

Regu K, David BV (1993). Taxonomic studies on Indian Aleyrodids of the tribe Aleurolobini (Aleyrodidae: Homoptera). Frederick Institute of Plant Protection & Technology (FIPPAT), Entomological Series 4: 1-79.

Riley CV, Howard LO (1893). The orange Aleyrodes. (*Aleyrodes citri* n. sp.). Insect Life. US Department of Agriculture, Washington DC 5: 219-226.

Russell LM (1947). A classification of the whiteflies of the new Tribe Trialeurodini (Homoptera: Aleyrodidae). Revista de Entomologia. Rio de Janeiro 18: 1-44.

Russell LM (1948). The North American species of whiteflies of the genus *Trialeurodes*. Miscellaneous Publications.United States Department of Agriculture 635: 1-85.

Russell LM (1962). New name combinations and notes on some African and Asian species of Aleyrodidae (Homoptera). Bull. Brooklyn Entomol. Soc. 57: 63-65.

Russell LM (1957). Synonyms of *Bemisia tabaci* (Gennadius) (Homoptera, Aleyrodidae). Bull. Brooklyn Entomol. Soc. 52: 122-123.

Russell LM (1960). A whitefly living on roses (Homoptera: Alevrodidae). Proceedings of the Royal Entomological Society of London (B) 29: 29-32.

Sampson WW (1943). A generic synopsis of the Hemiterous Superfamily Aleyrdoidea. Entomol. Am. 23: 173-223.

Sampson WW, Drews EA (1940). *Gymnaleurodes*, a new genus of Aleyrodidae from California (Homoptera). Pan-Pacif. Entomol. 16: 29-30.

Sampson WW, Drews EA (1941). Fauna Mexicana IV. A review of the Aleyrodidae of Mexico (Insecta, Homoptera). An. Esc. Cienc. Biol. Mex. 2: 143-189.

Schlee D (1970). Verwandtschaftsforschung an fossilen und rezenten Aleyrodina (Insecta: Hemiptera). Stuttgarter Beitrage zur Natiirkunde aus dem Sfaatlichen Museum filr Naturkunde in Stuttgart 213: 1-72.

Schrank FP von (1801). Fauan Bioca. Ingolstadt 2(1): 147.

Signoret V (1868). Essai monographique sur les aleurodes. Annls. Soc. Entomol. Fr. (4) 8: 369-402.

Signoret V (1868). Essai monographique sur les aleurodes. Annls. Soc. Entomol. Fr. (4) 8: 369-402.

Signoret V (1882). Séance du 14 décembre 1881. 4o Note. Annales de la Société Entomologique de France, 1, CLVIII.

Silvestri F (1911). Di una nuova specie di *Aleurodes* vivente sull' Olivo. Boll. Lab. Zool. gen. Agr. R. Scuola Agric. Portici 5: 214-225.

Silvestri F (1915). Contributo alla conoscenza degli insetti dell' `olivo dell' Eritrea e dell' Africa meridionale. Fam. Aleyrodidae. Boll. Lab. Zool. gen. agrar. R. Scuola Agric. Portici 9: 245-249.

Silvestri F (1934). Compendio di entomologia applicata 1. Portici, p.448

Singh K 1931). A contribution towards our knowledge of the Aleyrodidae (whiteflies) of India. Memoirs of the Department of Agriculture in India12: 1-98.

Sundararaj R, David BV (1991). Ten new species of *Dialeuronomada* Quaintance & Baker (Homoptera: Aleyrodidae) from India. Hexapoda 3: 27-47.

Sundararaj R, Dubey AK (2004). The whitefly genus *Martiniella* Jesudasan and David (Aleyrodidae: Hemiptera), with description of one new species from India. Entomol. 29(4): 357-360.

Sundararaj R, Dubey AK (2006). A review of the whitefly genus *Dialeurodes* Cockerell (Aleyrodidae: Hemiptera) with description of two new species from India. Journal of the Bombay Natural History Society 103(1): 62-67.

Takahashi R (1931). Some whiteflies from Formosa. (Part I). Trans. Nat. Hist. Soc. Formosa 21: 203-209.

Takahashi R (1932). Aleyrodidae of Formosa, Part I. Report. Department of Agriculture. Government Research Institute. Formosa 59: 1-57.

Takahashi R (1933). Aleyrodidae of Formosa, Part II. Report of the Department of Agriculture of the Government Research Institute, Formosa 60: 1-24.

Takahashi R (1934). Aleyrodidae of Formosa, Part III. Rep. Dep. Agric. Govt. Res. Inst. Formosa 63: 39-71.

Takahashi R (1936). Some Aleyrodidae, Aphididae, Coccidae (Homoptera), and Thysanoptera from Micronesia. Tenthredo 1: 109-120.

Takahashi R (1938). A few Aleyrodidae from Maritius and China (Hemiptera). Trans. Nat. Hist. Soc. Formosa 28: 27-29.

Takahashi R (1940). A new species of Aleyrodidae from Jugoslavia. Arb. rnorp. taxon. Entomol. Berl. 7: 148-149.

Takahashi R (1951a). Some species of Aleyrodidae (Homoptera) from from Madagascar, with a species from Mauritius. Memoires de l'Institiit Scientificfiie de Madagascar (A) 6: 353-385.

Takahashi R (1951b). Description of six interesting species of Aleyrodidae from Malaya (Homoptera). Kontyu 19: 1-8.

Takahashi R (1952). *Aleurotuberculatus* and *Parabemisia* of Japan (Aleyrodidae: Homoptera). Misc. Rep. Res. Inst. Nat. Res. Tokyo 25: 17-24.

Takahashi R (1957). Some Aleyrodidae from Japan (Homoptera). Insecta Matsumurana 21: 12–21.

Takahashi R, Mamet R (1952). Some species of Aleyrodidae from Madagascar (Homoptera) II. Mem. Inst. Scient. Madagascar (E) 1: 111-133.

Trehan KN (1939). Studies on the British Aleurodidae. Current Science 8: 266.

Trehan KN (1940). Studies on the British whiteflies (Homoptera: Aleyrodidae). Transactions of the Royal Entomological Society of London 90: 575-616.

Visnya A (1941). Vorarbeiten zur Kenntnis der Aleurodiden-Fauna von Ungarn, nebst systematischen Bemerkungen uber die Gattungen Aleurochiton, *Pealius* und *Bemisia* (Homoptera). Fragmenta Faunistica Hungarica. 4 (Suppl) 1-19.

Walker F (1852). List of the specimens of Homopterous insects in the collection of the British Museum. Supplement. pp. 369 London. 4: 909-1188.

Westwood JO (1840). An introduction to the modern classification of insects; founded on the natural habits and corresponding organization of different families. Longman, Orme, Brown and Green. London. p. 587

Westwood JO (1856). The new *Aleyrodes* of the greenhouse. Gardeners' Chronicle 1856: 852.

Zahradnik J (1956). Trois nouvelles especes des aleyrodides pour la faune tchecoslovaque. Sb. faun. Praci Entomol. Odd. Nar. Mus. Praze 1: 43-45.

Zahradnik J (1961). Nouvelles connaissances faunistiques et taxonomiques sur les aleyrodides de la Tchecoslovaquie (Homoptera, Aleyrodinea). Acta Faunistica Entomologica Musei Nationalis Pragae 7: 61-80.

Zahradnik J (1962). Donnees taxonomiques et faunistiques sur *Japaneyrodes* nov. gen. *similis europaeus* n. ssp. (Homoptera: Aleyrodinea). Acta Faunistica Enfomologica Miisei Nationalis Pragae 8: 13-19.

Zahradnik J (1963). Aleyrodina. in Die Tiem'elt Mitteleuropas (N.S.) 4: 1-19.

Zahradnik J (1991). Taxonomisches und Faunistisches über uropäische Mottenläuse (Aleyrodinea). Acta Universitatis Carolinae 35: 111-118.

Density-dependent phenotypic plasticity in body coloration and morphometry and its transgenerational changes in the migratory locust, *Locusta migratoria*

Amel Ben Hamouda[1,3*], Seiji Tanaka[2], Mohamed Habib Ben Hamouda[3] and Abderrahmen Bouain[1]

[1]Life Sciences Laboratory, Faculty of Sciences of Sfax, Route de la Soukra km 3.5 - B.P. n° 1171 – 3000, Sfax, Tunisia.
[2]Locust Research Laboratory, National Institute of Agrobiological Sciences at Ohwashi (NIASO), Tsukuba, Ibaraki 305-8634, Japan.
[3]Plant Protection laboratory, High Institute of Agronomy of Chott-Mériem, 4042 Chott Mériem, Sousse, Tunisia.

Migratory locust, *Locusta migratoria* (Orthoptera, Acrididae) changes phase in response to population density. By rearing nymphs from a solitarious (isolated-reared) and gregarious line at three different densities, we examined the effects of rearing density on body coloration and morphometry at the last nymphal instar, adult stage and hatchlings of the subsequent generation. Changes in density lead to phase transformation shown by a shift in the body coloration and size to either direction depending on population density. Nevertheless, the complete shift of solitarious locusts to gregarious phase cannot be acquired in the first generation of crowding and solitarious body coloration still appears even at high density (100 locusts/cage). In both phases, the shift of body colour in response to the variation of population density was more rapid than the morphometry. Adult rearing density affected also the progeny body colour and size. However, the parental prehistory and the environmental conditions of the offspring were also important to modify the phase characteristics. The most important gregarious characteristics of hatchlings (black body colour and large size) were observed if parents were maintained at the density of 40 per cage. These characteristics depend not only on parents phase state but also depend on the food abundance. We found a positive correlation between the darkness of body colour and size of hatchlings and this effect was more pronounced in gregarious line.

Key words: *Locusta migratoria*, rearing density, phase characteristics, progeny, parents' prehistory.

INTRODUCTION

The migratory locust, *Locusta migratoria,* and the desert locust, *Schistocerca gregaria,* both show density-dependent phase polyphenism, in which they transform reversibly between two extreme phases, solitary and gregarious, in response to population density (Chopard, 1938; Uvarov, 1966). This transformation involves graded changes in a number of characteristics, including behavior, colour, reproduction, development, morphometry and endocrine physiology (Albrecht, 1967; Dale and Tobe, 1990; Girardie, 1991; Pener, 1991). A series of intermediates (transient) exist between the two extreme phases during a transient period from one extreme phase to another. Individuals in low-density populations (the solitary phase) are characterized by a uniformly coloured body small eggs and hatchlings size, a relatively small ratio of fore-wing length to hind femur length and sedentary and solitary behaviors. Individuals in high-density populations (the gregarious phase) have a dark body colour, large eggs and hatchlings size, a large ratio of fore-wing length to hind femur length, and migratory and gregarious behaviours (Ben Hamouda et al., 2009; Dale and Tobe, 1990; Faure, 1932; Pener, 1991; Pener and Yerushalmi, 1998; Tanaka, 2006; Uvarov, 1921, 1966). Longer-term crowding of locusts results not only in increased behavioral tendencies for aggregation, but also in a change of morphology, both of which are

---

*Corresponding author. E-mail: ben.hamouda@yahoo.fr.

under hormonal control (Kraus and Ruxton, 2002). Maternal effects are sometimes interpreted as an important mechanism that allows a (presumably adaptive) phenotypic change in offspring depending upon the environmental cue perceived by the parents. In other words, female parents might adjust the phenotypes of their offspring in response to certain cues that predict the environment her offspring will encounter, in a way that enhances offspring fitness (Bernardo, 1996). One of the most important features of phase changes in locusts is the fact that offspring exhibit some of the phase characteristics of their mother. Hence, phase change is a cumulative process that is transmitted from one generation to the next (Albrecht et al., 1959; Faure, 1932; Papillon, 1960; Tanaka and Maeno, 2008; Uvarov, 1966). Such carry-over from one generation to the next of phase characteristics such as body colour, morphometry, body weight and mature oocyte numbers has been amply documented in the literature (Albrecht et al., 1959; Gunn and Hunter-Jones, 1952; Hunter-Jones, 1958; Papillon, 1970). The mechanism(s) of trans-generational transfer of behavioural phase state has been studied (Bouaïchi et al., 1995; Hägele and Simpson, 2000; Islam et al., 1994a, b; McCaffery et al., 1998; Simpson et al., 1999). Variation in these traits is continuous and not discrete, and transgenerational steps proceed toward the gregarious phase for several generations (Uvarov, 1966). Under both in the field and laboratory, crowding induces characteristics of the gregarious phase, whereas isolation promotes those of the solitary phase. However, phase transformation is a complex phenomenon. Some phase characteristics change within hours, whereas others take longer and show changes in the next stadium, or even in the next generation (Pener, 1991; Uvarov, 1966, 1977).

Most studies concerning the effects of density on phase characteristics and their transgenerational changes have been carried out in *S. gregaria*. Therefore, in the present study, the effects of population density on phase characteristics in last instar nymphs, adults and their progeny were examined in *L. migratoria*.

MATERIALS AND METHODS

Insects

Experiments were carried out with solitary and gregarious colony of *L. migratoria*. This strain was maintained for more than 6 years at the High Institute of Agronomy of Chott-Mariem, Tunisia. The colony was brought to the National Institute of Agrobiological Sciences at Ohwashi, since the summer of 2009 by permission from the Yokohama Plant Protection Station. Preliminary analysis of mitochondria DNA suggests that this strain belongs to the southern clade of this species that includes strains from Africa, France, Australia, Timor, Southern China and Southern Islands of Japan, and is distinctly different from the northern clade that includes strains of most areas of Japan and China (Tokuda and Tanaka, unpublished data). The crowd-reared colony was maintained at 30°C in large wood-framed cages (42 cm × 22 cm × 42 cm; 0.038 m³). Each cage was covered with nylon screen mesh except for the wood floor and the front sliding door, which was composed of a transparent acrylic plate. Locusts were fed cut grass inserted in

water jars and wheat bran. Grass was changed every 1 or 2 days. *Bromus* grass was grown in crop fields by the Field Management Department of the National Institute of Agrobiological Sciences at Ohwashi. Locusts lay egg pods in moist sand held in plastic cups (volume of 380 ml) which were incubated at 30°C.

Rearing methods

To examine the effect of rearing density on phase characteristics, hatchlings obtained from both crowd-reared and isolated-reared colonies were reared at densities of 20, 40 and 100 individuals per cage in a well-ventilated room (30°C) until their adult stage. Egg pods collected from adults were incubated at 30°C. Egg length was measured with an ocular micrometer in a binocular microscope and hatchlings were checked for their body weight and colour.

Scoring of locust body colour

To quantify the density effects on body colouration, locusts were photographed using a scanner (Epson GT-X770, Japan) connected to a computer using commercial software, Photoshop 7.0 (Adobe Systems Incorporated, San Jose, CA). Locusts were chilled on ice for 15 min and placed with one side down on the glass table of the scanner for photographing. The image type used was 48-bit colour at a resolution of 1200 d.p.i. for hatchlings and 600 d.p.i. for last instar nymphs and adults. Hatchling body colour was observed 6 to 12 h after hatching. Hatchlings were divided into three colour groups based on the method of Hamouda et al. (2009): whitish (W), grey (G) and black (B). The last instar nymphs body colour was categorized by five colours: four graded of beige body colour according to the darkness and without or with black patterns, respectively (B1) and (B2, B3 and B4) and green body colour (G). Adults were subdivided to five colours: beige without black patterns (B1), beige with black patterns (B2), black (BL), grey (G) and yellow (Y). Statistical differences were analysed by multiple comparison using Student–Newman–Kuels test (p ≤ 0.05) (SPSS.10 software).

Measurements of morphometric ratios

Electronic sliding callipers (accuracy 0.01 mm) were used to measure the following classical morphometric phase characteristics for last instars nymphs and adults in order to determine E/F, F/C and H/P ratios: E = length of forewing, F = length of hind femur, C = maximum head width, H = maximal height of pronotum and P = length of pronotum, (Joly, 1968; Pener and Yerushalmi, 1998; Uvarov, 1966). The statistical analysis was carried out using a multiple comparison using Student–Newman–Kuels test (p < 0.05) and discriminant analysis (SPSS.10 software).

RESULTS

Effects of different rearing densities on body colouration at the last nymphal instar and the adult stage in nymphs from gregarious and solitary lines

Figure 1 shows that rearing gregarious hatchlings in groups of 20 led to the appearance of solitary body colouration (B1) in 11% of the last instar nymphs (8% in the case of females (A) and 3% for males (B)). No solitary body colouration was observed at a density of 40 or

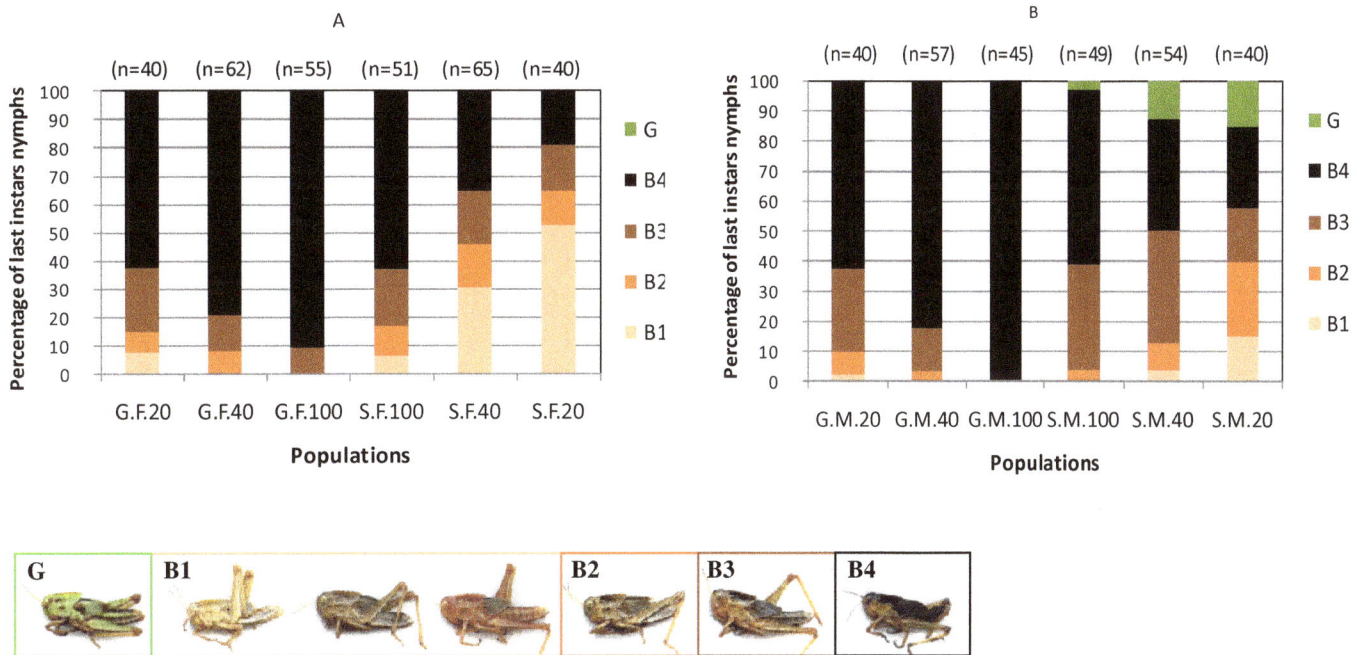

Figure 1. Effects of nymphal rearing density on body colouration at the last nymphal instar in *L. migratoria* individuals from a gregarious and solitarious line. (A) G.F/S.F: gregarious/solitarious females; (B) G.M/S.M: gregarious/solitarious males; 20, 40 and 100: the number of locusts/cage. Numbers in parentheses (n) indicate the number of individuals.

100 gregarious nymphs /cage. The intermediate body colouration (B2 and/or B3) appeared in different proportions at the three densities of gregarious females (Figure 1A). However, half of the total proportion of this colouration was provided in G.F.20. No intermediate body colouration was observed at the density of 100 gregarious males per cage (Figure 1B). Gregarious body colouration (B4) was observed at the three densities in nymphs from gregarious lines depending on population densities.

Figure 1 shows that even under crowded conditions, nymphs from solitary line conserved their solitary body colouration (B1 and/or G) at the last stage. The highest proportion of solitary body colouration (53% in the case of females (A) and 30% for males (B)) was observed at a density of 20 nymphs. At high density (100 individuals /cage), only 7 and 4% of individuals developed solitary body colouration for females and males, respectively. Intermediate body colouration was distributed almost equally at the three densities and the highest proportion of gregarious body colouration was found at a density of 100 nymphs /cage.

Figure 2A shows that in female adults, the darkness of body colour was correlated positively to an increase in the number of individuals per cage. In females from solitary lines, green colour did not appear when the locusts were reared in a group of 100. In males (Figure 2B), yellow colour, typical of gregarious phase, was observed at different densities in differential proportions. The

highest proportion of males with yellow body colour (75%) was obtained at a density of 40 locusts per cage. In adults from solitary lines (Figure 2B), a density of 100 produced the highest proportion of yellow males (74%), but also yielded individuals with green colour (8%).

Effects of different rearing densities on morphometric ratios at the last nymphal instar and the adult stage in nymphs from gregarious and solitary lines

The variation in rearing density did not affect the length of femur (F) and the height of pronotum (H) of nymphs from gregarious line (Figure 3). In nymphs from solitary line, rearing density resulted in significant differences in F and H; they increased considerably when the population density decreased, although this was observed only in females (Figure 3A and C). In the case of males, we found that two groups were statistically different by the height of pronotum; the first group of gregarious nymphs was characterized by a low H and the second contains solitary nymphs with high H (Figure 3D).

Discriminant analysis was performed on different population densities for E/F, F/C and H/P ratios. In the case of females (Figure 4A), the first discriminant factor exhibited 90.8% of the total variability. It separates between two groups; the first includes gregarious females while solitary ones were contained in the second group.

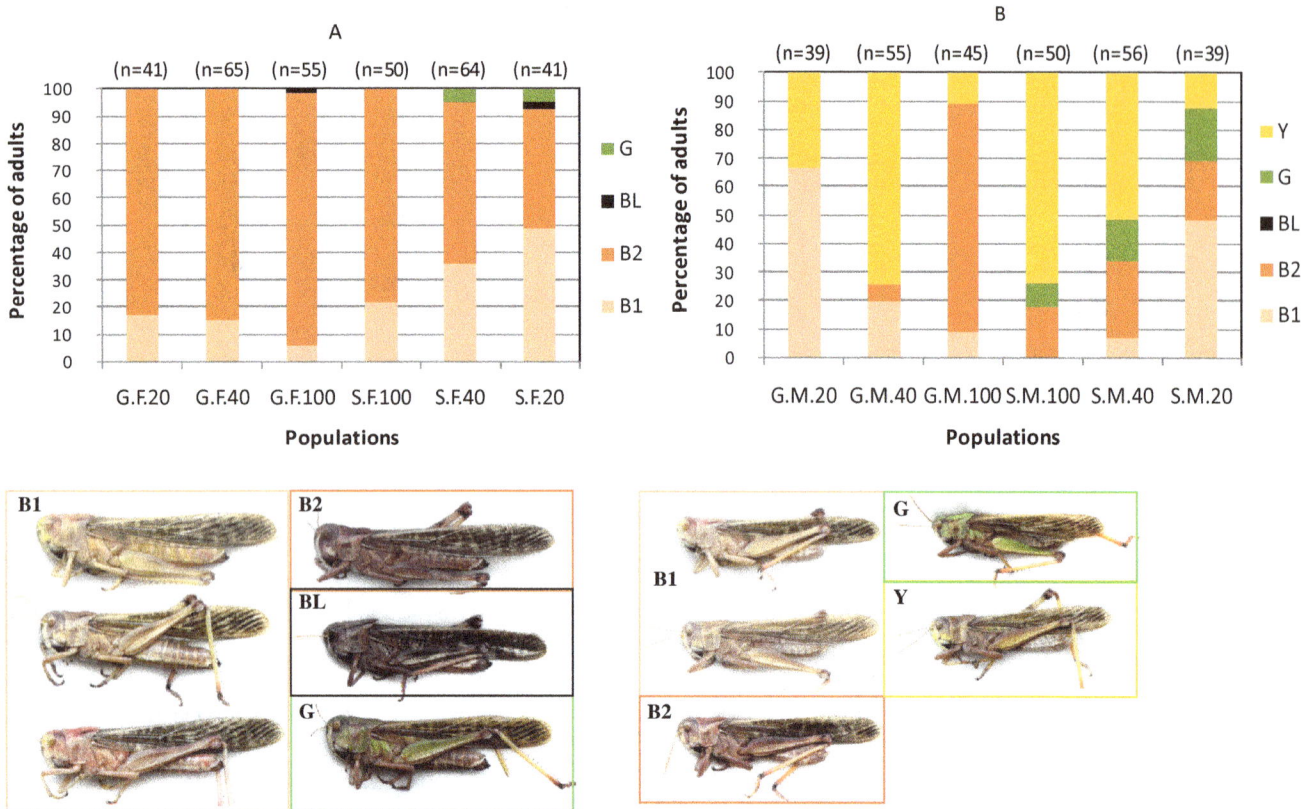

Figure 2. Effects of nymphal rearing density on body colouration at the adult stage in *L. migratoria* individuals from a gregarious and solitarious line. (A) G.F/S.F: gregarious/solitarious females; (B) G.M/S.M: gregarious/solitarious males; 20, 40 and 100: the number of locusts/cage. Numbers in parentheses (n) indicate the number of individuals.

This factor was correlated negatively to the F/C ratio, on the side of the solitary groups. This means that solitary females have the highest F/C ratio. The same factor was positively correlated with E/F ratio towards gregarious females. In the case of males (Figure 4B), the first discriminant factor explains 90% of the total variation of the complete data sets. It was related positively to F/C on the side of solitary groups and negatively to E/F on the side of gregarious males. It appears that morphometrical segregation was established according to the origin of populations studied (gregarious or solitary phase).

Effects of parental rearing density on hatchlings phase characteristics

Egg length was compared among eggs laid by gregarious female adults maintained at different rearing densities (Figure 5). The mean egg length at a density of 20 locusts /cage was 5.71±0.4 mm; (n = 595), which was significantly shorter than those produced at a density of 100 (5.85±0.3 mm; n = 437) and 40 locusts /cage (6.12±0.4 mm; n = 442). Females which were derived from solitary line and reared at a density of 40 locusts

/cage produced the smallest eggs (5.17±0.3 mm; n = 492).

Concerning the body weight of hatchlings, Figure 6 shows that the density of 40 gregarious locusts /cage yielded the largest hatchlings (1.51±0.2; n = 139) in comparison with the density of 20 (1.32±0.2; n = 122) that showed the smallest hatchlings and the density of 100 (1.39±0.2; n = 155). It seems that the density of 40 locusts /cage was the best to induce large hatchlings. The difference was not significant between solitary locusts maintained at 20 and 40 locusts /cage which had 1.19±0.2; n = 95 and 1.21±0.1; n = 165 respectively. The smallest hatchlings were obtained from a density of 100 (1.14±0.1; n=113) but statistically, the difference was not significant.

In the same purpose, it seems that the density of 40 individuals/cage induces the highest proportion of black hatchlings (Figure 7). For both gregarious and solitary phase, there was a positive correlation between hatchlings body weight and their body colouration except for hatchlings from solitary females maintained at 100 (Table 1).

No correlation was found between eggs length and mother's age at deposition except in gregarious females

A

F (mm)   (n=40)   (n=57)   (n=45)   (n=49)   (n=54)   (n=40)

Populations

B

F (mm)   (n=40)   (n=57)   (n=45)   (n=49)   (n=54)   (n=40)

Populations

C

H (mm)   (n=40)   (n=62)   (n=55)   (n=51)   (n=65)   (n=49)

Populations

D

H (mm)   (n=40)   (n=57)   (n=45)   (n=49)   (n=54)   (n=40)

Populations

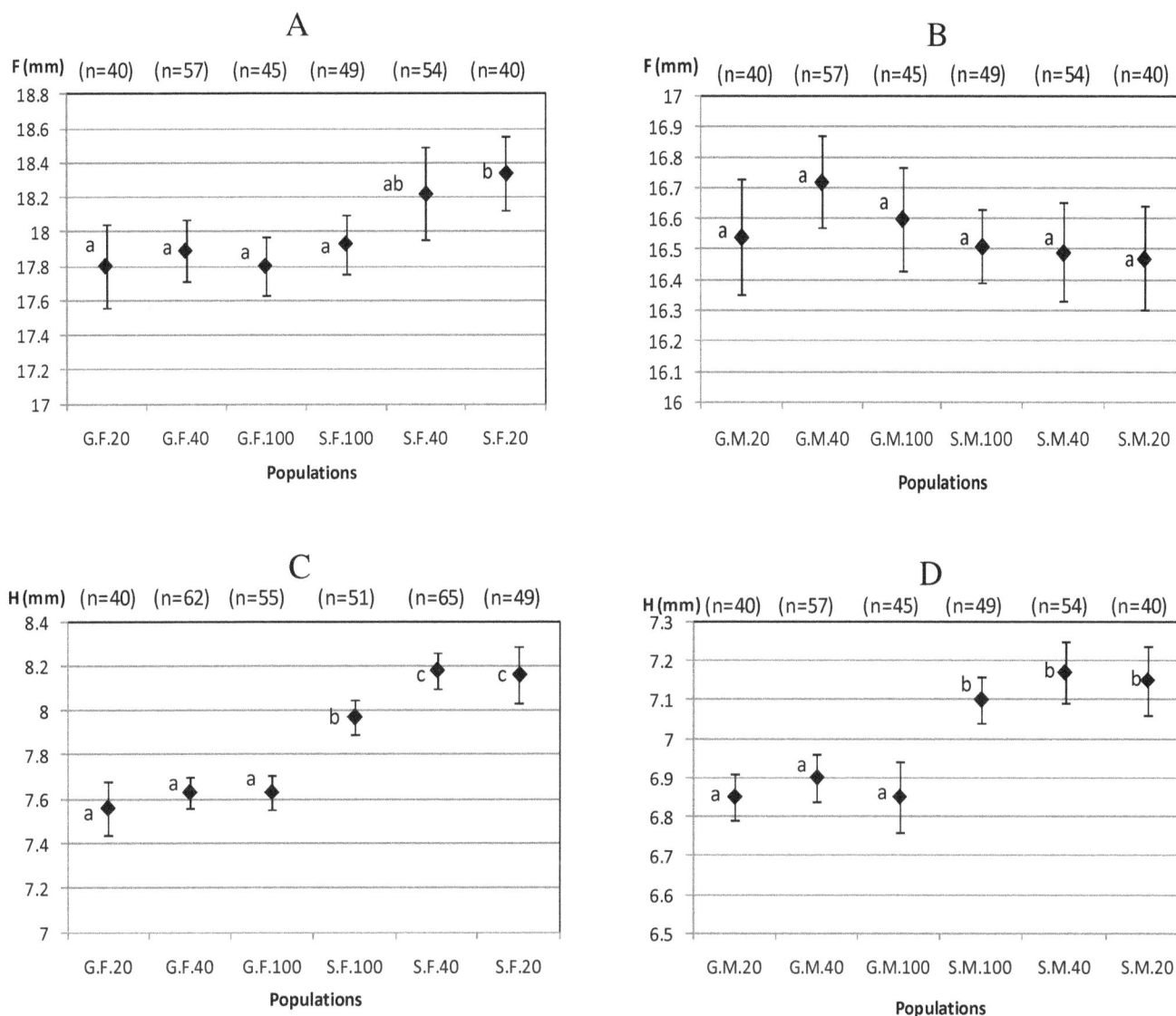

Figure 3. Effect of crowding density of hatchlings on gregarious and solitarious last instar nymphs morphometry. (A): F for females; (B): F for males; (C): H for females; (D): H for males; G.F/S.F: gregarious/solitarious females; G.M/S.M: gregarious/solitarious males; H: maximal height of pronotum; F: length of hind femur. Numbers in parentheses (n) indicate the number of individuals. Different letters in each panel indicate a significant difference among H and F values at $p \leq 0.05$ by Student–Newman–Kuels test.

maintained at the densities of 20 ($r = -0.438$; $p \leq 0.05$) and 40 ($r = -0.171$; $p \leq 0.05$) that tended to produce smaller eggs as the female adults grew older (Figure 8). Figure 9A shows that gregarious females maintained at low density (20 individuals/cage) produce small hatchlings with age ($r = -0.263$; $p \leq 0.05$). While gregarious females maintained at 100 individuals/cage and solitary females maintained at 40 locusts/cage (Figure 9B) tend to produce bigger hatchlings, respectively ($r = 0.225$; $r = 0.361$; $p \leq 0.05$).

DISCUSSION

In the present study, we examined the effect of population density by rearing gregarious and solitary hatchlings at three conditions: 20, 40 and 100 individuals /cage. The target of this study was to establish a system by which we could understand the relationship between the population density, phase changes and the trans-generational changes of phase characteristics in the migratory locust, *L. migratoria*. Our experiments confirmed the finding that the shift of various characteristics from one phase to another was density dependant and this effect was transgenerational in both phases. However, most studies concerning this phenomenon were done on *S. gregaria*. Consequently, we are interested in analyzing more of the subsequent occurrence on *L. migratoria*.

As shown previously by Hamouda et al. (2009) and

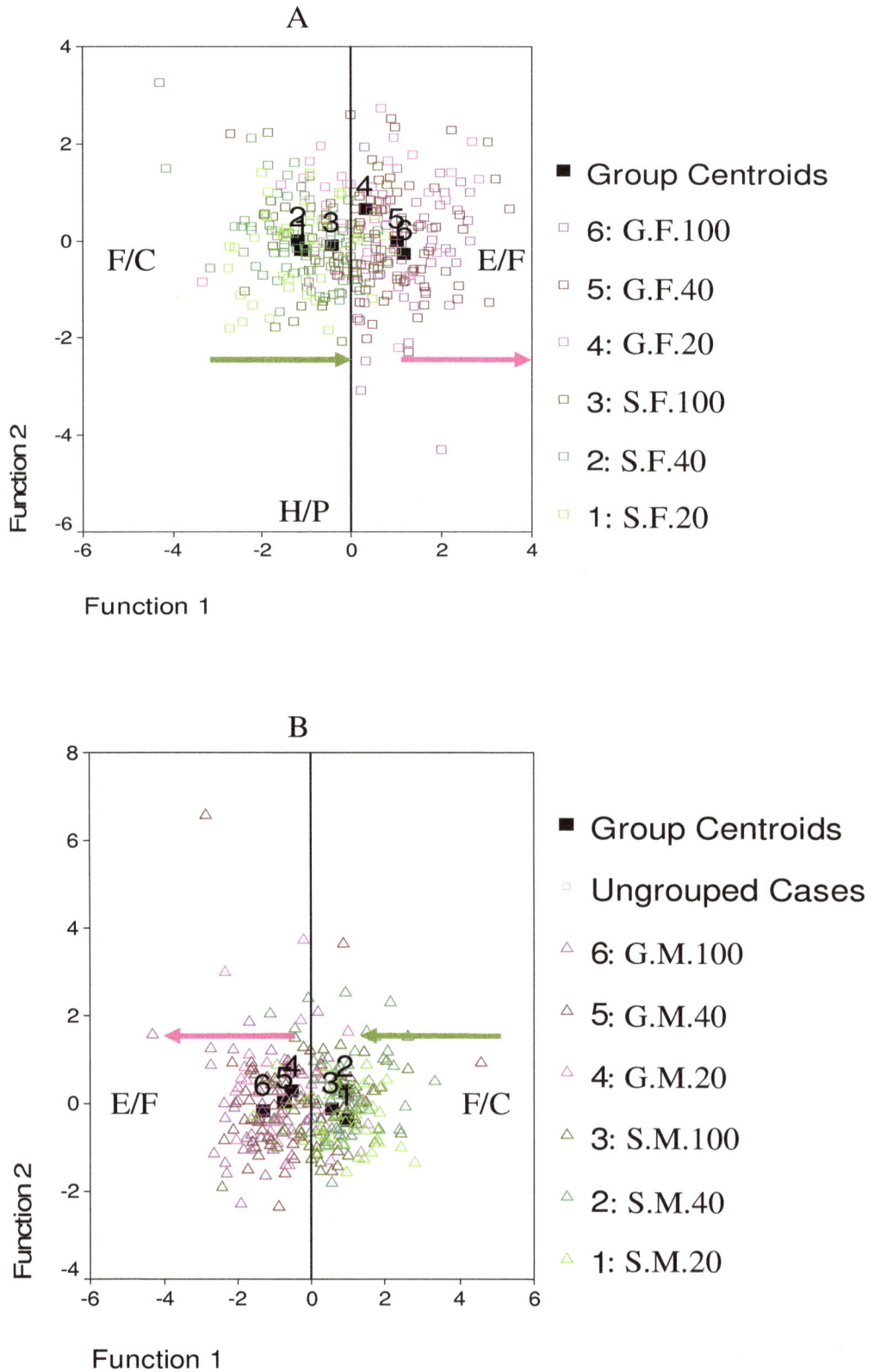

Figure 4. Main variables distribution measured in response to different population densities of females (A) and males (B) harvested in the factorial plan defined by the discriminant analysis. G.F/S.F: gregarious/solitarious females; G.M/S.M: gregarious/solitarious males. 20, 40 and 100: the number of locusts/cage. Arrows show the direction of the evolution of gregarious and solitarious populations in response to the increase of the number of locusts/cage.

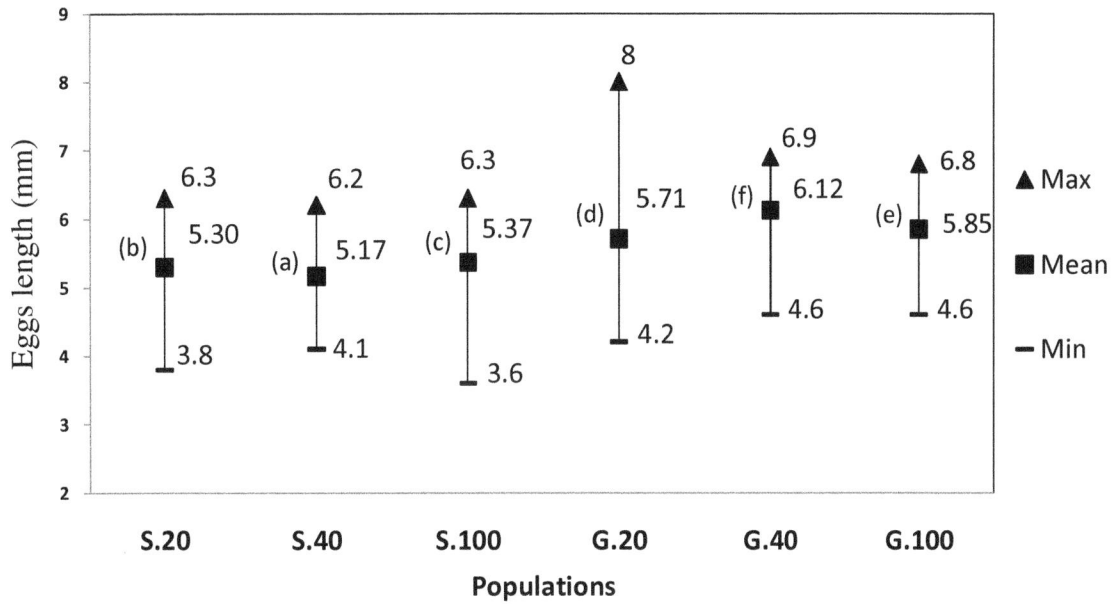

Figure 5. Effect of parental densities on solitarious and gregarious eggs length. Different letters in each panel indicate a significant difference among eggs length at p ≤ 0.05 by Student–Newman–Kuels test.

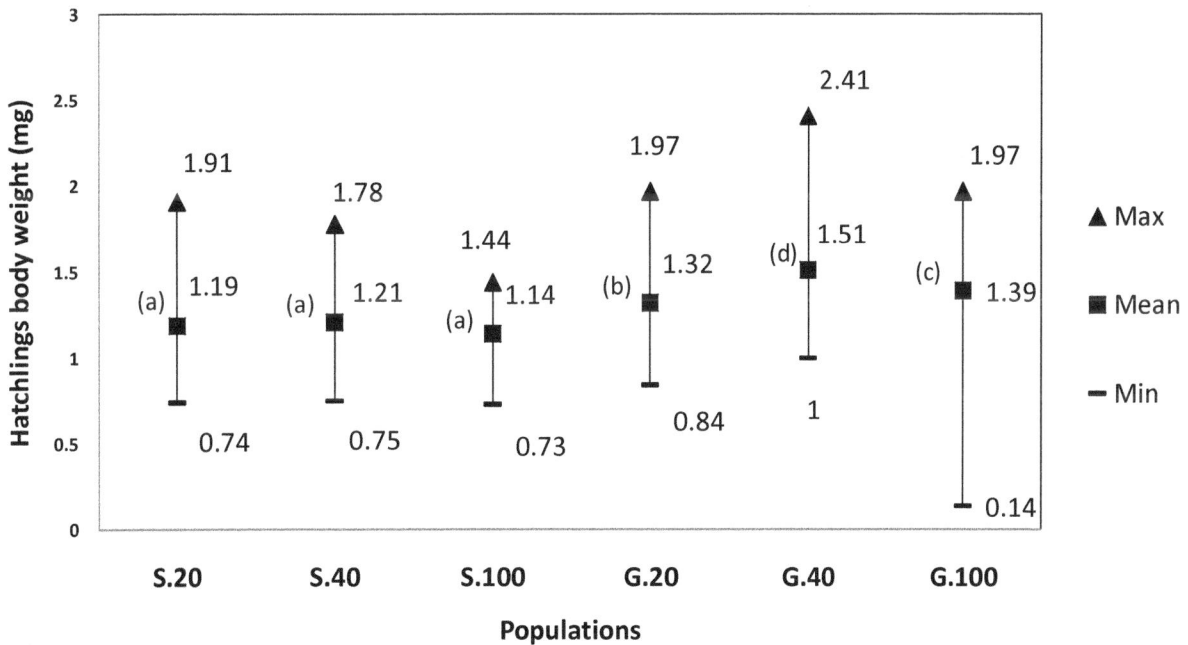

Figure 6. Effect of parental densities on solitarious and gregarious hatchlings body weight. Different letters in each panel indicate a significant difference among eggs length at p ≤ 0.05 by Student–Newman–Kuels test.

Gullan and Cranston (2010) by splitting a single locust egg pod into two: rearing the offspring at low densities induces solitary locusts, whereas their siblings reared under crowded conditions develop into gregarious locusts. The body colouration was one of the most rapid phase polyphenism characters that changes depending on the population density, and the development of black pigmentation in response to change in density is widespread in grasshoppers (Applebaum and Heifetz, 1999, Pener, 1991, Rowell, 1967, 1971).

In the present experiments, solitary nymphs that were whitish at the early stages tended to exhibit an orange body colour with black patterns in last nymphs stadium. Indeed, at a density of 20 locusts /cage, 20% of females

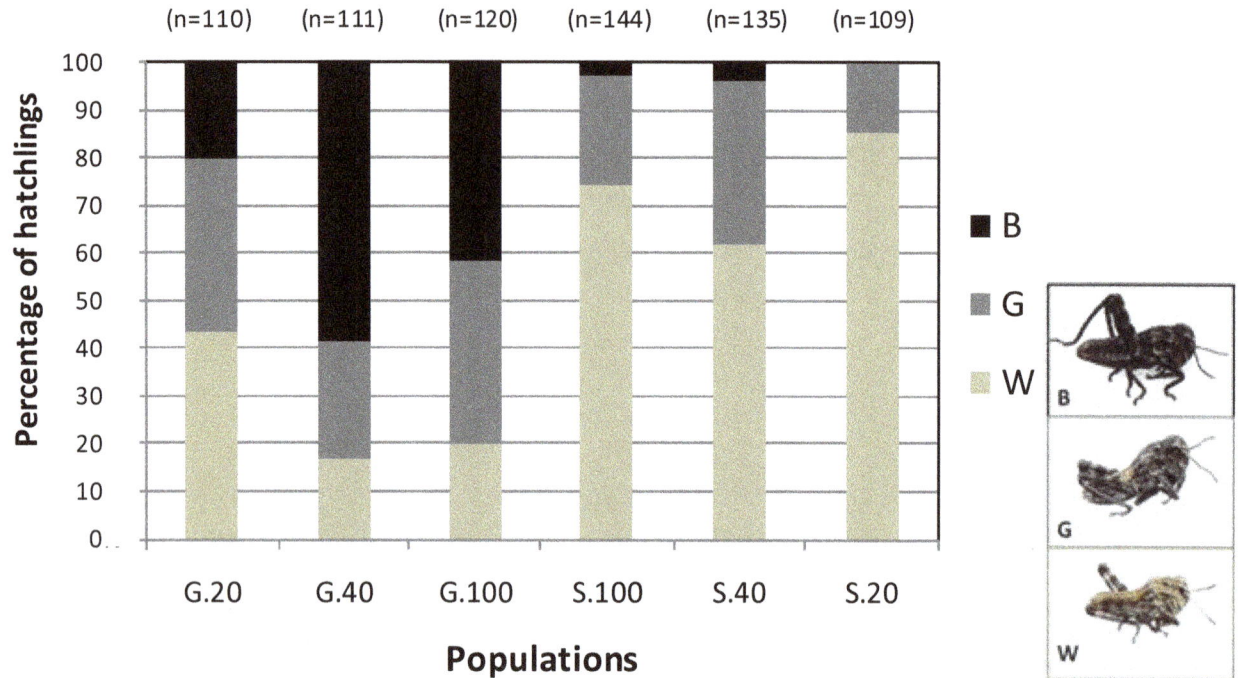

Figure 7. Effect of parental densities on solitarious and gregarious hatchlings body colour. B: black; G: grey; W: whitish body colour of hatchlings. Numbers in parentheses (n) indicate the number of hatchlings.

Table 1. Matrix structure of F/C, E/F, and H/P ratios in discriminant analysis for gregarious and solitarious populations maintained at different densities for each sex.

Morphometrical ratio		Discriminant factors	
		1	2
	F/C	-0.885*	0.071
Females	E/F	0.783*	0.579
	H/P	0.349	-0.716*
	F/C	0.958*	0.147
Males	E/F	-0.631*	0.464
	H/P	-0.238	0.517

* Largest absolute correlation between each morphometrical ratios and any discriminant factor.

and 28% of males shifted their body colour from solitary to the gregarious phase. From these observations, we concluded that in laboratory conditions, a density of 20 solitary nymphs per a volume of 0.038 m^3, in other words, 526 solitary nymphs per 1 m^3 were sufficient to induce the gregarious body colouration in some nymphs since the first generation of crowding.

The response of body colouration to the increase of population density was observed on other species of locusts; in the desert locust, *S. gregaria*, a wide spectrum of body-colour variation has been found depending on the population density (Gunn and Hunter-Jones, 1952; Hunter-Jones, 1958; Maeno and Tanaka, 2006; Stower, 1959; Tanaka and Maeno, 2010). Duck (1944) found that

solitary larvae of *Schistocerca obscura* were green, but when reared two or more per cage were brown with more or less dense black markings. Larvae of *Spodoptera littoralis* which were reared in crowded condition produced darker coloured larvae and when they were reared in isolation they were pale in colour (Hodjat, 1970). 7 and 4% of solitary body colour, in females and males respectively, were observed when solitary nymphs were maintained at the density of 100. This result indicates that one generation of crowding solitary nymphs at high density (100 nymphs/ cage) was insufficient to the complete shift to the gregarious body colouration and the sensitivity of solitary nymphs to crowding was variable among individuals.

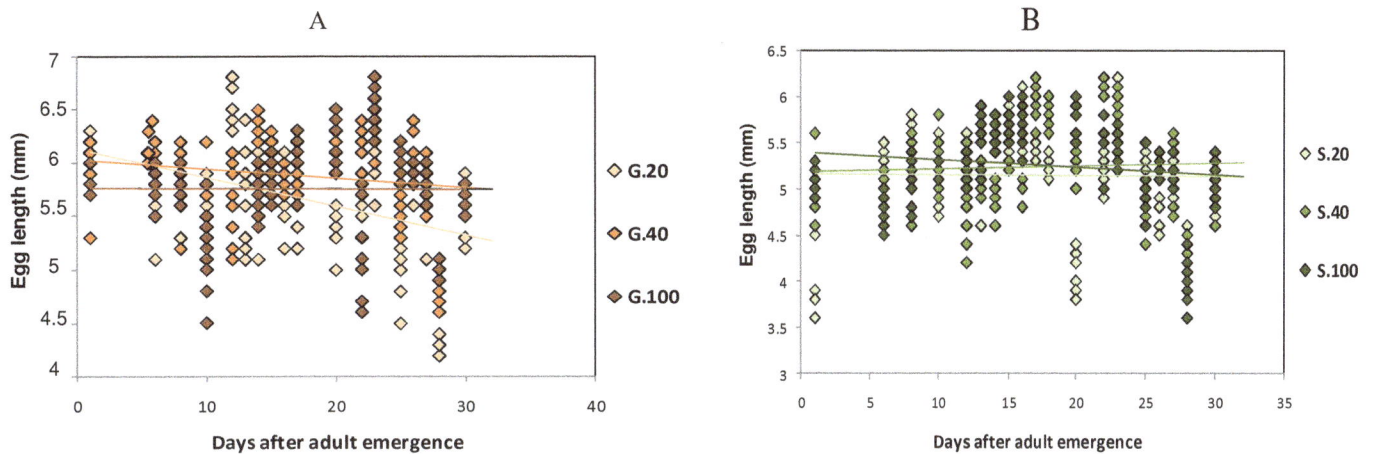

Figure 8. Temporal variation in eggs length of gregarious (A) and solitarious (B) females under the variation of population density.

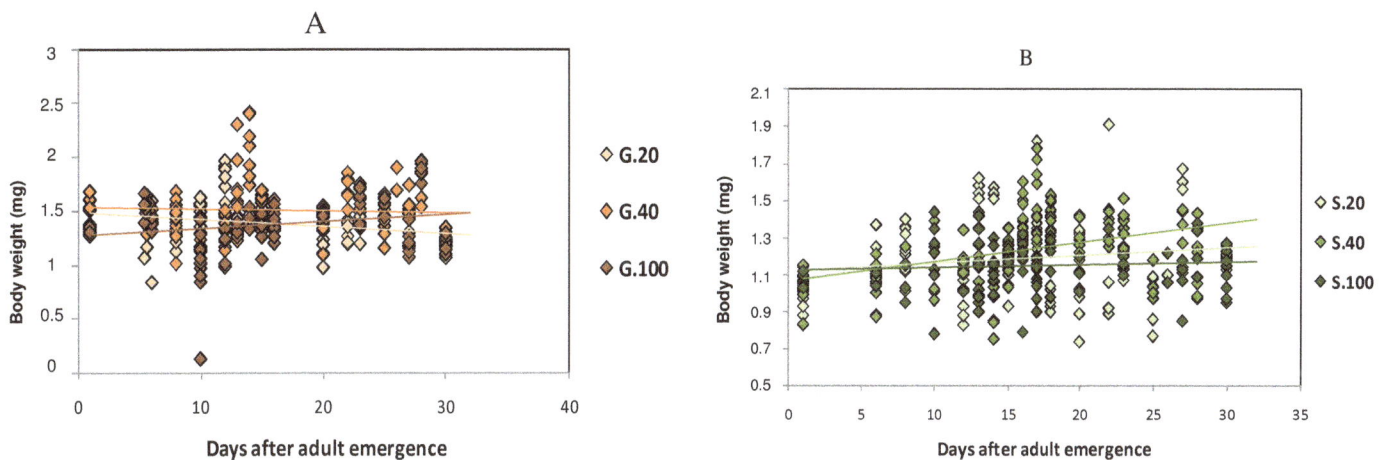

Figure 9. Temporal variation in hatchlings body weight of gregarious (A) and solitarious (B) females under the variation of population density.

Rearing gregarious hatchlings at a low density (20 individuals /cage) resulted in the appearance of some individuals (11%) with solitary body colouration at the last nymphal instar. It means that maintaining a number of 20 gregarious nymphs per cage induces the solitarioussness of some nymphs but not the whole number. Furthermore, the shift to the solitary colouration was observed since the second nymphal stadium. We suppose that the body colour was influenced by the environmental conditions experienced during the previous stadium. This was in concordance with notes claimed by Cockrell (1933) in which he claimed that body colour seems to be inheritance when the first stage larva from *gregaria* would show the dark colouration. But this was due to the "locustine" in the egg, and disappeared in the second stage if the insect was reared alone. This it was observed also in *S.gregaria* (Hunter-Jones, 1958; Injeyan et al., 1979; Pener, 1991; Pener and Simpson, 2009; Uvarov, 1966).

At 40 gregarious nymphs/cage, solitary colouration

disappears completely. These results mean that the shift from gregarious phase to the solitary one starts when the density of gregarious nymphs decreases less than 20 into a volume of 0.038 m^3. On other hand, rearing at high density (100 individuals/cage) consistently produced high proportion of gregarious body colouration in the late nymphal stage. These results were consistent with previous studies related to the body colour polyphenism expressed in response to the population density (Faure, 1932; Pener, 1991; Stower, 1959; Uvarov, 1966) and with studies made by Maeno and Tanaka (2008) on *S. gregaria*.

In solitary *L. migraotria* adults females, the proportion of gregarious body colouration (B2+BL) increased by 32% and the proportion of yellow (Y) males by 61% when the density of nymphs increased from 20 to 100 solitary individuals/cage. At a density of 100 individuals /cage, the green body colouration did not exceed 8% in solitary males, while it was 18% when the locusts were reared at a density of 20 individuals /cage. It was clearly suggested that the body colouration was positively correlated to the rearing density and the phase state of locusts cannot still stable when individuals were exposed to the variation in density.

At a density of 20 solitary individuals /cage, and if we start by a 100% solitary body colouration at hatchling stage, 54% of solitary colouration in females and 67% in the case of males, remain at the adult stage. At 100 locusts, 22% of solitary colour persists in females and only 8% in the case of males. It seems that at low density, the response of body colouration of solitary adults to crowding was more rapid in the case of females than males. Contrariwise, at high density, the response of males was more pronounced.

Morphometric characteristics seem less labile than the body colouration to the variation of the population density and no significant changes were observed for H and F parameters in gregarious nymphs at all rearing densities tested. This might because the individuals already had reached a gregarious phase by virtue of being reared under crowded conditions for many generations prior to the start of distributed them into different densities. We concluded that one generation of rearing gregarious nymphs at a density of 20 per cage was insufficient to allow them to change their morphometry to the solitary phase.

In the case of solitary nymphs, a significant difference was observed at a density of 100 females/cage; their F and H lengths were decreased significantly showing the shift from solitary to the gregarious phase when they were maintained at the density of 100. On the contrary, males could not change their morphometrical state since the first generation of crowding even at very high rearing density. It was clear that solitary females were more sensitive to crowding than males.

Discriminant analysis of morphometrical ratios of adults showed that, mainly, the switch of body size depends essentially on population densities; there was a clear discrimination between the gregarious and the solitary populations. Inside each group, the segregation was performed depending on the population density; in solitary females and males, as the number of locusts per cage increases, as the solitary populations brought closer to the gregarious locusts and their F/C ratio decreases. In the case of gregarious populations, as the number of individuals per cage increases, they rolled away from solitary populations and their E/F ratio increases.

It has been proved that adult density influences the progeny characteristics in locusts (Hunter-Jones, 1958; Injeyan and Tobe, 1981; Maeno and Tanaka, 2008; Norris, 1950, 1952; Pener, 1991; Uvarov, 1966). However, most experiments were carried out on *S. gregaria*. What is the relationship between the prehistory of parents and the impact of rearing density on progeny? The present study demonstrated that maternal crowding influences progeny size in *L. migratoria* and the mother's rearing conditions shifts the phase state of her hatchlings accordingly: gregarious females tend to produce smaller

eggs and hatchlings and high proportion of whitish hatchlings at low rearing density (G.20). But at high density, food competition as one of the rearing conditions parameters may control the progeny size. This was observed at a density of 100 gregarious individuals /cage since the eggs and hatchlings size becomes smaller and the proportion of black hatchlings reduced by 16.89% from the density of 40 to the density of 100 individuals /cage. It seems that the suitable density of gregarious locusts to produce big eggs and large and black hatchlings was that of 40 individuals /cage. This means that the transmission of gregarious phase characteristics to the progeny in gregarious locusts was influenced not only by the population density, but also by the characteristics of rearing conditions, especially the food abundance.

Solitary females reared at a high density (100 individuals /cage) tend to produce larger eggs (5.37 mm) than the two others solitary lines but these eggs were smaller than the gregarious ones which among them the smallest mean was 5.71 mm. From these solitary eggs hatched small and whitish hatchlings. It seems that at high population density even they tended to produce larger eggs, solitary mothers yield small and whitish hatchlings. It was clear that in solitary phase, the phase characteristics of the progeny depends more to the prehistory and the phase state of parents than the population density. We concluded from this that solitary locusts are very sensitive to crowded conditions and food abundance appears to be important in the determination of eggs and hatchlings size and colour. This parameter may represent a limiting factor to produce large-sized hatchlings. Consequently, rearing density may play a role in the induction of the enlargement or the miniaturization of eggs and hatchlings indirectly by affecting their parents' size. In fact, it has been hypothesized by Tanaka and Maeno (2010) that crowding stimulus perceived is likely to be transmitted to the brain which in turn may cause some factor controlling egg size to be released. One possibility is that this factor is released in response to high adult density and acts on the ovary to increase egg size. Alternatively, it is released in response to low adult density and stimulates the ovary to reduce egg size. It has been shown also that adult females of *Locusta* from larvae reared in isolation weighed 1.5 g while others from larvae reared in crowds weighed only 1.2 g (Chapman, 1998). These findings indicate clearly that size of the parents may be influenced by rearing density in which their eggs and hatchlings size depend.

As shown in Table 1, there was a strong positive correlation between the darkness of body colour and size of hatchlings produced by gregarious line. The same conclusion was founded previously in the case of *S. gregaria* by Maeno and Tanaka (2009). Additionally, set apart the effect of phase, the larger the hatchling body size the larger the degree of darkening and this correlation is consistently observed in the case of gregarious as that of solitary phase. We found in 729 gregarious and

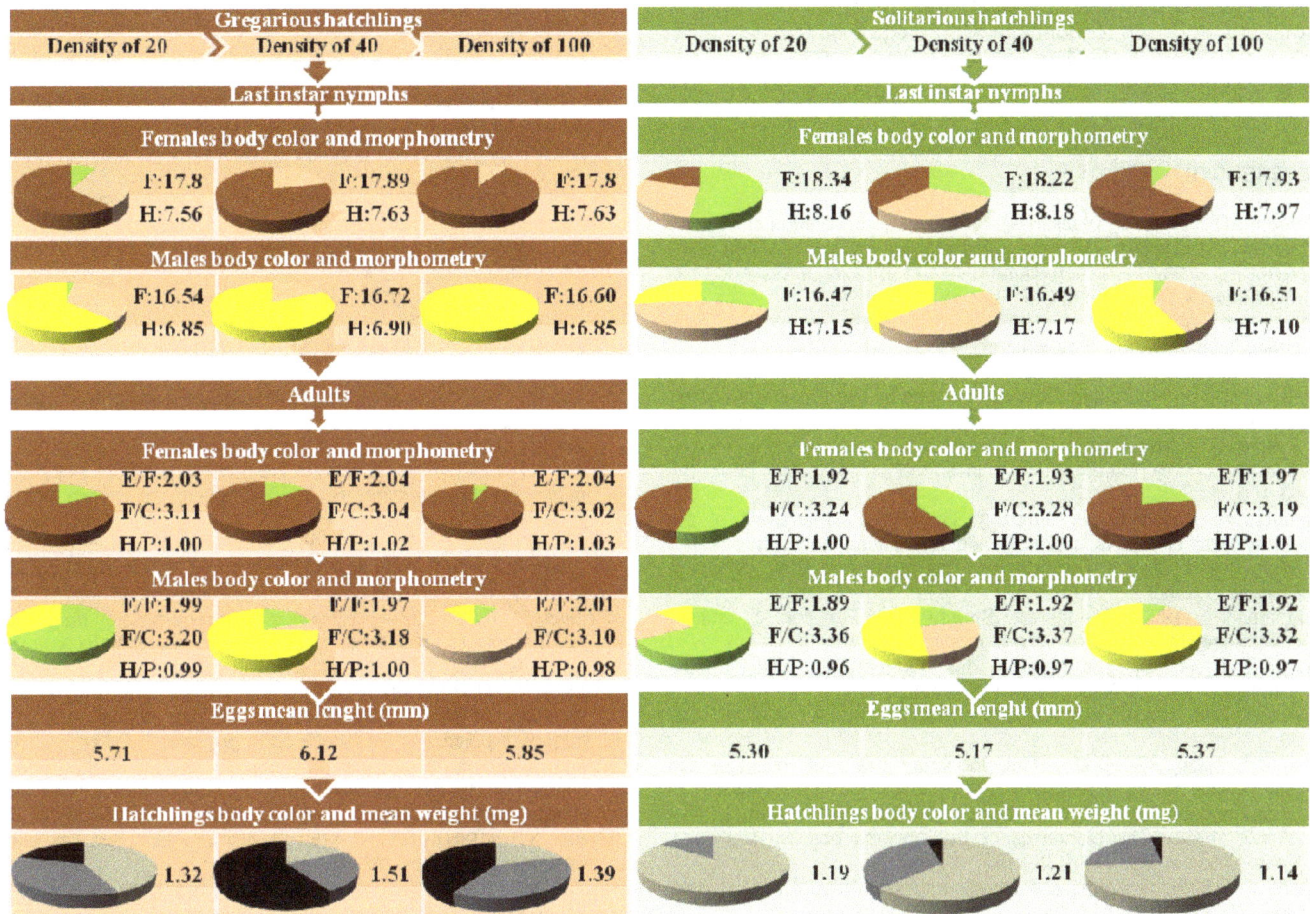

Figure 10. A synopsis showing the effect of the variation of population density on body color and morphometry of last instar nymphs, adults and their progeny. F: length of hind femur; H: maximal height of pronotum; E/F, F/C, H/P: morphometrical ratios of adults; ▭; solitarious body color; ▭ transient body color; ▬ females gregarious body color; ▭ : males gregarious body color; ▭ ; whitish, ▬ : grey and ▬ : black body color of hatchlings.

solitary hatchlings observed, significant differences in body weight between whitish, grey and black hatchlings. Among those hatchlings, the mean average of black hatchling (1.58 mg) was 1.35 times bigger than the mean average of whitish hatchling (1.17 mg). Whereas, grey hatchlings have a medium mean average of body size (1.34 mg). These observations were in concordance with those of Tanaka and Maeno (2008) in the case of *S. gregaria*; they found also that the correlation was observed even within the same egg pod that produces a mixture of green and black hatchlings.

Our results are very important for the study of locust phase polyphenism, especially the migratory locust, *L. migratoria*. In fact, a change of transgenerational of some phase characteristics in response to population density will help us to predict locust outbreaks. Figure 10 shows clearly the response of each stage to the variation of population density and the process of the transgenerational of the changes to the progeny.

ACKNOWLEDGEMENTS

We thank Ms. Totsuka, Ms. Yokota, Ms. Ikeda, Ms. Higuchi and for laboratory assistance. We are grateful to Dr. Koutaro Maoeno and Dr. Harano Ken-Ichi for kind advice and encouragement at the Locust Research Laboratory, National Institute of Agrobiological Sciences at Ohwashi, Tsukuba. The grass used was raised by Field Management Section of NIASO. This study was supported by the "Japanese Association of University Woman" Research Fellowships.

Abbreviations: W, Whitish body colour of hatchlings; **G,** grey body colour of hatchlings; **B,** black body colour of hatchlings; **B1, B2, B3** and **B4,** beige body colour of the last instar nymphs according to the darkness; **G,** green body colour of the last instar nymphs; **B1,** beige body colour of adults without black patterns; **B2,** beige body colour of adults with black patterns; **BL,** Black body colour of adults; **G,** grey body colour of adults; **Y,** yellow body colour of adults; **E,** length of forewing; **F,** length of hind femur; **C,** maximum head width; **H,** maximal height of pronotum; **P,** length of pronotum; **G.F,** gregarious females; **S.F,** solitarious females; **G.M,** gregarious females; **S.M,** solitarious male

REFERENCES

Albrecht FO, Verdier M, Blackith RE (1959). Maternal control of ovariole number in the progeny of the migratory locust. Nature. 184: 103-104.

Albrecht FO (1967). Polymorphisme phasaire et biologie des Acridiens migrateurs. Ed. Masson et Cie, Paris, p. 194.

Applebaum SW, Heifetz Y (1999). Density-dependent physiological phase in insects, Ann. Rev. Entomol, 44: 317-341.

Ben Hamouda A, Ammar M, Ben Hamouda MH, Bouain A (2009). The role of egg pod foam and rearing conditions on the phase state of the Asian migratory locust *Locusta migratoria migratoria* (Orthoptera, Acrididae). J. Insect. Physiol., 55: 617-623.

Bernardo J (1996). Maternal effects in animal ecology. Am. Zool., 36: 83-105.

Bouaïchi A, Roessingh P, Simpson SJ (1995). An analysis of the behavioural effects of crowding and re-isolation on solitary-reared adult desert locusts (*Schistocerca gregaria*) and their offspring. Physiol. Entomol., 20: 199-208.

Chapman RF (1998). The Insects: Structure and Function. 4th edition. Cambridge University Press, New York, p. 788.

Chopard L (1938). La biologie des Orthoptères. Encyclopédie entomologique. Ed. Paul Lechevalier, Paris,p. 478.

Cockerell TDA (1933). The Phases of African Locusts. Am. Nat., 67: 93-96.

Dale JF, Tobe SS (1990). The endocrine basis of locust phase polymorphism. In: Chapman, R.F., Joern, A. Ed. Biology of Grasshoppers. John Wiley and Sons, New York, pp. 393-414.

Duck LG (1944). The bionomics of *Schistocerca obscura* (Fabr). J. Kans. Entomol. Soc., 17: 105-119.

Faure JC (1932). The phases of locusts in South Africa. Bull. Entomol. Res., 23: 293-405.

Girardie A (1991). Régulation endocrinienne du développement, de la reproduction et du polymorphisme phasaire. La lutte anti-acridienne. Ed. AUPELF-UREF, John Libbey Eurotext, Paris, pp. 119-127.

Gullan PJ, Cranston PS (2010). Insects: An Outline of Entomology, 4th edition. Blackwell Science, 584p.

Gunn DL, Hunter-Jones P (1952). Laboratory experiments on phase differences in locusts. Anti-locust. Bull., 12: 1-29.

Hägele BF, Simpson SJ (2000). The influence of mechanical, visual and contact chemical stimulation on the behavioural phase state of solitary desert locusts (*Schistocerca gregaria*). J. Insect. Physiol., 46: 1295-1301.

Hodjat SH (1970). Effects of crowding on colour, size and larval activity of *Spodoptera littoralis* (Lepidoptera: Noctuidae). Entomol. Exp. Appl., 13: 97-106.

Hunter-Jones P (1958). Laboratory studies on the inheritance of phase characters in locusts. Anti-Locust Bull., 29: 1-32.

Injeyan HS, Tobe SS, Rapport E (1979). The effects of exogenous juvenile hormone treatment on embryogenesis in *Schistocerca gregaria*. Can. J. Zool., 57: 838-845.

Injeyan HS, Tobe SS (1981). Phase polymorphism in *Schistocerca gregaria*: reproductive parameters. J. Insect. Physiol., 27: 97-102.

Islam MS, Roessingh P, Simpson SJ, McCaffery AR (1994a). Effects of population density experienced by parents during mating and oviposition on the phase of hatchling desert locusts, *Schistocerca gregaria*. Proc. R. Soc. Lond., 257: 93-98.

Islam MS, Roessingh P, Simpson SJ, McCaffery AR (1994b). Parental effects on the behaviour and colouration of nymphs of the Desert Locust *Schistocerca gregaria*. J. Insect. Physiol., 40: 173-181.

Joly P (1968). Endocrinologie des insectes. Ed. Masson et Cie, Paris, p. 18.

Krause J, Ruxton GD (2002). Environmental effects on grouping behavior: Behavioural changes induced by crowding: the desert locust. *In*, Living in groups. Oxford Series in Ecology and Evolution. Ed, Oxford University Press Inc., New York, p. 128.

Maeno K, Tanaka S (2006). Effects of hatchling body colour and rearing density on body colouration in last stadium nymphs of the desert locust, *Schistocerca gregaria* (Forskal) (Orthoptera: Acrididae). Physiol. Entomol., 32: 87-94.

Maeno K, Tanaka S (2008). Maternal effects on progeny size, number and body colour in the desert locust, *Schistocerca gregaria*: density- and reproductive cycle- dependent variation. J. Insect. Physiol., 54: 1072-1080.

Maeno K, Tanaka S (2009). Artificial miniaturization causes eggs laid by crowd-reared (gregarious) desert locust to produce green (solitary) offspring in the desert locust, *Schistocerca gregaria*. J. Insect. Physiol., 55: 849-854.

McCaffery AR, Simpson SJ, Islam MS, Roessingh P (1998). A gregarizing factor present in the egg pod foam of the desert locust *Schistocerca gregaria*. J. Exp. Biol., 201: 347-363.

Norris MJ (1950). Reproduction in the African migratory locust (*Locusta migratoria migratorioides* R. & F.) in relation to density and phase. Anti-Locust Bull., 6: 1-48.

Norris MJ (1952). Reproduction in the desert locust (*Schistocerca gregaria* Forskål) in relation to density and phase. Anti-Locust. Bull., 13: 1-49.

Papillon M (1960). Etude préliminaire de la répercussion du groupement des parents sur les larves nouveau nées de *Schistocerca gregaria* Forsk. Bull. Biol. Fr. Belg., 94: 203-263.

Papillon M (1970). Influence du groupement des adultes sur leur fécondité et sur le polymorphisme de leur descendance chez le criquet pèlerin, *Schistocerca gregaria* Forsk. Colloques Internationaux du Centre National de la Recherche Scientifique L'influence des Stimuli Externes sur la Gamétogenèse des Insectes, Paris, CNRS, pp. 71-86.

Pener MP (1991). Locust phase polymorphism and its endocrine relations. Adv. Insect. Physiol., 23: 1-79.

Pener MP, Yerushalmi Y (1998). The physiology of locust phase polymorphism: an update. J. Insect. Physiol., 44: 365-377.

Pener MP, Simpson SJ (2009). Locust phase polyphenism: an update. Adv. Insect. Physiol., 36: 1-286.

Rowell CHF (1967). Corpus allatum implantation and green/brown polymorphism in three African grasshoppers. J. Insect. Physiol., 13: 1401-1412.

Rowell CHF (1971). The variable colouration of the acridoid grasshoppers. Adv. Insect. Physiol., 8: 145-198.

Simpson SJ, McCaffery AR, Hägele BF (1999). A behavioural analysis of phase change in the desert locust. Biol. Rev. Camb. Philos. Soc., 74: 461-480.

Stower WJ (1959). The colour patterns of "hoppers" of the desert locust (*Schistocerca gregaria* Forskål). Anti-Locust Bull., 32: 1-75.

Tanaka S (2006). Corazonin and locust phase polyphenism. Appl. Entomol. Zool., 41: 179-193.

Tanaka S, Maeno K (2008). Maternal effects on progeny body size and colour in the desert locust, *Schistocerca gregaria*: examination of a current view. J. Insect Physiol., 54: 612-618.

Tanaka S, Maeno K (2010). A review of maternal and embryonic control of phase dependent progeny characteristics in the desert locust. J. Insect Physiol., 56: 911-918.

Uvarov BP (1921). A revision of the genus *Locusta*, L. (= *Pachytylus*, Fieb.), with a new theory as to periodicity and migrations of locusts. Bull. Entomol. Res. 12: 135-163.

Uvarov B (1966). Grasshoppers and Locusts,. Cambridge University Press, Cambridge, Vol. 1.

Uvarov B (1977). Grasshoppers and Locusts. Centre for Overseas Pest Research, Vol. 2.

3

Damage assessment and management of cucurbit fruit flies in spring-summer squash

R. Sapkota*, K. C. Dahal and R. B. Thapa

Institute of Agriculture and Animal Science Chitwan, Nepal.

Cucurbit fruit fly, *Bactrocera cucurbitae* (Coquillett), is one of the most important pests of cucurbits, and squash (*Cucurbita pepo* Lin.) is highly prone to damage by this pest in Nepal. Because of the difficulties associated with the control of this pest by chemical insecticides, farmers experienced great losses in cucurbits. Therefore, a participatory field experiment was conducted under farmer field conditions to assess losses and to measure the efficacy of different local and recommended management options to address the problem of it in squash var. Bulam House (F_1). The experiment consisted of six different treatments including untreated control, and there were four replications. All the treatments were applied 40 days after transplanting. Cucurbit fruit fly preferred young and immature fruits and resulted in a loss of 9.7% female flowers. Out of total fruits set, more than one-fourth (26%) fruits were dropped or damaged just after set and 14.04% fruits were damaged during harvesting stage, giving only 38.8% fruits of marketable quality. Application of locally made botanical pesticide 'Jholmal' was found superior in terms of fruit size (895 g), quality and yield (62.8 t/ha), and reduced fruit fly infestation in squash as compared to other treatments. Although, 'Jholmal' preparation is easy and its application is effective for the management of cucurbit fruit fly, it involves more labor cost and frequent application is a tedious process. Future efforts should be made to find the ways to reduce the cost of its application to make vegetable cultivation more profitable.

Key words: *Bactrocera cucurbitae*, *Cucurbita pepo*, pesticide, melon fly, food baits.

INTRODUCTION

The cucurbits such as cucumber, bitter gourd, sponge gourd, ridge gourd, bottle gourd, snake gourd, ash gourd, chayote, pointed gourd, and pumpkins are some of the major vegetables grown across Nepal. Several biotic factors limit the production and productivity of cucurbits, of which cucurbit fruit fly (*Bactrocera cucurbitae* Coquillett) has been the most prominent pest over the last several decades in Nepal (Manjunathan, 1997; GC and Mandal, 2000). Depending on the environmental conditions and susceptibility of the crop species, the extent of losses varies between 30 to 100% (Gupta and Verma, 1992; Dhillon et al., 2005a, b, c; Shooker et al., 2006). The field experiments on assessment of losses caused by cucurbit fruit fly in different cucurbits been reported 28.7 - 59.2, 24.7 - 40.0, 27.3 - 49.3, 19.4 - 22.1, and 0 - 26.2% yield losses in pumpkin, bitter gourd, bottle

gourd, cucumber, and sponge gourd, respectively, in Nepal (Pradhan, 1976). Considering previous facts and reports, it is apparent that >50% of the cucurbits are either partially or totally damaged by fruit flies and are unsuitable for human consumption. Although, several management options, such as hydrolyzed protein spray, para-pheromone trap, spraying of ailanthus and cashew leaf extract, neem products, bagging of fruits, field sanitation, food baits, and spray of chemical insecticides (Pawar et al., 1991; Zaman, 1995; Neupane, 1999, 2000; Akhtaruzzaman et al., 2000; GC and Mandal, 2000; Satpathy and Rai, 2002; Dhillon et al., 2005c; Palaniappan and Annadurai, 2006; Jacob et al., 2007) have been in use for the management of cucurbit fruit fly, some of them either fail to control the pest and/or are uneconomic and hazardous to non-target organisms and the environment (Manjunathan, 1997; Singh and Singh, 1998; Neupane, 2000; Dhillon et al., 2005c). In mid hill district of Nepal, farmers attempted different methods of management, like indigenous (70%), chemical (32%),

*Corresponding author.E-mail: merekha_iaas@yahoo.com

Picture 1. Adult female of cucurbit fruit fly.

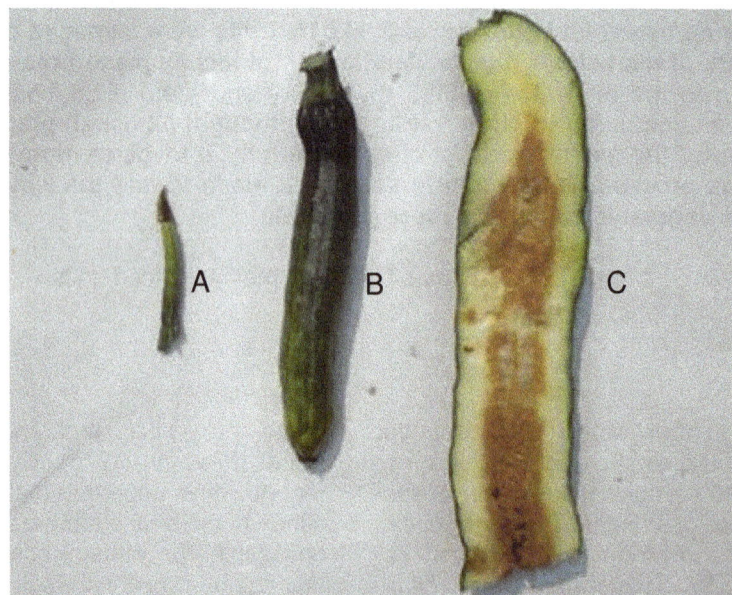

Picture 2. Different stages of squash fruit damaged by cucurbit fruit fly. **A.** Pre-set damage, **B.** Post-set damage, **C.** Harvested damage.

mechanical (80%) and combination of two or more methods (68%) to combat the problems of fruit fly.

(Sapkota, 2009). Considering the hazardous impact of chemicals on non-target organisms and the environment, present studies were undertaken to assess the losses caused by *B. cucurbitae* and efficacy of different control measures aiming to develop an eco-friendly and sustainable pest management system in cucurbits (1).

MATERIALS AND METHODS

Pits of size 45 × 45 × 45 cm were dug at spacing of 1 × 1 m as per the recommendation of Rajan and Markose (2001). Then, twenty - eight days old seedlings of popular squash var. Bulam House (F₁) were transplanted on 25th January, 2008 in a pit with standard dose of manure and fertilizer 40 t FYM + 120:80:60 NPK kg/ha. Field experiment was arranged under six farmers' fields spaced at more than 500 m apart as adopted by Nasiruddin et al. (2002) in randomized complete block design (RCBD) with four replications. The treatments consisted of: 1) Cue-lure [5 drops of cue-lure (4-p-acetoxyphenyl - 2-butanone) and 10 drops of malathion treated cotton-wool wick, recharged at 15 days interval]; 2) Indigenous food bait (Fermented rice 200 g + 5 ml molasses + 4 g borax and 1 ml malathion, replaced at 4 days interval); 3) Dichlorvos (NUVAN® at 2 ml/lit water, sprayed at weekly interval); 4) Banana pulp bait (Over ripe banana 500 g + 10 ml molasses + 10 g borax and 2.5 ml malathion, replaced at 4 days interval); 5) Botanical pesticide 'Jholmal' - leaf extract with cow urine, fresh cow dung and selective spices [Half kg leaves of each: neem (*Azadirachta indica* A. Juss.), ashuro (*Adhatoda vasica* Vasaka), tulsi (*Osimum sanctum* Lin.), tomato (*Lycopersicon esculentum* Mill), titepati (*Artemisia vulgaris* Mugwort), bojho (*Acorus calamus* Calamus), marigold (*Tagetes* sp.), khirro (*Sapium insigne* Royle), chrysanthemum (*Chrysanthemum* sp.), simali (*Vitex negundo* Lin.), and 100 g of each garlic (*Allimum sativum* Lin.)*, chilli (*Capsucum annuum* Lin.), and ginger (*Zingiber officinale* Roscoe)]; and vi) Untreated control. Treatments were imposed on the same day just before the initiation of flowers that is 40 days after transplanting (DAT). Micronutrients (Multiplex® at 2.5 ml/litre water) were applied at 20, 40, and 60 DAT in all the experimental plots, and other management operations like irrigation, weeding, hoeing etc., were managed by the farmers as per recommendations. Sixteen interior plants (4 plants per treatment per replication) were tagged for observations in each treatment. The observations were recorded on pre-set damage or ovary damage, post-set damage (PSD), and harvested damage (HD) at three-day intervals starting from flowering till last harvest. Observations were also recorded on losses of squash fruits due to other factors than fruit fly. Total number of set fruits per plant was calculated by adding the total post-set damage and total harvested fruits per plant.

Pre-set damage: Unopened female flowers (ovary) damaged by cucurbit fruit fly.

Post-set damage (PSD): Just after set to immature fruits (<100 g) damaged by cucurbit fruit fly.

Harvested damage (HD): Unmarketable fruits (≥100 g) damaged by cucurbit fruit fly recorded at harvest.

$$\text{Total marketable fruits (\%)} = \frac{\text{Number of harvested marketable fruits}}{\text{Total number fruits set}} \times 100$$

$$\text{HD\%} = \frac{\text{Number of harvested fruit fly damaged fruits}}{\text{Total number fruits set}} \times 100$$

$$\text{PSD\%} = \frac{\text{Number of PSD fruit fly damaged fruits}}{\text{Total number fruits set}} \times 100$$

$$\text{Harvested marketable fruit (\%)} = \frac{\text{Number of harvested marketable fruits}}{\text{Sum of total harvested fruit number}} \times 100$$

$$\text{Marketable fruit weight (\%)} = \frac{\text{Weight of harvested marketable fruits}}{\text{Sum of total harvested fruit weight}} \times 100$$

Similarly, percentage of unmarketable fruits and their weights were also calculated separately. The average weight of a marketable fruit/plant/treatment was calculated dividing the total marketable weight by the sum of the total marketable fruit number of the respective plant. Total yield per hectare (t/ha) was also computed considering the per plant yield (g/plant) as an output of 1 m² area.

The data were subjected to analysis of variance (ANOVA) using MSTATC statistical package (MSTATC, 1986), and the treatment means were compared using Duncan Multiple Range Test at P = 0.05.

RESULTS AND DISCUSSION

Cucurbit fruit fly damage at different stages

Average number of ovary (Pre-set) damage due to cucurbit fruit fly infestation per plant differed significantly (Table 1). Ovary damage was significantly lower in 'Jholmal', cue-lure, rice food bait and banana pulp bait treated plots than the control, however, ovary damage in control was at par with that of chemical treated plot. More than one (1.2) ovary of each plant was damaged by cucurbit fruit fly before anthesis. Out of total female flowers, 9.7% flowers did not open due to the infestation by cucurbit fruit fly. It notify that, besides genetic and environmental factors, significant variation in cucurbit fruit fly damage in the ovaries played an important role in fruit set, and yield of cucurbits.

Out of total set fruits, more than one-fourth (26%) fruits were dropped or damaged just after set, while 14.04% fruits were damaged during harvesting by the cucurbit fruit fly. Significantly higher numbers of fruits were damaged in control (32.5%) than 'Jholmal' (18.6%) and banana pulp bait (24.1%) treated plots. It was noted that the number of fruits damaged during harvesting was economically critical as compared to unopened flower damage and just after set damage. The cucurbit fruit fly favored early stages of fruits, and the infested fruits failed to develop properly and dropped-off from the plant. Earlier studies on fruit flies have also reported that the adult females preferred unopened flowers and young fruits for egg laying (Weems and Heppner, 2004; Dhillon et al., 2005a, b; Ronald and Kessing, 2007).

Fruit set and damage

Out of total squash fruits set, 40% were damaged by cucurbit fruit fly, and 21.2% losses in fruiting bodies due to other biotic and abiotic factors, of which 66% were lost due to hailstorm and remaining 34% due to rotting, blossom end shrinkage, abnormal growth, caterpillar infestation etc. (Table 2). Remaining 38.8% fruits were of marketable size and quality. The fruit damage due to other factors than fruit fly in 'Jholmal' treated plots was significantly lower, and numbers of marketable fruits (59.7%) significantly higher than that in other treatment plots.

There were a total of 10 harvests in insecticide treated plots as compared to a total of 12 harvests in other treatments. The total fruit set was also significantly lower in insecticide treated plot (8.94) than that in other treatment plots, suggesting that the use of chemical

Table 1. Fruit fly damage in unopened flowers (ovary), and post set and harvested fruits of squash under farmers' field conditions during spring-summer, Lamjung, 2008.

S.N.	Treatments	Cucurbit fruit fly damage		
		Ovary (No.)	Post-set (%)	Harvested (%)
1	Cue-lure	0.70[b]	25.2[abc]	12.38[bc]
2	Rice food bait	1.02[b]	26.1[abc]	11.72[bc]
3	Chemical treatment	1.35[ab]	29.5[ab]	17.65[ab]
4	Banana pulp bait	1.22[b]	24.1[bc]	10.06[c]
5	Leaf extract 'Jholmal'	0.65[b]	18.6[c]	10.59[bc]
6	Control	2.27[a]	32.5[a]	21.82[a]
Grand mean		1.20	26.0	14.04
LSD at P = 0.05		0.95	7.55	6.63

Values following different letters in a column are significant at P = 0.05.

Table 2. Total fruits set, marketable fruits and fruits damaged due to fruit fly and other factors in squash under farmers' field condition during spring - summer, Lamjung, 2008.

S.N.	Treatments	Fruit damage (%) by		Marketable fruits (%)	Total fruit set/plant (No.)
		Cucurbit fruit fly	Other factors		
1	Cue-lure	37.6[bc]	26.8[a]	35.6[b]	12.38[a]
2	Rice food bait	37.8[bc]	23.5[ab]	38.7[b]	12.50[a]
3	Chemical treatment	47.2[ab]	23.0[ab]	29.8[b]	8.94[b]
4	Banana pulp bait	34.1[c]	29.1[a]	36.7[b]	11.50[a]
5	Leaf extract 'Jholmal'	29.2[c]	11.1[c]	59.7[a]	11.81[a]
6	Control	54.3[a]	13.8[bc]	31.9[b]	11.81[a]
Grand mean		40	21.2	38.8	11.49
LSD at P = 0.05		10.28	10.09	8.83	2.302

Values following different letters in a column are significant at P = 0.05.

insecticides is inferior option in terms of fruit set and protection against cucurbit fruit fly damage in squash over the other control measures. Lower number of set fruit in chemically treated plot might be due to chemical sensitivity in flowers resulting in poor fruit set, early plant maturity, and less numbers of crop harvests as compared to other treatments. Since, the fruit fly maggots feed inside the fruiting bodies, it is difficult to control this pest with insecticides. Neupane (2000) also concluded that use of chemical pesticides to control fruit fly was only burden to the environment and increased cost of production. Similarly, Dhillon et al., (2005c) also pointed out difficulties to control this pest with insecticides.

Damaged fruit yield

Out of total harvested fruits, over two-third (67.1%) were marketable, while the remaining 25.4% and 7.5% unmarketable fruits were damaged by cucurbit fruit fly, and

other biotic and abiotic factors, respectively (Table 3).Unmarketable fruits due to cucurbit fruit fly damage in terms of numbers and weights were significantly lower in 'Jholmal', banana pulp bait, cue-lure, and rice food bait treated plot as compared to that in insecticide treated and untreated control plots.

The numbers of marketable fruits were significantly higher in 'Jholmal' treated plots (84.11%) as compared to that in other treatment plots. Furthermore, the 'Jholmal' contains nutrient enriched supplements for fruit growth and development, and the decomposed plant materials might have repelled or destroyed the successful life cycle of cucurbit fruit fly resulting in reduced fruit damage by this pest. Earlier studies have also reported that the application of 'Jholmal' increased the quality and yield of vegetables (Khatiwada and Pokhrel, 2004), neem derivatives repel insect pests of cucurbits (Budhathoki et al., 1993), and spraying of ailanthus and cashew leaf extract reduced cucurbit fruit fly attack (Jacob et al., 2007).

Table 3. Numbers and weights of harvested fruits of squash under different treatment conditions in the farmer's fields during spring-summer, Lamjung, 2008.

S.N.	Treatments	Harvested fruits (%)			
		Unmarketable by cucurbit fruit fly		Marketable	
		Number	Weight	Number	Weight
1	Cue-lure	21.8[b]	10.2[b]	62.2[b]	84.6[a]
2	Rice food bait	20.3[b]	8.7[b]	67.6[b]	88.4[a]
3	Chemical treatment	36.8[a]	16.2[b]	63.2[b]	83.8[ab]
4	Banana pulp bait	19.2[b]	8.3[b]	69.7[b]	88.7[a]
5	Leaf extract 'Jholmal'	15.1[b]	6.6[b]	84.11[a]	92.7[a]
6	Control	38.9[a]	25.3[a]	55.9[b]	73.5[b]
Grand mean		25.4	12.6	67.1	85.6
LSD at P = 0.05		11.69	8.77	12.96	9.46

Values following different letters in a column are significant at P = 0.05.

Conclusion

The present studies conclude that the cucurbit fruit fly causes significant damage in squash preferably in young and immature stages. The cucurbit fruit fly causes about 50% (10% flower and 40% set fruits) losses in squash yield under farmers field conditions in uncontrolled situations. Application of locally made botanical pesticide "Jholmal" offers superior yield in terms of fruit size and quality, and reduced fruit fly infestation in squash. Although, 'Jholmal' is easy to prepare locally and is effective for the management of fruit fly, it requires more frequent applications owing to more labor cost. It is also concluded that spraying of chemical insecticide is worthless in fruit fly management options. Therefore, future efforts should be made to find ways to reduce the cost of application of bio-pesticides like 'Jholmal' to make vegetable cultivation a profitable business, and to protect environment and life.

ACKNOWLEDGEMENTS

The authors are grateful to Dr. M. K. Dhillon, ICRISAT, Andhra Pradesh, India for his inspiration and technical support to research and to bring the manuscript in this form. The authors are also thankful to Directorate of Research, IAAS, Rampur and World-Vision International Nepal Lamjung, Area Development Program for the financial support to this study.

REFERENCES

Akhtaruzzaman M, Alam MZ, Ali-Sardar MM (2000). Efficiency of different bait sprays for suppressing fruit fly on cucumber. Bull. Inst. Trop. Agric. (Kyushu University). 23: 15-26.

Budhathoki K, Gurung GB, Lohar DP (1993). Vegetable crops: Indigenous knowledge and technology in the western hills of Nepal. Seminar Paper 1992/93, Lumle Agric. Res Centre, Kaski, Nepal.

Dhillon MK, Naresh JS, Singh R, Sharma NK (2005a). Evaluation of bitter gourd (Momordica charantia L.) genotypes for resistance to

melon fruit fly, Bactrocera cucurbitae. Indian J. Pl. Prot. 33(1): 55-59.

Dhillon MK, Naresh JS, Singh R, Sharma NK (2005b). Influence of physico-chemical traits of bitter gourd, Momordica charantia L. on larval density and resistance to melon fruit fly, Bactrocera cucurbitae (Coquillett). J. Appl. Ent. 129(7): 393-399.

Dhillon MK, Singh R, Naresh JS, Sharma HC (2005c). The melon fruit fly, B. cucurbitae: A review of its biology and management. J. Insect Sci. 5: 40-60 [Online]. Available on: http://www. Insectscience.org. [Retrieved on: 7th Oct., 2007].

GCYD, Mandal CK (2000). Integrated management of fruit fly (Bactrocera cucurbitae) on bitter gourd (Momordicha charantia L.) during the summer of (1998/99). IAAS Res. Rep. (1995-2000): 171-175.

Gupta D, Verma AK (1992). Population fluctuations of the maggots of fruit flies Dacus cucurbitae Coquillett and D. tau (Walker) infesting cucurbitaceous crops. Adv. Pl. Sci. 5: 518-523.

Jacob J, Leela NK, Sreekumar KM, Anesh RY, Heema M (2007). Phyto-toxicity of leaf extracts of multipurpose tree against insect pests in bitter gourd (Momordica charantia) and brinjal (Solanum melongena). Allelopathy J. 20(2): 1-2.

Khatiwada B, Pokhrel BP (2004). Botanical pesticides 'Jholmal' for organic agriculture. Ecocentre Tech. Bull. 1(2): 1-2.

Manjunathan TM (1997). A report on the integrated pest management (IPM) consultancy for Lumle Agriculture Research Centre (LARC). Occasional Paper No. 97/2. LARC, Kaski, Nepal.

MSTATC (1986). A micro-computers program for the design, manage-ment and analysis for agronomy research experiments. East Lansing, Michigan State Univ., USA.

Nasiruddin M, Alam SN, Khorsheduzzaman M, Jasmine HS, Karim ANMR, Rajotte E (2002). Management of cucurbit fruit fly, Bactrocera cucurbitae, in bitter gourd by using pheromone and indigenous bait traps and its effect on year-round incidence of fruit fly [Online]. Available on: http//www.oired.vt.edu/ipmcrsp/communications/ ammrepts/annrepoz/Bangladesh/bang-topic 10 pdf. [Retrieved on: 18th Feb. 2008].

Neupane FP (1999). Field evaluation of botanicals for the management of cruciferous vegetable insect pests. Nepal J. Sci. Tech. 2: 95-100.

Neupane FP (2000). Integrated management of vegetable insects. CEAPRED, Bakhundol, Lalitpur, Nepal. 172 p.

Palaniappan SP, Annadurai K (2006). Organic farming: Theory and practices. Scientific Publishers, Jodhpur, India. 257p.

Pawar DB, Mote UN, Lawande KE (1991). Monitoring of fruit fly population in bitter gourd crop with the help of lure trap. J. Res. (Maharashtra Agric. Univ). 16: 281.

Pradhan RB (1976). Relative susceptibilities of some vegetables grown in Kathmandu valley to D. cucurbitae Coq. Nep. J. Agric. 12: 67-75.

Rajan S, Markose BL (2001). Summer squash. In: Thamburaj Y and Singh N (eds.) Textbook of Vegetables, Tuber Crops and Spices. ICAR, New Delhi, India. pp. 286-289.

Ronald FIM, Kessing JM (2007). Bactrocera cucurbitae (Coq.)

[Online]. Available on: http//www. extento.hawaii.edu/ kbase/crop/Type/bactro_c.htm. [Retrieved on: 15th Sept. 2007].

Sapkota R (2009). Damage assessment and field management of cucurbit fruit fly (*Bactrocera cucurbitae* Coquillett) in squash during spring summer season of mid hill Nepal. Thesis, M. Sc. Ag., Tribhuvan University/IAAS, Rampur, Nepal.

Satpathy S, Rai S (2002). Luring ability of indigenous food baits for fruit fly B. cucurbitae (Coq.). Indian J. Ent. Res. 26(3): 249-252.

Shooker P, Khayrattee F, Permalloo S (2006). Use of maize as a trap crops for the control of melon fly, *B. cucurbitae* (Diptera:Tephritidae) with GF-120. Bio-control and other control methods [Online]. Available on: http\\www.fcla. edu/FlaEnt/fe87 p354.pdf. [Retrieved on: 20th Jan. 2008].

Singh S, Singh RP (1998). Neem (Azadirachta indica) seed kernel extracts and Azadirachtin as oviposition deterrents against the melon fly (B. cucurbitae) and oriental fruit fly (*B. dorsalis*). Phytoparasitica. 26(3): 1-7.

Weems HVJR, Heppner JB (2004). Melon fly, *Bactrocera cucurbitae* Coq. (Insecta: Diptera: Tephritidae). Florida Department of Agriculture and Consumer Services, Division of Plant Industry, and T.R. Fasulo, University of Florida. Univ. of Florida Pub., EENY-199.

Zaman M (1995). Assessment of the male population of the fruit flies through kairomone baited traps and the association of the abundance levels with the environmental factors. Sarhad J. Agric. 11: 657-670.

4

Virulence of *Beauveria bassiana* against Sunn pest, *Eurygaster integriceps* Puton (Hemiptera: Scutelleridae) at different time periods of application

Abdul Nasser Trissi[1], Mustapha El Bouhssini[2*], Mohammad Naif Al Salti[1], Mohammad Abdulhai[4], Margaret Skinner[3] and Bruce L. Parker[3]

[1]Aleppo University, Faculty of Agriculture, Aleppo, Syria.
[2]International Center for Agricultural Research in the Dry Areas, P. O. Box 5466, Aleppo, Syria.
[3]Entomology Research Laboratory, University of Vermont, Burlington, VT, USA. 05405-0105, USA.
[4]General Commission for Scientific Agricultural Research, Aleppo Center, Aleppo, Syria.

The virulence of *Beauveria bassiana* to overwintering Sunn pest adults was determined in the laboratory using micro-application techniques. Five microliter of 2×10^3, 2×10^4, 2×10^5, 2×10^6, 1×10^7, 2×10^7, 1×10^8 conidia/ml^{-1} were used in November, January and March. Applications were made to the mesosternum of adults. Mortality was recorded every 2 days for 16 days. Isolates were efficacious to Sunn pest at the higher concentrations. The cumulative corrected mortality after 12 days varied from 2.3% (SPT22) at 2×10^3 conidia/ml^{-1}, when treatment was conducted in November, to 100% (SPSR2) at 1×10^8 conidia/ml^{-1} in March. The LC$_{50}$ for SPSR2 decreased from 1.1×10^7 to 3.7×10^3 conidia/ ml^{-1} when treatment was conducted in November and March, respectively. The 50% lethal time (LT$_{50}$) varied from 17.07 days for GHA (the commercial fungal isolate – in *BotaniGard*), to 9.31 days for SPSR2 in November. The most efficacious isolate was SPSR2.

Key words: Sunn pest, *Eurygaster integriceps*, *Beauveria bassiana*, LC$_{50}$, LT$_{50}$.

INTRODUCTION

Sunn pest, *Eurygaster integriceps* Puton, is a major pest of wheat in Eastern Europe and the near and Middle East (Parker et al., 2011). The vegetative stage of wheat and maturing grains are affected by adults and nymphs. Sunn pest feeds on leaves, stems and grains (Hariri et al., 2000). Through feeding on the grain, the insects inject a prolyl endoprotease (Darkoh et al., 2010). This enzyme causes extensive breakdown of gluten, which greatly reduces the baking quality of the dough (Hariri et al., 2000).

Control of Sunn pest mostly relies on the use of chemical insecticides; around US$ 40 million is spent each year on chemical pesticides (El-Bouhssini et al., 2009). The continuous use of chemical insecticides for

control has resulted in serious management problems such as resistance to the insecticides, pest resurgence, elimination of beneficial insects and toxicity to humans and wildlife (Hendrawan and Ibrahim, 2006). These problems and the requirement for pesticide-free foods have increased pressure to find alternative management strategies (Mahdneshin et al., 2009). Integrated pest management (IPM), which utilizes ecological factors such as parasitoids, predators and microbial control agents are very attractive alternatives to the conventional chemicals used in the management of plant pests and diseases (Hanh et al., 2007). Entomopathogenic fungi that parasitize insects are valuable weapons for biocontrol and play an important role in promoting IPM (Cooke, 1977). *Beauveria bassiana* (Blasamo) Vullemin is the most prevalent fungus attacking *E. integriceps* populations in countries where this insect is a problem (Parker et al., 2003). Various strains of *B. bassiana* have been studied and have shown potential for inclusion in an IPM

*Corresponding author. E-mail: m.bohssini@cgiar.org.

Table 1. Isolate designation, source and country of origin of *B. bassiana* isolates.

Isolate	Host	Country of origin
GHA	*Bemisia* sp.	USA
SP22	Litter	Turkey
SP566	Sunn pest	Iran
SPSR2	Sunn pest	Syria

program for Sunn pest (Edgington et al., 2007). The objective of the research herein was to determine the virulence of some *B. bassiana* isolates collected from different regions against overwintering Sunn pest under laboratory conditions.

MATERIALS AND METHODS

Sunn pest adults were collected from overwintering sites under the litter around pine trees from the International Center for Agricultural Research in the Dry Areas (ICARDA), Aleppo, Syria at three different time periods in early and late winter and spring (November, 2008; January and March, 2009). In each date, Sunn pest was collected one day before treatment and stored in plastic ventilated containers in a refrigerator at 5°C until used. Four Isolates of *B. bassiana*, SP22, GHA (the active fungal ingredient of commercially available *BotaniGard*), SP566 and SPSR2 were obtained from the fungal culture collection maintained at ICARDA; isolate designation, source and country of origin of initial isolates are given in Table 1.

For each of the treatments, five culture plates were prepared by spreading 100 μl of 1×10^6 conidia/ ml^{-1} onto quarter strength Sabouraud dextrose agar supplemented with 0.25% technical yeast extract (SDAY/4- neopeptone 2.5 g/L, dextrose 10 g/L, agar 15 g/liter, and yeast 2.5 g/L; all Difco, Becton-Dickinson) (Parker et al., 2003). Plates were held for 14 days at 22 ± 2°C and 65 ± 5% rH, to maximize spore production. Conidia were then harvested by flooding each plate with 10 ml^{-1} of sterile distilled water (SDW) containing 0.01% (v/v) Tween 80 (Sigma) (Liu et al., 2003) and dislodging the conidia into suspension by stirring with a glass rod. All samples were vortexed 3 min to break up the conidial chains or clumps. Conidia were separated from hyphae and substrate materials by filtration of the suspension through two layers of cheese-cloth. The concentrations of fungal conidia in suspensions were determined using a haemocytometer. Viability of conidia was determined by spreading conidial suspensions onto SDAY/4 in Petri dishes and incubated at 25 ± 2°C. The percentage of germinated conidia was quantified after incubation for 24 h by examining a minimum of 200 conidia from each of three replicate plates. Conidia with germ tubes equal to at least half the conidial length were considered germinated. Suspensions with >90% germination were used for treatments.

Adults were treated with *B. bassiana* isolates at 2×10^5, 2×10^6, 1×10^7, 2×10^7, 1×10^8 conidia/ ml^{-1}, and were repeated at the three time periods for adults collection. Concentrations were prepared in SDW containing Tween 80 (0.01% v/v). Ten overwintered adults (5 ♂ and 5 ♀) per isolate and concentration were used. To secure the insects for topical application individuals were placed on a strip of scotch tape dorsal side down, and 5 μl of conidial suspension applied/insect to the mesosternum. Control insects were treated with SDW containing Tween.

After the application was dry (20 min) insects were transferred to wheat growing in pots (7 cm diameter and 8.5 cm height)

surrounded by clear plastic cages and incubated at 22 ± 2°C and 65 ± 5% rH for 14 days. A split-plot design with five replicates was used. Mortality was counted 4, 6, 8, 10 and 12 days post application. Sunn pest adults were considered dead if they failed to move following slight probing. Dead insects from each treatment were surface sterilized and kept separately in Petri dishes containing sterile paper toweling moistened with 0.10% streptomycin sulfate and 0.02% penicillin G (Lacey and Brooks, 1997). Dishes were then incubated at 22 ± 2°C and 65 ± 5% rH for 2 weeks to observe fungal outgrowth.

Statistical analyses

Cumulative mortality was corrected for natural mortality using Abbott's formula (Abbott, 1925), and normalized using an arcsine transformation, and then analyzed statistically using ANOVA. Means were separated using Fisher's Unprotected LSD at $P = 0.05$. The computations were done using GenStat Ed:10 (Payne et al., 2007). Probit analysis was used to estimate LC_{50} of the isolates with 95% confidence limits (CL) and LT_{50} values (SPSS, 1999).

RESULTS

Germination varied from 90 to 98% (unpublished data). Data for percentage corrected mortality of Sunn pest caused by four isolates of *B. bassiana* at different concentrations during three time periods of application are presented in Figure 1. Sunn pest mortality varied significantly depending on isolate (df=3, 36; F= 70.23; P<0.001), conidial concentration tested (df=24, 288; F= 26.56; P<0.001) and time period of application (df=2, 12; F= 209.62; P<0.001). Mortality of Sunn pest began ~5 days after application, and then increased slowly. It is clear that the increase in percent mortality was pronounced between the first and third time period of application for all isolates (df=6, 36; F= 16.58; P<0.001). Differences in mortality between the low conidial concentrations (2×10^3, 2×10^4 and 2×10^5 conidia/ml^{-1}) were not statistically significant for any of the isolates tested when treatments were made in November and January, but mortality was significantly different in March (df=6, 81; F= 44.66; P<0.001), between all concentrations. Moreover, significant differences (df= 48, 288; F=1.64; P= 0.007) were observed between the different concentrations, especially for high concentrations (2×10^6, 1×10^7, 2×10^7, 1×10^8 conidia/ml^{-1}), for all isolates at the three time periods of application. The mortality caused by SPSR2 was higher than the other isolates tested with a maximum Sunn pest mortality of 100% at 10^8 and 2×10^7 conidia/ml^{-1} when Sunn pest were treated in March. Whereas, the mortality was around 64 and 36% when Sunn pest were treated at the same concentrations, respectively, in November. When adults were treated with SP566, mortality was 100 and 90% with 10^8 and 2×10^7 conidia/ml^{-1}, respectively, in March. While, mortality was around 49 and 30% at the same concentrations, respectively, when treatment was made in November. Mortality of SP22 was 38, 42 and 80% with 10^8 conidia/ml^{-1}, when Sunn pest was treated in November, January and March,

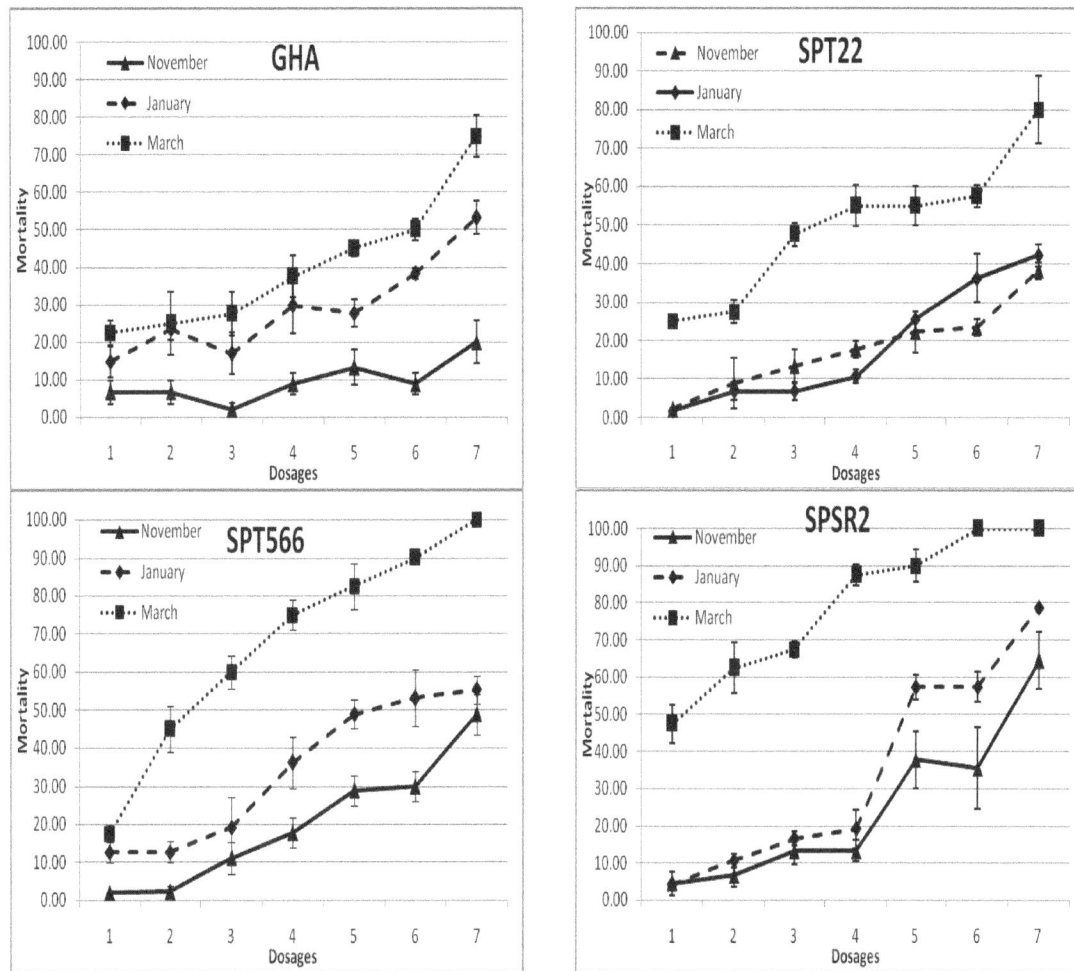

Figure 1. Percentage corrected mortality of Sunn pest caused by four isolates of *B. bassiana* at different concentrations during three time periods of application.

respectively. The lowest mortality was observed with GHA.

The LC_{50} estimates ranged from 3.7×10^3 to 1.5×10^{13} conidia/ ml^{-1}, depending on isolates and time periods of application (Table 2). Sunn pest were more sensitive to infection by *B. bassiana* when treatment was made in March (time of Sunn pest migration to wheat fields). The lowest LC_{50} was 3.7×10^3 (2.0×10^3- 1.1×10^4), conidia/ ml^{-1}, when Sunn pest were treated in March with SPSR2, compared with 1.1×10^7 (6.9×10^5- 2.1×10^7) conidia/ ml^{-1}, when they were treated in November. The highest LC_{50} 1.5×10^{13} (2.4×10^{10} - 8.2×10^{18}) was observed with GHA in November.

The LT_{50} values varied from 5.48 to 30.02 days Depending on isolate, concentration, and time of application (Table 3). The shortest LT_{50} (5.48 days) was obtained with SPSR2 at 10^8 conidia/ml^{-1} in March. The highest LT_{50} (30.02 days) was obtained with GHA at 10^7 conidia/ml^{-1} in November. Sunn pest was killed faster when treatment was made in March for all isolates at the three concentrations tested.

DISCUSSION

Although all isolates caused mortality to Sunn pest adults, they showed different levels of virulence. These findings corroborate previous studies that isolate-specific differences in virulence toward a single species of insect occurs (Haji Allahverdi Pour et al., 2008; Tsuda et al., 1996; Todorova et al., 1994). Under the same experimental conditions, the same insect host can be resistant to certain isolates of *B. bassiana*, while being susceptible to other isolates of the same pathogen (Tanada and Kaya, 1993; Todorova et al., 2002).

SPSR2 had significantly higher mortality (100%), and the lowest LT_{50} (Table 3) to Sunn pest adults than the other isolates, while GHA had the lowest mortality (75%) at the highest concentration (1×10^8 conidia/ml^{-1}) when treatment was made in March. This is in agreement with Tanada and Kaya (1993) that isolates from the same hosts cause higher mortality to that host than those from other hosts. However, Haji Allahverdi Pour et al. (2008) showed that soil isolates had higher virulence to Sunn

Table 2. LC_{50} (conidia ml^{-1}) values with 95 fiducial limits and probit analysis parameters for adult Sunn pest at different time periods of application.

Isolates	Time periods of application	LC_{50} (conidia/ ml^{-1})	95% fiducial limits		Intercept (a)	Slope (b)	x^2 value	p-value
			Lower	Upper				
GHA	November	1.5×10^{13}	2.4×10^{10}	8.2×10^{18}	3.139	0.141	0.89	0.989
	January	2.0×10^{8}	4.6×10^{7}	1.0×10^{9}	2.787	0.266	0.67	0.995
	March	7.9×10^{6}	3.6×10^{6}	2.6×10^{7}	3.177	0.264	0. 73	0.994
SP22	November	6.7×10^{7}	3.2×10^{7}	2.2×10^{8}	1.848	0.403	0.32	1
	January	4.4×10^{7}	2.8×10^{7}	7.8×10^{7}	0.647	0.551	0.2	1
	March	9.02×10^{5}	4.2×10^{5}	2.1×10^{6}	3.301	0.285	0.7	0.995
SP566	November	4.8×10^{7}	2.5×10^{7}	1.3×10^{8}	1.685	0.431	0.25	1
	January	1.8×10^{7}	7.6×10^{6}	5.0×10^{7}	2.843	0.297	0.61	0.996
	March	7.8×10^{4}	4.1×10^{4}	1.2×10^{5}	2.544	0.502	0.34	0.999
SPSR2	November	1.1×10^{7}	6.9×10^{6}	2.1×10^{7}	1.621	0.48	0.25	1
	January	3.5×10^{6}	2.9×10^{6}	5.5×10^{6}	0.746	0.116	0.18	1
	March	3.7×10^{3}	2.0×10^{3}	1.1×10^{4}	3.665	0.375	0.28	0.977

Table 3. LT_{50} values in days for Sunn pest adults treated with different isolates of *B. bassiana* at 1×10^{7}, 2×10^{7} and 1×10^{8} conidia ml^{-1} at different time periods of application.

Isolate	Time period of application	Concentration		
		1×10^{7}	2×10^{7}	1×10^{8}
		LT_{50} (days)		
GHA	November	30.02	17.68	17.07
	January	15.38	15.05	11.77
	March	11.92	11.68	9.95
SP22	November	16.40	14.88	10.75
	January	14.29	13.83	11.27
	March	9.31	9.52	8.22
SP566	November	15.08	13.16	11.78
	January	11.44	11.98	10.13
	March	7.91	7.41	5.22
SPSR2	November	13.29	12.12	9.31
	January	8.54	8.25	7.29
	March	7.12	6.34	5.48

pest adults than the isolates from the insect source.

Results of mortality showed that Sunn pest was more susceptible to infection by *B. bassiana* toward the end of the overwintering period (March), where percent mortality reached 100%. Sunn pest adults do not feed during the overwintering period, but food reserves are depleted and thus may increase the susceptibility from January to March. Kouassi et al. (2003) revealed that development of fungal pathogens within hosts can be influenced

indirectly by host food. Furthermore, Hajek and St. Leger (1994) demonstrated that successful development of *B. bassiana* within hosts is based on overcoming the host hemocyte response.

The differences observed in LC_{50} estimates with the four different isolates tested could reflect genetic, physiological differences between isolates or factors such as toxins or characteristics of the host (Butt et al., 1992; Todoraova et al., 2002). Our experiments were conducted using adults collected from overwintering sites. Because the adults spend about 9 months under the litter in these sites (Parker et al., 2011), we believe that these insects would be more vulnerable to fungi than adults in the field (summer populations). This may explain why Haji Allahverdi Pour et al. (2008) demonstrated that field-collected nymphs were more susceptible to infection by B. bassiana than new generation adults (summer populations). The variations observed between LC50's can be explained by variations in the virulence of a single isolate to related species of the host insect (Sabbahi et al., 2008).

Results of LT$_{50}$ indicated that SPSR2 is more effective than the other isolates; this isolate killed Sunn pest faster than SP22 as reported by Parker et al. (2003). The increase of insecticidal activity of conidia against Sunn pest through application time periods (as referenced above) may be related to the reduction of fat body in insects through the overwintering period.

In conclusion, adult Sunn pest are susceptible to some B. bassiana isolates. The most promising of these isolates was SPSR2, isolated from Sunn pest adults in wheat fields in the summer. This isolate should be tested further in an IPM program for Sunn pest.

REFERENCES

Abbott WS (1925). A method of computing the effectiveness of an insecticide. J. Econ. Entomol. 18:265-267.

Butt TM, Barrisever M, Drummond J, Schuler TH, Tillemans FT, Wilding N (1992). Pathoginicity of the entomopathogenus Hyphomycetes fungus Metarhizum anisopliae against the chrysomelid beetles Psylliodes chrysocephata and Phaedon cochleariae. Biocontrol. Sci. Technol. 2:327-334.

Cooke R (1977). The biology of symbiotic fungi. John Willey and Sons, New York.

Darkoh C, EL Bouhssini M, Baum M, Clack B (2010). Characterization of a prolyl endoprotease from Eurygaster integriceps Puton (Sunn Pest) Infested Wheat. Arch. Insect. Biochem. 74(3):163-178.

Edgington S, Moore D, EL-Bouhssini M, Sayyadi Z (2007). Beauveria bassiana for the control of Sunn pest (Eurygaster integriceps) (Hemiptera: Scutelleridae) and aspects of the insects daily activity relevant to a mycoinsecticide. Biocontrol. Sci. Technol. 17:63-79.

EL Bouhssini M, Street K, Joubi A, Ibrahim Z, Rihawi F (2009). Sources of Wheat resistance to Sunn pest, Eurygaster integriceps Puton, in Syria. Genet. Resour. Crop. Eviron. 56:1065-1069.

Hajek AE, St. Leger RJ (1994). Interaction between fungal pathogens and insect host. Annu. Rev. Entomol. 39:293-322.

Haji Allahverdi Pour H, Ghazavi M, Kharazi-Pakdel A (2008). Comparison of the virulence of some Iranian isolates of Beauveria bassiana to Eurygaster integriceps (Hem.: Scutelleridae) and production of the selected isolate. J. Entomol. Soc. Iran. 28:13-26.

Hanh Vu VIL, Hong S, Kim K (2007). Selection of Entomopathogenic Fungi for Aphid control. J. Biosci. Bioeng. 104:498-505.

Hariri G, Williams PC, Jaby EL-Haramein F (2000). Influence of Pentatomidae insects on the physical dough properties and two-layered flat-bread baking quality of Syrian wheat. J. Cereal Sci. 31:111-118.

Hendrawan S, Ibrahim Y (2006). Effect of dust formulations of three entomopathogenic fungal isolates against Stiphilus oryzae (Coleoptera: Curculionidae) in rice grain. J. Biosci. 17:1-7.

Kouassi M, Coderre D, Todorova SI (2003). Effect of plant type on the persistence of Beauveria bassiana. Biocontrol. Sci. Technol. 13:415-427.

Lacey LA, Brooks MW (1997). Initial handling and diagnosis of diseased insects. Pages 1-15. In: Manual of Techniques in Insect Pathology. L.A. Lacey (Editor). Academic Press, New York p. 409.

Liu H, Skiner M, Parker BL (2003). Bioassay method for assessing the virulence of Beauveria bassiana against tarnished plant bug, Lygus lineolaris (Him., Miridae). J. Appl. Entomol. 127:299-304.

Mahdneshin Z, Safaralizadah MH, Ghosta Y (2009). Study on the efficacy of Iranian isolates Beauveria bassiana (Blasamo) Vullemin and Metarhizum anisopliae (Metsch.) Sorokin against Rhyzopertha dominica F. (Coleoptera: Bostrichidae). J. Biol. Sci. 9:170-174.

Parker BL, Amir-Maafi M, Skinner M, Kim J, EL-Bouhssini M (2011). Distribution of Sunn Pest, Eurygaster integriceps Puton (Hemiptera: Scutelleridae), in overwintering sites. J. Asia-Pacific Entomol. 14:83-88.

Parker BL, Skinner M, Costa SD, Goulli S, Reid W, EL Bouhssini M (2003). Entomopathogenic fungi of Eurygaster integriceps Puton (Hemiptera: Scutelleridae): collection and characterization for development. Biol. Control 27:260-272.

Payne RW, Murray DA, Harding SA, Baird DB, Soutar DM (2007). GenStat for Windows (10th Edition) Introduction. VSN International, Hemel Hempstead.

Sabbahi R, Merzouoki A, Guertin C (2008). Efficacy of Beauveria bassiana (Bals.) Vuill. Against the tarnished plant bug, Lygus lineolaris L. in strawberries. J. Appl. Entomol. 132:124-134.

SPSS (1999). SPSS for Windows user's guide release 10. 1st Edn. SPSS inc., Chicago.

Tanada Y, Kaya HK (1993). Insect pathology. Academic Press. New York, p. 665.

Todorova SI, Cloutier C, Cote JC, Coderre D (2002). Pathogenicity of six isolates of Beauveria bassiana (Balsamo) Vuillemin (Deuteromycotina, Hyphomycetes) to Perillus bioculatus (F) (Hem., Pentatomidae). J. Appl. Entomol. 126:182-185.

Todorova SI, Cote JC, Martel P, Coderre D (1994). Heterogeneity of two Beaveria bassiana strains revealed by biochemical test. Protein profiles and bio-assays on Leptinotarsa decemlineata (Col., Chrysomelidae) and Coleomegilla maculata lengi (Col., Chrysomelidae). J. Appl. Entomol. 120:159-163.

Tsuda K, Yoshioka T, Tsutsumi T, Yamanaka M, Kawarabata T (1996). Pathogenicity of some entomogenous fungi to brown-winged green bug, Plautia stali (Hemiptera: Pentatomidae). Jpn. J. Appl. Entomol. Z. 40:318-321.

5

Larvicidal activity of the saponin fractions of *Chlorophytum borivilianum santapau* and *Fernandes*

S. L. Deore* and S. S. Khadabadi

Government College of Pharmacy, Kathora Naka, Amravati- 444604, (M.S), India.

The present communication deals with the laboratory studies carried out to ascertain the larvicidal properties of *Chlorophytum borivilianum* Sant. and Fernand. saponin extracts for the mosquito species *Anopheles stephensi*, *Culex quinquefasciatus* and *Aedes aegypti*. Methanolic extract (ME), crude saponin extract (CSE) and purified saponin fractions (PSF) were used as test solutions. Concentration to kill 50% larvae and concentration to inhibit emergence of 50% adult that is LC_{50} and EC_{50} values respectively were calculated. All extracts found to be larvicidal but PSF was found more effective.

Key words: Larvicidal, *Chlorophytum borivilianum*, saponin.

INTRODUCTION

Chlorophytum borivilianum Sant. and Fernand. belonging to family Liliaceae is a very well known plant for its aphrodisiac as well as immunomodulatory properties (Oudhia, 2001). Roots of the plant are used both in Ayurveda and Unani system to treat oligospermia, arthritis, diabetes and dysuria (Wealth of India 1996). In earlier studies, Antiviral (Siddiqui YM, 2005), Anticancer (Arif JM, 2005), immunomodulatory (Singh et al., 2004), anti-diabetic (Govindrajan et al., 2005), antistess (Gopalkrishna et al., 2006), aphrodisiac (Thakur et al., 2006), antimicrobial (Deore et al., 2007) and anti-inflammatory (Deore et al., 2008) activities of root extracts have been evaluated. Roots of this plant contain carbohydrates, phenolic compounds, saponins and alkaloids (Deore et al., 2008).

As the saponins reported to have insecticidal and larvicidal actions (Sparg et al., 2004), the present study was conducted to ascertain the larvicidal properties against larvae of three species of mosquito (*Anopheles stephensi*), *Culex quinquefasciatus* and *Aedes aegypti*. The most common mosquito larvicides used currently are organophosphates, insect growth regulators and microbial larvicides. Current research focuses on microbials such as *Bacillus thuringiensis* and *Bacillus sphaericus* as well as herbal larvicidal, oviposition inhibiting, repellent or insect growth regulatory effects.

Such products contain a multitude of active ingredients with different modes of action, which lessens the chance of resistance developing in mosquito populations (Rajkumar et al., 2005).

MATERIALS AND METHODS

Plant materials, extraction and isolation

C. borivilianum Sant. and Fernand. roots were purchased from local cultivator and a specimen sample was deposited at Department of Botany, Vidarbh institute of Science and Humanities, Amravati. The roots were washed dried and powdered and defatted by petroleum ether. Thereafter, extracted with methanol for 3 h with mild heating. Methanol extract was concentrated and methanol extract (ME) was obtained. In order to get the crude saponins, extract was again dissolved in methanol and acetone was added (1:5 v/v) to precipitate the saponins as described by Yan et al. (1996). The precipitate was dried under vacuum. The whitish amorphous powder, thus obtained is named as a crude saponin extract (CSE). To get the pure saponin fraction (PSF), certain amount of CSE was fractionated by applying to silica gel-60 (230 400) mesh. Column chromatography and eluted successfully with chloroform-methanol-water (70:30:10) as described by Favel et al. (2005). Eluted fractions combined to give PSF.

Test mosquitoes

Laboratory-reared III instars mosquito larvae of *A. stephensi, C. quinquefasciatus* and *A. aegypti* was provided by the Head of Zoology Department, Vidarbh Institute of Science and Humanities, Amravati.

---

*Corresponding author. E. mail: sharudeore_2@yahoo.com.

Table 1. Mean LC$_{50}$ values of different fractions of *C. borivilianum Sant.* and *Fernand.* tuber extracts.

Mosquito species	LC$_{50}$ (PPM)			
	ME	CME	PSF	Malathion
A. stephensi	8066.67	5300	4000	0.00547
C. quinquefasciatus	6300	4833.33	3850	0.00477
A. aegypti	5733.33	4250	3916.67	0.00503

Table 2. Mean EC$_{50}$ values of different fractions of *C. borivilianum* Sant. and Fernand. tuber extracts.

Mosquito species	EC$_{50}$ (PPM)			
	ME	CME	PSF	Malathion
A. stephensi	7219	4305	4872	0.0024
C. quinquefasciatus	5201	3988	2981	0.0036
A. aegypti	4721	3201	2290	0.0022

METHODOLOGY

The larvicidal bioassay followed the World Health Organization (WHO) standard protocols (World Health Organization (1981). Instructions for determining the susceptibility or resistance of mosquito larvae to insecticides. WHO/VBC, 81:807.) with slight modifications. Each of the concentrations of different extracts of *C. borivilianum* (0.1 - 0.5%) was transferred into sterile glass Petri dishes (9 cm diameter/150 ml capacity). 10 third instar larval form of *S. aegypti* were separately introduced into different Petri dishes containing graded concentrations and the mortality was recorded for 48 h of the exposure period. Dead larvae were identified when they failed to move after probing with a needle in the siphon or cervical region. The experiments were replicated three times and conducted under laboratory conditions at 25 - 30°C and 80 - 90% relative humidity. Similar types of bioassay were conducted with different concentration of each extracts of the *C. borivilianum* and with a chemical insecticide, Malathion, on third instar larval forms of selected mosquito larvae. A negative control was run in tap water. 6 replicates were run under the same microclimatic conditions. The lethal concentration was determined and compared with Malathion.

The effects of the treatments were monitored by counting the number of dead larvae each day. For the LC$_{50}$ the data of 48 h was uses because till that time no pupa was observed even in control treatments. During the course of experiment, a food based on the baby food was provided to the larvae. In another series of experiments, observations on the emergence and larval duration of larvae that were reared at sublethal doses of the active fractions of the treatments were made and the emergence of the 50% of the test larvae (EC$_{50}$values) was determined. Experiment was carried out in triplicate and results are expressed Table 1 and 2.

RESULTS AND DISCUSSION

Mosquitoes transmit several public health problems, such as malaria, filariasis, dengue and Japanese encephalitis; causing millions of deaths every year (World Health Organization, 1981). Mosquitoes in the larval stage are attractive targets for pesticides because they breed in water and, thus, are easy to deal with them in this habitat. The use of conventional chemical pesticides has resulted in the development of resistance, undesirable effects on non-target organisms and fostered environmental and human health cosncerns (Vatandoost et al., 2001, Severini et al., 1993). Plants are rich source of bioactive organic chemicals and offer an advantage over synthetic pesticides as these are less toxic, less prone to development of resistance (World Health Organization, 1970) and easily biodegradable (Forget, 1989).

The secondary compounds of plants make up a vast repository of compounds with a wide range of biological activities. Most studies report active compounds as steroidal saponins (Hostettmann et al., 1995). Saponins are freely soluble in both organic solvents and water, and they work by interacting with the cuticle membrane of the larvae, ultimately disarranging the membrane, which is the most probable reason for larval death (Wiesman et al., 2005).

In 48 h experimental period, the crude extracts of tuber of *C. borivilianum* Sant. and Fernand. has been found to possess larvicidal and adult emergence inhibition activity against the mosquito *A. stephensi*, *C. quinquefasciatus* and *A. aegypti*. Among the three mos-quito species tested, A. aegypti was the most sensitive followed by *C. quinquefasciatus* and *A. stephensi* in case of all extracts. The biological activity of the extracts might be due to the saponins and alkaloids exist in plants, these compounds may jointly or independently contribute to produce larvici-dal and adult emergence inhibition activity. The larvicidal efficacy of *C. borivilianum* Sant. and Fernand. is not comparable to well established insecticidal plant species or synthetic insecticides such as Malathion but it can be suggested that its use for control of this mosquitoes. Future scope needs to isolate responsible single constituent/s should be identified and utilized, if possible, in preparing a commercial product / formulation to be used as a mosquito repellent.

Conclusion

In conclusion, *Chlorophytum borivilianum* Sant. and Fernand offers promised as a potential bio control agent against A. *Aegypti, C. quinquefasciatus* and *A. stephensi* particularly in its markedly larvicidal effect. The extract or isolated bioactive phytochemical from the plant could be used in stagnant water bodies which are known to be the breeding grounds for mosquitoes.

ACKNOWLDGEMENTS

We are thankful to the Head of Zoology Department, Vidarbh Institute of Science and Humanities, Amravati for providing the larvae of *A. stephensi, C. quinquefasciatus* and *A. aegypti*.

REFERENCES

Arif JM (2005). Effects of Safed Musli on Cell Kinetics and Apoptosis in Human Breast Cancer Cell Lines, International Conference on Promotion and Development of Botanicals with International Co-Ordination Exploring Quality, Safety, Efficacy and Regulation Organized by School of Natural Product Study, Jadavpur University, Kolkata , India Feb 25-26.

Deore SL, Khadabadi SS (2007) In vitro antimicrobial studies of *Chlorophytum borivilianum* (Liliaceae) root extracts Asian. J. Microbiol. Biotech. Environ. Sci., 9 (4): 807-809.

Deore SL, Khadabadi SS (2008) Anti-Inflammatory and Antioxidant Activity of *Chlorophytum Borivilianum* Root Extracts. Asian J. Chem., 20 (2), 983-986.

Favel A, Kemertelidze E, Benidze M, Fallague K, Regli P (2005). Antifungal activity of steroidal glycosides from *Yucca gloriosa* L. Phytother. Res., 19(2): 158-161.

Forget O (1989). Pesticides, necessary but dangerous poisons. The IDRC Reports, 18:7-13.

Gopalkrishna B, Patil SH (2006). Preliminary phytochemical investigation and in-vitro Anti-stress activity of *Safed musli - Chlorophytum borivilianum*. Indian Drugs, 43(11):878-880.

Govindrajan R, Sreevidya N, Vijaykumar M, Thakur M, Dixit VK (2005) In-vitro antioxidant activity of ethanolic extract of *Chlorophytum borivilianum*. Nat. Prod. Sci. 11(3): 165-169.

Hostettmann K, Marston A (1995). Saponins (Chemistry and Pharmacology of Natural Products). Cambridge: Cambridge University Press, pp.132.

Oudhia P (2001). My experiences with wonder crop Safed Musli. In Sovenier. International seminar on medicinal plants & quality standardization VHERDS, Chennai, India June 9-10.

Rajkumar S, Jebanesan (2005). Larvicidal and adult emergence inhibition effect of *Centella asiatica* -- Brahmi (Umbelliferae) -- against mosquito Culex quinquefasciatus (Diptera: Culicidae) Afr .J. Biomed. Res., 8:31-33.

Severini C, Rom R, Marinucci M, Rajmond M (1993). Mechanisms of insecticide resistance in field populations of *Culex pipiens* from Italy. J. Am. Mosq. Control Assoc., 9:164-168.

Siddiqui YM (2005). Potent Antiviral Effect of Safed Musli on Poliovirus Replication, International Conference on Promotion and Development of Botanicals with International Co-Ordination Exploring Quality, Safety, Efficacy and Regulation Organized by School of Natural Product Study, Jadavpur University, Kolkata , India, Feb 25-26.

Singh P, Gupta P, Singh R (2004). Safed Musli as a Divya Aushad, Chem. Biol. Interface, Synergestic New Frontiers, New Delhi, India: Nov. 21-26.

Sparg SG, Light ME, Staden J (2004). Biological activities and distribution of plant saponins. J. Ethnopharmacol., 94(2-3), 219-243.

Thakur M, Dixit VK. (2006). Effect of *C. borivilianum* on androgenic and sexual behaviour of male rats. Indian Drugs, 43:300-306.

Vatandoost H, Vaziri M (2001). Larvicidal activity of neem extract (*Azadirachta indica*) against mosquito larvae in Iran. Pestol., 25:69-72.

Wealth of India-raw material, (1996). Vol-3, (Ca-ci), CSIR New Delhi, pp. 482-483.

Wiesman Z, Chapagain BP (2005). Larvicidal effects of aqueous extracts of *Balanites aegyptiaca* (desert date) against the larvae of *Culex pipiens* mosquitoes. Afr. J. Biotechnol., 4:1351-1354.

World Health Organization (1970). Insecticide resistance and vector control. XVII Report of WHO expert Committee on Insecticides. World Health Organ Tech. Rep. Ser., 443:279.

World Health Organization (1981). Instructions for determining the susceptibility or resistance of mosquito larvae to insecticides. WHO/VBC., 81:807.

Yan W, Ohtani K, Kasai R, Yamasaki K (1996). Steroidal saponin from fruits of *Tribulus terrstris*. Phytochem., 42(5): 1417-1422.

6

Population dynamics of the leaf curl aphid, *Brachycaudus helichrysi* (Kalt.) and its natural enemies on subtropical peach, *Prunus persica* cv. Flordasun

R. K. Arora, R. K. Gupta* and K. Bali

Division of Entomology, Faculty of Agriculture, Sher-e-Kashmir University of Agricultural Sciences and Technology-J, Jammu-180 002, India.

Population dynamics of leaf curling aphid, *Brachycaudus helichrysi* (Kalt.) and natural enemies associated with the pest were studied. The initial occurence of the alate forms of *B. helichyrsi* on peach was noticed as the first alates (1.33/10 leaves) in the 3rd week of October, 2004. In the aphid colonies, alates appeared (2.33 alates/10 leaves) in the 3rd week of February and their percent population increased gradually in proportions. However, the winged aphids totally disappeared from 2nd week of December to 2nd week of February. There was a decline in the aphid population from 3rd week of April (12.62%) and onwards both apterous and alate forms disappeared from the host plants. During 1st week of February, the overwintering nymphs developed into apterous adults. There was a rapid increase in aphid population (170.67/10 leaves) and peak population (592.00/10 leaves) was recorded in 3rd week of March when the mean atmospheric temperature and relative humidity were 16.2°C and 74.44% respectively. A positive correlation was observed between the population build up of the pest and combined effect of atmospheric temperature and relative humidity ($R_{1.23}$ = 0.7051). The independent effect of humidity ($r_{13.2}$ = 0.7741) and atmospheric temperature ($r_{12.3}$ = 0.5551) was also found positively correlated with population build up of the pest. Overwintering stage of the pest from December - January revealed that nymphs produced in the last week of November overwintered in the axils and bases of dormant buds of peach tree. The population of overwintering nymphs was more in January (3.0/10 buds) than in December (0.67/10 buds). *Coccinella septempunctata* L., *Leis dimidiata* F., *Coelophora sauzeti* Muls., *Ischiodon scutellaris* (F.), *Paragus* (*Paragus*) *serratus* (Fabr.) and *Paragus* (*Pandasyophthalmus*) *tibialis* (Fallen) were recorded as aphidophagous predators and *Diaeretiella rapae* M'Intosh and *Aphidius* sp. as parasitoids of the pest.

Key words: Population dynamics, *Brachycaudus helichrysi* (Kalt.), natural enemies, peach cv. Flordasun

INTRODUCTION

The peach, *Prunus persica* (L.) Balsch is grown around the world between 25° and 45° latitudes above and below the equator (Childers, 1975) and its commercial cultivation is in vogue in a number of countries like the U.S.A, Italy, France, Japan, Argentina, Australia, Mexico, Korea, Germany, New Zealand, Turkey, Canada, Chile, India etc.

In India, peach cultivation is more or less confined to the mid hill zones of the Himalayas extending from Jammu and Kashmir to Khasi hills at an attitude of 1500 - 2000 m above mean sea level. Low chilling cultivators are also grown in sizeable area in the sub-mountainous regions and eastern parts extending from Punjab, Haryana, Delhi to Western Uttar Pradesh. Limited cultivation also exists in the hills of the South and in the North eastern region of the India (Ghosh, 1976).

However, in Jammu and Kashmir, peach cultivation is restricted to the temperate and intermediate agro climatic zones. Inspite of early bearing, high yield potential of this

*Corresponding author. E-mail: rkguptaentoskuast@gmail.com

fruit crop and concerted efforts of both the government and some non-governmental organizations to motivate farmers to accept its commercial cultivation, the area and production have not increased substantially due to the fact that peach is subjected to the ravages of the leaf curling aphid, *Brachycaudus helichyrsi* (Kaltenbach), which is one of the major limiting factors for its successful cultivation under sub-tropical conditions (Verma and Singh, 1990). The pest is highly polyphagous, infesting about 175 species belonging to 115 genera into 49 plant families in India (Ghosh and Verma, 1988; Thakur et al., 1995). This pest is also a vector of *Plum Pox virus* which further aggravates the situation and imparts a sickly appearance to the host plant (Maison and Massonie, 1982). Keeping in view the economic importance of this pest, this research work was initiated to study its population dynamics and natural enemies associated with it.

MATERIALS AND METHODS

The studies were carried out in the laboratory of Division of Entomology and peach orchard under Division of Pomology and Post Harvest Technology, Sher-e-Kashmir University of Agricultural Sciences and Technology, Udheywalla, Jammu

Initial occurence of the alates of *B. helichyrsi* on the host and population build-up of the pest

For recording the initial occurence of the alate forms of *B. helichyrsi* on peach trees and subsequent population build up during the season, three peach trees were randomly selected and marked in the orchard where no treatment was applied. From each tree, 10 twigs were selected randomly from all geographical directions, besides one from the central whorl and from each twig 10 buds/ leaves were observed for all forms *i.e.* alate, apterous and nymphs with naked eye and hand lens (x10) at weekly intervals. The buds were observed from December and January, 2004 while the leaves were observed during the rest of the year. All forms of aphid were counted. The observations on leaves were recorded weekly starting from first week of September to last week of August in the following year. The data, thus generated were analysed statistically as per the methods advised by Panse and Sukhatme (1978).

Simple and multiple correlations of the aphid population were also worked out with weather parameters. The data for weather parameters viz., atmospheric temperature and relatively humidity were obtained from the Meteorology station of Water Management Research Centre (ICAR), Jammu, India.

Overwintering of *B. helichyrsi*

Weekly observations were also recorded on the axils and bases of the dormant buds under microscope in the laboratory to see the overwintering egg, nymphal or adult stages.

Natural enemies of *B. helichyrsi*

The maggots of syrphid and grubs of coccinellids were collected from the aphid infested peach trees and were reared in the laboratory by providing them aphids as daily feed till the adults emerged.

Similarly, the parasites were collected from the aphid mummies in the laboratory. The adults of the predators and para-sites thus obtained were got identified by the taxonomy section of Division of Entomology, IARI, New Delhi.

To record the population of predators and parasites, three untreated peach trees were selected and kept under observation. Population of each predator was determined on the randomly selected branches/ bunches at weekly interval commencing from 1st week of March to 4th week of April for the period under study. Similarly, for recording the population of parasites, a random sample of 100 aphids was kept under observations in the laboratory to see the intensity of parasitism based on the formula given Root and Skelsey (1969).

$$\text{Parasitism (\%)} = \frac{\text{Total aphid mummies}}{\text{Total live aphids + Total aphid mummies}} \times 100$$

RESULTS AND DISCUSSION

Initial appearance of the alate forms of *B. helichyrsi*

The first alates (1.33/10 leaves) of *B. helichyrsi* appeared in the 3rd week of October of respective years. The population of winged aphids continued to increase and maximum population (3.67 alates/10 leaves) was recorded in 3rd week of November. Thereafter, the population of winged aphids declined until disappear from 2nd week of December to 2nd week of February. Their per cent population increased gradually in proportions to the apterous population till 2nd week of April (16.01). There was a decline in the aphid population from 3rd week of April (12.62%) and onwards both apterous and alate forms disappeared from the host plant. This decline in aphid population was attributed to increase in temperature and appreciable activities of the biotic agents.

Studies conducted by (Madsen and Bailey, 1958; Sandhu and Khangura, 1977; Ghosh and Raychaudhuri, 1981; Verma and Singh, 1990; and Gupta and Thakur, 1993) had revealed that the alate forms of *B. helichyrsi* migrated from the alternate hosts to the primary host (peach) during autumn between October November for laying eggs or nymphs. As observed in the present study. The slight variation in time of their initial occurrence might have been due to the difference in agroclimatic conditions, altitude of the area and cultivar of the host plant etc. However, the alate forms were not observed till 2nd week of February probably due to extreme cold weather in this northern part of the country. These observations fall in line with that of Madsen and Bailey (1958) who also did not observe the alate forms during winter months. During 3rd week of February with rising temperature, these forms again started occuring (2.33/10 leaves) and their maximum per cent population (16.01) was recorded in the 2nd week of these observations are in close conformity to the findings of Verma and Singh (1990) who also observed the of winged forms in the aphid colonies

Table 1. Population of alate and apterous morphs of *B. helichyrsi* (Kalt.) on peach cv. Flordasun.

| Month | Week | Population* load | | | Per cent population | | Aphid counted from |
		Alate	Apterous	Total	Alate	Apterous	
October	II	0.00	0.00	0.00	0.00	0.00	Leaves
	III	1.33	0.00	1.33	100.00	0.00	Leaves
	IV	2.00	0.33	2.33	85.84	14.16	Leaves
November	I	3.00	3.00	3.67	81.74	18.26	Leaves
	II	3.33	3.33	4.33	76.90	23.10	Leaves
	III	3.67	3.67	5.67	64.72	35.28	Leaves
	IV	2.00	2.00	5.00	40.00	60.00	Leaves
	V	1.00	1.00	3.33	30.03	69.97	Leaves
December	I	0.33	0.33	1.33	24.81	75.19	Leaves
	II	0.00	0.00	1.00	0.00	100.00	Buds
	III	0.00	0.00	0.67	0.00	100.00	Buds
	IV	0.00	0.00	0.67	0.000	100.00	Buds
January	I	0.00	0.00	1.67	0.00	100.00	Buds
	II	0.00	0.00	2.00	0.00	100.00	Buds
	III	0.00	0.00	2.00	0.00	100.00	Buds
	IV	0.00	0.00	3.00	0.00	100.00	Buds
February	I	0.00	0.00	5.67	0.00	100.00	Leaves
	II	0.00	0.00	14.67	0.00	100.00	Leaves
	III	2.33	2.33	170.67	1.36	98.64	Leaves
	IV	5.67	5.67	301.33	1.88	98.12	Leaves
March	I	11.33	11.33	393.00	2.88	97.12	Leaves
	II	26.67	26.67	472.33	5.65	94.35	Leaves
	III	34.33	34.33	592.00	5.80	94.20	Leaves
	IV	19.33	19.33	175.33	11.02	88.98	Leaves
	V	14.00	14.00	101.33	13.82	86.18	Leaves
April	I	13.33	13.33	87.00	15.32	84.68	Leaves
	II	8.33	8.33	52.00	16.01	83.99	Leaves
	III	4.67	4.67	37.00	12.62	87.38	Leaves
	IV	0.00	0.00	0.00	0.00	0.00	Leaves

* 10 buds/ 10 leaves, ** Over-wintering population

in the month of June which subsequently migrated to the summer alternate hosts.

Population builds up of *B. helichyrsi* during the season

With the commencement of sap flow in the trees during spring (1st week of February), the overwintering nymphs developed into apterous adults and started multiplying in the curled leaves (Table 1). There was a rapid increase in aphid population (170.67/10 leaves) and the peak population (592.00/10 leaves) was recorded in 3rd week of March when the mean atmospheric temperature and relative humidity were 16.2°C and 74.44%, respectively, (Figure 1). From 4th week of March onwards, aphid population started declining (175.33 aphids/10 leaves) and this tendency continued till 3rd week of April (37.00/10 leaves). In the 4th week of April, when the mean atmospheric temperature and relative humidity were 26.7°C and 58.62%, respectively, there was complete absence of aphid population.

During the course of recording the data, it was observed that with the onset of spring overwintering nymphs developed into apterous adults who reproduced tremendously within the curled leaves. The population of apterous aphid was more than that of alate forms. Alate aphids were more active, seen moving on the leaves, while the apterous were sedentary and remained confined to feeding sites.

The perusal of data (Table 2) revealed a positive correlation between the population build up of the pest and combined effect of atmospheric temperature and relative humidity ($R_{1.23} = 0.7051$). The independent effect of humi-

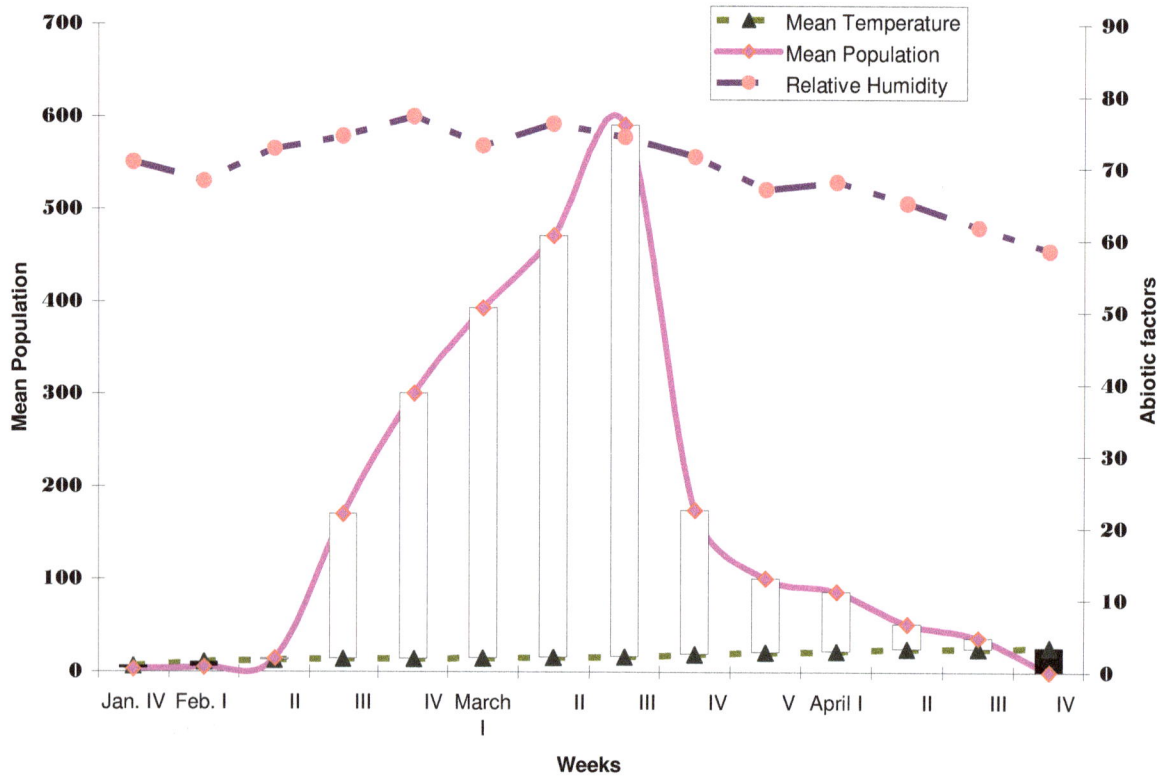

Figure 1. Influence of abiotic factors on the population builds up of *Brachycaudus helichrysi* (Kalt.).

Table 2. Seasonal population build-up of *B. helichyrsi* (Kalt.) on peach cv. Flordasun.

Month	Week	Mean aphid population/10 leaves			Mean temperature(°C) (X_1)	Mean relative humidity (%) (X_2)
		Nymphs	**Adults**	**Total (Y)**		
January	IV	3.00	-	3.00	6.6	70.85
	I	4.00	1.67	5.67	10.9	68.20
February	II	11.67	3.00	14.67	12.7	72.70
	III	120.00	50.67	170.67	13.9	74.50
	IV	202.33	99.00	301.33	14.2	77.20
	I	263.00	130.00	393.00	14.9	73.14
	II	325.33	147.00	472.33	15.7	76.19
March	III	387.00	205.00	592.00	16.2	74.44
	IV	48.33	127.00	175.33	19.0	71.60
	V	23.00	78.33	101.33	21.4	67.07
	I	11.00	76.00	87.00	22.3	68.13
	II	3.33	48.67	52.00	24.9	65.20
April	III	0.00	37.00	37.00	25.2	61.80
	IV	0.00	0.00	0.00	26.7	58.62

Multiple correlation co-efficient $(R_{1.23})$ = 0.7051*, Partial correlation co-efficient $(r_{12.3})$ = 0.5551*, $(r_{12.3})$ = 0.7741*, Coefficient of determination (R_2) = 0.4972, Significant at P ≤ 0.05

dity ($r_{13.2}$ = 0.7741) was also found positively correlated with population build up of the pest. The analysis of independent effect of atmospheric temperature on the population fluctuation of aphid reflected a positive correlation ($r_{12.3}$ = 0.5551).

Population of the pest started multiplying from the fundatrix with the commencement of the flow of sap in the host plant. Observations recorded by Gupta and Thakur (1993) support the present findings since they also visualized that egg hatching or Fundatrix development rate depended on the initiation of the sap flow. A rapid increase in the pest population, synchronized with increase in temperature and humidity was observed in the 3rd week of February. The peak population (592.00 aphids/ 10 leaves) was recorded in the 3rd week of March with mean atmospheric temperature and relative humidity being 16.2°C and 74.44 %, respectively. These observations suggest that among abiotic factors, temperature and relative humidity play predominant role in determining the population of soft bodied sucking insects. These findings fall in line with those of Ghosh and Raychaudhuri (1981) and Verma and Singh (1990) who also recorded rapid multiplication of this pest parthenogenetically resulting in huge population build-up within a short period. The peak pest occurrence varies due to variation in elevation as observed by (Sharma et al., 1968) who recorded pest activity in mid February at higher elevations and by the end of January at lower elevations. The data recorded in the present study revealed that rising temperature and decreasing humidity contributed to the declining of pest population which reached its low ebb (37.00/10 leaves) in the 3rd week of April at mean atmospheric temperature and relative humidity of 25.2°C and 61.80% respectively. This change in weather conditions also led to the production of winged forms within the aphid colonies which subsequently migrated to the summer alternate hosts and completely disappeared from trees. Further, a positive correlation was observed between aphid population build up and combined effect of mean atmospheric temperature on the population fluctuation of aphid reflected a positive correlation ($r_{12.3}$ = 0.5551). However, Ram and Pathak (1987) reported humid climate and stable temperature responsible for its population build up of B. helichrysi.

Mode of attack

The nymphal instars and apterous adults, having succivorous feeding habits, caused considerable damage by desapping the host plant (Plate 1). These stages of the pest drain considerable quantities of sap from the buds, blossoms and newly emerged leaves. Consequently, the foliage gets distorted and leaves become profusely crumpled, undersized fruits, pre mature fruit fall and ultimately poor fruit yield is the resultant effect. The resultant effect

of desapping was distorted foliage, profusely crumpled leaves, premature fruit drop, shriveled and under sized fruits and untimely low fruit yield. The present observations fall in line with those reported by Gupta et al. (1986) and Kapoor et al. (1989), who also reported the similar type of damage caused by this pest.

Overwintering of B. helichyrsi

Observations regarding the overwintering stage of the pest from December, to January (Table 1) revealed that nymphs produced in the last week of November over wintered in the axils and bases of dormant buds of peach tree. The population of overwintering nymphs was more in January than in December. The mean minimum number of over-wintering nymphs (0.67/10 buds) was during 3rd and 4th week of December and their maximum population (3.0/ 10 buds) was in the last week of January. The pest remained under overwintering conditions from 2nd week of December till last week of January as has also been recorded by Bennett (1955), Gupta and Thakur (1993) and Thakur et al. (1995).

Associated natural enemies of B. helichyrsi

Various predators and parasitoids found associated with B. helichyrsi are presented in Table 3. Coccinella septempunctata L., Leis dimidiata F., Coelophora sauzeti Muls., Ischiodon scutellaris (F.), Paragus (Paragus) serratus (Fabr.) and Paragus (Pandasyophthalmus) tibialis (Fallen) were recorded as aphidophagous predators and parasite, Diaeretiella rapae M'Intosh and Aphidius sp. as parasitoids of the pest (Plate 2, 3, 4 and 5).

C. septempunctata appeared in the 1st week of March whereas, L. dimidiate, C. sauzeti and I. scutellaris in the 2nd week of March (Table 4). Their population started building up and the maximum population was recorded in the 4th week of March which synchronized with the post peak aphid activity.

Adults and grubs of coccinellids and maggots of the syrphid flies were found feeding on aphids. The coccinellids captured the aphids by their legs and punctured them on the abdominal region by their powerful jaws and finally sucked out body contents. In some, complete swallowing of the prey was observed. The population of C. septempunctata was than all other predators, followed by I. scutellaris (2.50 per bunch) and maximum parasitism (7.25%) by D. rapae and Aphidius sp. was recorded in the 4th week of March. On wild weed, Ageratum, the maggots of two small species of Diptera namely P. serratus and P. tibialis were also preying on the aphids during November. Thalji (1981), Voicu et al. (1987) and Verma and Singh (1989) also reported the association of various predacious coccinellid beetles and syrphid maggots with B. helichyrsi. Predation of this pest by various

Table 3. Natural enemies complex of *B. helichyrsi* (Kalt).

Natural enemies	Order	Family
Coccinella septumpunctata Linn.	Coleoptera	Coccinellidae
Leis dimidiate F.	Coleoptera	Coccinellidae
Coleophora sauzeti Muls.	Coleoptera	Coccinellidae
Ischiodon scuilellaris (F.)	Diptera	Syrphidae
Paragus (Paragus) Serratus (Fab.)	Diptera	Syrphidae
Paragus (Pandasyophthalmus) tibialis (Fallen)	Diptera	Syrphidae
Diaeretiella rapae M'Intosh	Hymenoptera	Braconidae
Aphidius sp.	Hymenoptera	Braconidae

Table 4. Incidence and abundance of natural enemies associated with *B. helichrysi* (Kalt.) on peach.

Natural enemies	Weekly mean population*												
	March					April				May			
	I	II	III	IV	V	I	II	III	IV	I	II	III	IV
Coccinella Septumpunctata L.	0.75	1.75	2.50	3.25	3.00	2.75	2.50	2.25	1.75	1.00	0.50	-	-
Leis dimidiate F.	-	1.25	1.75	2.00	1.50	1.75	1.50	0.75	-	-	-	-	-
Coelophora sauzeti Muls.	-	0.50	0.75	1.00	1.00	0.75	0.50	0.25	-	-	-	-	-
Ischiodon scuilellaris F.	-	0.75	2.00	2.50	2.25	1.75	1.50	0.50	-	-	-	-	-
1. *Diaeretiella rapae* M'Intosh													
2. *Aphidius* sp.	3.25**	4.25	6.75	7.25	6.50	5.25	1.25	0.75	-	-	-	-	-

Mean of four observations per bunch/branch/tree. Per cent parasitism by parasitoid.

coccinellid species has also been reported by Bhagat et al. (1988), Chakrabarti et al. (1995), Bharadwaj (1995), Semyanov et al. (1996) and Sharma (2001). Observations regarding the association of hymenopteran parasites are confirmed by the findings of Verma and Singh (1989); Carver et al. (1993); Bhagat and Mir (1995) and Singh et al. (1995) who reared *Aphidius* spp. and *D. rapae* from *B. helichyrsi*. These natural enemies have been observed playing significant role in suppressing the pest population (Semyanov et al., 1996).

In the present investigations, the maximum per cent parasitism by *D. rapae* and *Aphidius* sp. was 7.25 which is almost similar to the findings of Verma and Singh (1989), who recorded 8.05% parasitism by *Aphidius matricariae* During the course of present studies, maggots of small dipteran species were found preying upon aphids on wild *Ageratum*. Predation on aphids by these species have not been reported from Jammu and Kashmir, as such they represent new records from this region. However, Agarwala et al. (1983) have reported the preda-tion of aphids by *Paragus* species from the West Bengal and Sikkim.

REFERENCES

Agarwala BK, Dutta S, Raychaudhuri DN (1983). An account of syrphid (Diptera: Syrphidae) predators of aphids available in Darjeeling District of West Bengal and Sikkim. Pranikee 4: 238-244.

Bennett SH (1955). The biology, life history and methods of control of the leaf curling plum aphid, *Brachycaudus helichyrsi* (Kalt). J.Hortic. Sci: 30: 252-259.

Bhagat KC, Masoodi MA, Koul VK (1988). A note on the

occurrence of a coccinellid predator in Kashmir. Current Research, University of Agricultural Sciences, Banglore 17(5) : 49-51.

Bhagat KC, Mir NA (1995). Aphidid parasitoids (Hymenoptera) of aphids (Homoptera) of Jammu- new records, host range and biological notes. J. Aphid 5(1&2): 90-96.

Bharadwaj SP (1995). Biological management of Brachycaudus helichyrsi (Kalt.) infesting almond in dry temperature region of Himachal Pradesh. J. Aphido (1&2): 26-31.

Carver M, Hart PJ, Wellings PW (1993). Aphids (Hemiptera: Aphididae) and associated biota fron The Kingdom of Tonga, with respect to biological control. Pan-pacific-Entomologist 63(3): 250-260.

Chakrabarti S, Debnath N, Ghosh D (1995). Bioecology of Harmonia eucharis (Mulsant) (Coleoptera: Coccinellidae) An aphidiophagous predator in Western Himalayas. Entomon 20(3&4): 191-196.

Childers NF (1975). Modern Fruit Science. Horticulture Publication, Rutgers University, Brunshrick NJ.

Ghosh LK, Verma KD (1988). A new host record of Brachycaudus helichyrsi (Kaltenbatch) in India (Homoptera: Aphididae). J. Aphid. 2 (1&2): 66-68.

Ghosh MR, Raychaudhuri DN (1981). Aphids (Homoptera: Aphididae) infesting Rosaceous fruit plants in Darjeeling district of West Bengal, Sikkim Entomon. 6 (1): 61-68.

Ghosh SP (1976). Research report, Fruit research workshop held at Ranikhet, India pp 13-19.

Gupta BP, Joshi R, Tripathi GM (1986). Control of peach leaf curling aphis, Brachycaudus helichyrsi (Kalt.) with granular formulations of some insecticides. Progressive Hortic. 18(1&2): 129-131.

Gupta PR, Thakur JR (1993). Sexual generation and overwintering of the peach leaf curling aphid, Brachycaudus helichyrsi (Kalt.) in Himachal Pradesh, India. Ann. Appl. Biol. 122 (2): 215-221.

Kapoor TR, Kashyap RK, Daulta BS (1989). A note on the effect of time of insecticidal application for the control of peach leaf curl aphid in peach (Prunus persica Linn). Haryana J. Horticl. Sci. 18(3&4): 239-241.

Madsen HF, Bailey JB (1958). Biology and control of leaf curl plum aphid in Northern California. J. Econ. Entomol. 51(2): 226-229.

Maison P, Massonie G (1982). First observation on the specificity of the resistance of peach to aphid transmission of plum pox virus. Agronomie 2(7): 681-683.

Panse VG, Sukhatme PV (1978). Statistical method for agricultural workers. Indian Council of Agricultural Research Publications, New Delhi p.108.

Ram LK, Pathak KA (1987). Occurrence and distribution of pest complex of some tropical and temperate fruits in Manipur. Bull. Entomol. 28(1): 12-18.

Root BR, Skelsey (1969). Biotic factors involved in crucifer aphid outbreaks following insecticides application. J. Econ. Entomol. 62(1): 223-233.

Sandhu GS, Khangura JS (1977). Chemical control of peach leaf curl aphid, Brachycaudus helichyrsi (Kaltenbach). Pestol. 4: 9-11.

Semyanov VP, Polesny F, Muller W, Olszak RW (1996). Lady beetles (Coleoptera: Coccinellidae) of Leningrad region orchards (fauna, biology and their role in the pest population dynamics).Bull. OILB-SROP. 19(4): 208-211.

Sharma N (2001). Studies on the predatory complex of peach leaf curling aphid, Brachycaudus helichrysi (Kalt) with special reference to the bug, Anthocoris minki (Dhorn). M.Sc.Thesis, Dr.YS Parmar Univ.Horti. & Forest, Nauni Solan.

Sharma PL, Attri BS, Shandil RL, Bhalla OP (1968). Biology and control of peach leaf curl aphid, Brachycaudus (Anuraphis) helichyrsi Kalt, Indian J. Entomol. 30: 289-294.

Singh NS, Paonam MS, Singh TK (1995). Aphid parasitoids (Hymenoptera: Aphididae) from Nagaland, North East India. J. Aphidol. 5 (1&2) : 61-63.

Thakur JR, Chauhan U, Dogra GS (1995). Autumn population and setting behaviour of peach leaf curl aphid, Brachycaudus helichyrsi (Kalt) on early varieties in Himachal Pradesh. J. Aphidol. 5(1&2): 77-85.

Thalji R (1981).Natural enemies of the aphid Brachycaudus helichyrsi (Kalt) (Homoptera: Aphididae) a pest of sunflower in Vojvodina. Zastita Bilja. 32(2): 147-153.

Verma KL, Singh M (1989). Evaluation of different predators and parasites as biological control agents on peach leaf curl aphid. Pesticides 23(1): 29-31.

Voicu MC, Sapunaru T, Nagler C, Buzdea S (1987). The role of predatory insects in the reduction of the population of aphid pests on sunflower. Cercetari-Agronomice in Moldova. 20(2): 135-139.

7

Electromagnetic radiation (EMR) clashes with honeybees

Sainudeen Pattazhy

Department of Zoology, S.N. College, Punalur, Kerala, India. Email-sainudeenpattazhy@hotmail.com

Apiculture has developed into an important industry in India as honey and bee-wax have become common products. Besides, honey bees do great environmental service by pollinating flowers. Bee keeping proves worthwhile from a monetary point of view as honey and wax command rewarding profits. In an average colony, there may be between 20000 to 31000 bees consisting normally of a queen and a few hundred drone. 90% of the population is made up of the workers. Recently, a sharp decline in population of honey bees has been observed throughout the Indian subcontinent resulting in devastating loses. For example, Kerala province has seen around 60% plunges in its commercial bee populations. Although the bees are susceptible to diseases and attack by natural enemies like wasps, ants and wax moth, constant vigilance on the part of the bee keepers can overcome these adverse conditions. The present plunge in population was not due to these reasons, it was caused by man.

Key words: Honey bees, bee-wax, population,

FULL TEXT

Bees and other insects has survived and evolved complex immune system on this planet over a span of millions of years. It is not logical that they would now suddenly die out now due to diseases and natural parasites. This suggests that another factor has been introduced into their environment that disrupts their immune system. This man made factor is the mobile towers and mobile phones.

The public is not being informed of the threat due to deliberate attempts on the part of mobile phone makers to mask the direct causal relationship. Over the past several months, a cadre of scientists, funded by the deep pockets of the mobile phone industry, has suggested that viruses, bacteria, and pesticides are to be blame for the unprecedented honey bee decline. Rather than critically assessing the problem, the industry is dealing with it as a political and public relation problem thus manipulating

Abbreviations: CCD, Colony collapse disorder; **BSNL,** bharat sanchar nigam ltd; **RFR,** radio frequency radiation; **KMC,** Kootenai medical center; **EMR,** electromagnetic radiation; **SARS,** severe acute respiratory syndrome; **WHO,** world health organization.

perception of the appropriate remedy. Sadly, this deceptive practice is business as usual for the mobile phone industry.

If the reason behind the population decrease were biological or chemical, there would be a pattern of epidemic spread. Observers would be able to trace the spread of bee disappearance from a source similar to the spread of severe acute respiratory syndrome (SARS) a few years ago. This pattern did not occur, however mobile towers and mobile phones meet the criterion.

New experiments suggest a strong correlation between population decline and cellular equipment. In one experiment, a mobile device was placed adjacent to bee hives for 10 min, for a short period of only 5 to 10 days. After few days, the worker bees never returned home.

The massive amount of radiation produced by towers and mobile phones is actually frying the navigational skills of the honey bees and preventing them from returning back to their hives. The thriving hives suddenly left with only queens, eggs and hive bound immature worker bees. Thus, electromagnetic radiation (EMR) exposure pro-vides a better explanation for Colony Collapse Disorder (CCD) than other theories. The path of CCD in India has followed the rapid development of cell phone towers, which cause atmospheric EMR.

Insects and other small animals would naturally be the first to obviously be affected by this increase in ambient radiation since naturally they have smaller bodies and hence less flesh to be penetrated by exposure to microwaves. The behavioral pattern of bees alters when they are in close proximity to mobile phones and towers. The vanished bees are never found, but thought to die singly far from home. Bee keepers observed that several hives have been abruptly abandoned. If towers and mobile phones increase, the honey bees might be wiped out in ten years. Radiation of 900 MHz is highly bioactive, causing significant alternation in the physiological function of living organisms.

Some countries have sought to limit the proliferation of mobile towers with strict rules. But in India, no such rules have been formulated or implemented. Given the proliferation of mobile phone towers and their vital role in communications, solutions to the problem will not be as simple as eliminating the towers. One possibility is shielding. Experiments confirmed that light aluminium does effectively block microwave radiations from hitting the hive directly. Since wrapping mobile phone and towers in aluminium foil would prevent communication, it would be better for mobile companies to pay bee keepers to protect their bee colonies to some extent from mobile phones and towers radiations with alumunium shielding. While this option could be easily implemented and has low costs, it proves less than optimal because worker bees will still be exposed to radiation as they fly to and from the hive.

Another solution would be granting local communities the ability to control whether or not to install mobile towers. On one hand, community members would be able to exert some control over their environment and determine whether the benefits outweigh the costs and risks. On the other hand, it is highly susceptible to manipulation by powerful influences, especially since the bee keepers have significantly less influence, power and wealth than the mobile phone companies.

However, Indians could risk losing even this right to self determination if the cellular providers can impose a country wide mandate prohibiting regulation against them, similar to the Telecommunications Act of 1996 in the United States. The Act prohibited local governments from making sitting decisions based on the perceived health impacts of wireless facilities. Indian advocates are concerned that such regulations might be upheld in India as they were in the United States in order to "eliminate service gapes in its cellular telephone service area."

In Kerala, there are about 600,000 bee hives, and over 100,000 workers are engaged in apiculture. A single hive may yield 4-5 kg of honey. Moreover, the destruction of bee hives could be a major environmental disaster. Honeybees are responsible for pollinating over 100 commonly eaten fruit and vegetable crops, and without bees, the food system would be in serious trouble. Rural village dependent on locally grown foods would be most vulnerable. The need of the hour is to check unscientific proliferations of mobile phone towers. More research is essential on how to protect the bee hives from the electromagnetic exposure, but perhaps more to study the impacts on humans.

Recently, the Bharat Sanchar Nigam Ltd. (BSNL) has suggested that the mobile phone towers that have been erected across the Kerala State do not cause health problems. Although the BSNL contends that it is safe, yet the government agencies involved in regulating radio frequency radiations have yet to prove that the towers were harmless. Keeping on these as the focal point, a field study has been undertaken by the Kerala Environmental Researchers Association (KERA) In Kollam Taluk areas. The study was conducted in more than 2000 houses situated within a kilometer of the towers in Kollam Taluk. In Kollam Taluk alone, there were more than 80 towers. In many places, more than three of them were present within half a kilometer radius. This unprecedented proliferation and construction of several towers across the state has raised the question of potential adverse health effects of microwave radiations emitted from these towers. The study showed that more than 40% of the people living in the vicinity of the towers, especially those of the middle age group and children, complained of eye problems, oblivion, sleep disorders, headache, etc. The people said that this predicament has emerged two years after the installation of the towers.

All mobile phone towers emit microwave radiations, which is in the radio frequency radiation (RFR), part of the spectrum of electromagnetic waves. Though RFR like ultra-violet (UV) and infra-red light, is a source of non-ionizing radiation, these radiations, together with ionizing EMR such as X- rays and gamma rays make up the electromagnetic spectrum. Radio frequency of the electromagnetic waves ranged from 100 kilo hertz (KHz) to 300 Giga hertz (GHz). RFR is a source of thermal energy and in adequate doses, has all the known effects of heating on biological systems, including burns and cataracts in the eyes. Human and animal studies in America indicate that radio frequency fields can cause harmful effects because of excessive heating of internal tissues. For most of the range of RFR, the skin does not easily detect the heating caused by these fields. The heating effect of RFR can become a problem in individuals with metallic implants such as rods in bones and electromagnetic interference can interact with cardiac pace makers. Acute high dose exposure to RFR may cause injury to the eyes. The cornea and lens are particularly susceptible to frequency of the 1- 300 GHz range, and formation of lesions in the retina is also possible. Long-term exposure to low level RFR has induced a variety of effects in the nervous system and components of the immune system of small animals. However, significance of these in humans is still not clear. A research study in Britain has suggested that RFR may act as a cancer promoter in animals. International organizations such as the world health organization (WHO) recognized the biological effects from exposure in the RF range that may affect health. In the Freiberger report, over 3000 German doctors have linked wireless phones and cell tower radiation to dramatic increase in disorders of learning, concentration and behavior among their patients.

Dr. Henry Lai, a leading radiation and biomedical researcher working at the University of Washington, stated that numerous medical studies show serious health effects can occur at irradiation levels far below current exposure standards. Anderson points to a December 2006 article in *Coeur d' Alene Magazine* which disclosed that the Kootenai Medical Center (KMC), serving as a magnet hospital for the five northern Idaho countries, is now inundated with cancer conditions of all types. The article reported that an average of 210 patients are cared for daily at KMC's North Idaho Cancer Center, and that more than 100 new cancer patients join the ranks of the "Big C" club every month.

According to the article, crushing is the cancer case load in North Idaho, that other cancer centers are being rapidly expanded in post falls and sand point. "One hospital worker told us that experienced medical personnel talk among themselves as having never before seen so much cancer among young people," Anderson said. "Our city has been increasingly saturated with microwave radiations from wireless tower and roof top transmitters since about the mid 1990s. Scientists say cancers can have a latency period of around 10 years, which computes with our area's growing cancer epidemic, she observed.

Dr. Lai's research group reported findings that microwave radiations can actually be physically addictive. The cell phone "high" is triggered by endorphins released into the brain when microwaves enter through the ear. Wireless industry adds, and promotions continually prod kids to buy new glitzy wireless hardware for watching TV, downloading music and texting. Kids know nothing about wireless health hazards because the industry is not required to warn them that at least 17 epidemiological studies show cell phone usage greatly increases their risk of developing brain cancer. "It's a real problem," said Anderson. "The more kids get hooked on wireless toys, the more towers are needed to service those toys, and if, they have their schools and play grounds irradiated by nearby transmitters, they are getting a double whammy."

CONCLUSION

Despite a growing number of warnings from scientists, the Government has done nothing to protect people and the environment. Steps must be taken to control the near thickly populated areas, educational institutions, hospitals, etc. Sharing of towers by different companies should be encouraged, if not mandated. To prevent overlapping high radiations fields, new towers should not be permitted within a radius of one kilometer of existing towers.

More must also be done to compensate individuals and communities who are put at risk. Insurance covering diseases related to towers, such as cancer, should be provided for free to people living in 1 km radius around the tower. Independent monitoring of radiation levels and overall health of the community and nature surrounding towers is necessary to identify hazards early. Communities need to be given the opportunity to reject cell towers, and national governments need to consider ways of growing their cellular networks without constantly exposing people to radiation.

Perhaps most importantly, bee keepers and humans have an inherent right to live without being exposed to radiation that affects their natural behavior and increases their risks of developing biological deformities, like cancer.

REFERENCES

Aday WR (1975). Introduction: Effects of electromagnetic radiation on the nervous System". Ann. NY Acad. Sci., 247: 15-20.

Goldsmith JR (1996). Epidemiological studies of radio-frequency radiation". Current status and areas of concern. Sci. Total Environ., 180: 3-8.

Leffell DJ (2000)."The scientific basis of skin cancer". J. Am. Acad. Dermatol., 42(1-2): 18-22.

Maes A, Collier MD, Verschaeve L (1995). Cytogenetic effects of Microwaves from mobile communication frequencies M Electro. Magnetobil., 954(14): 91-98.

Repacholi MH, Basten A, Gebski V, Noonand Finnie J, Harris AW (1997). "Lymphomas in Emupim I transgenic mice exposed to pulsed 900 MHz Electromagnetic field" Reat. Res., 147(5): 631-640.

Ross, Adey W (1998). "Cell Membranes, The Electromagnetic Environment and Cancer promotion". Neurochem. Res., 13(7): 671-677.

Wallaczek J (1992)."Electromagnetic Field effects on cells of the Immune System" The role of calcium signaling. F/ASEB J., 6: 3176-3185.

8

Indigenous knowledge of termite control: A case study of five farming communities in Gushegu District of Northern Ghana

Dokurugu Maayiem[1], Baatuuwie Nuoleyeng Bernard[1]* and Aalangdong Oscar Irunuoh[2]

[1]University for Development Studies, Faculty of Renewable Natural Resources, Department of Forestry and Forest Resources Management, P.O. Box TI 1882, Tamale
[2]University for Development Studies, Faculty of Renewable Natural Resources, Department of Range and Wildlife Management, P.O. Box TI 1882, Tamale

The study was conducted in five communities selected at random in the Gushegu-Karaga district of northern region of Ghana. The objective was to identify suitable and sustainable indigenous methods adopted by resource poor farmers for termite control. Semi-structured questionnaires were administered to 20 farmers in each of the selected communities who practiced indigenous termite control methods. A total of 100 farmers were interviewed. There was one female and the remaining 99 were males who had applied various indigenous treatments on their crop fields against termite infestation. The study recorded a total of 24 termite species, which varied in presence at each locality, with a few serious pest species damaging agricultural products such as maize, yam, millet, and other natural resources in the area. Five termite prevention and control methods were identified: (i) burial of plant and animal materials, (ii) application of wood ash, (iii) application of a mixture of salt and Shea butter residue, (iv) planting of elephant grass and (v) 'banchi' methods. Planting of elephant grass was found to be the most common method used by the farmers, while burial of plant and animal materials was found to be the most effective method of termite control in the area. Despite their well known role as pests, termites are considered important in the area because they provide necessary ecosystem services.

Key words: Infestation, pest species, damage, banchi' methods, elephant grass, wood ash.

INTRODUCTION

Termite infestation is prevalent worldwide especially in the tropics where distribution, extent of spread, problems and constraints results in livelihood threats (Dennis, 1987; Fenemore and Prakash, 2006), particularly among rural small scale farmers (Sileshi et al., 2008). The ever growing interest in sustainable agriculture and food security on the African continent highlights the need for a more balanced approach to termite control (Sileshi et al., 2008) that will prevent serious ecological damage and loss of ecosystem services provided by termites whilst using the available resources without exhausting them

(Logan et al., 1990).

Termites are abundant and diverse throughout the world (Donald and Dweight, 1970); with about 660 species out of the total of 2600 species found in Africa (Eggleton, 2000). In Ghana, 86 species are found, which belong to 38 genera, comprising of mound building and dry wood termites (Forsyth, 1966). In Ghana, some species (e.g. *Macrotermes, Microtermes* and *Odontotermes* species) cause widespread damage to crop seedlings whilst others (e.g. *Ancistrotermes, Allondoter-mes* and *Pseudacanthotermes* species) cause localized damage to forest trees, rangelands, food crops and other natural resources (UNESCO, 1997). Damage caused by termites is greater during periods of drought than during

*Corresponding author. E-mail: tuuwiebernard@yahoo.com.

the periods of regular rainfall (Logan et al., 1990; Nyeko and Olubayo, 2005).

The problem of termite infestation can have several effects such as agronomic, economic, or social constraints. The agronomic influence includes the role of termites as pests and ecosystem engineers; whereas, the economic aspect involves the destructive tendencies of termites due to their foraging activities on plants and wood products which cause economic hardship to individual producers (Fenemore and Prakash, 2006).

Reliable information on economic losses may not be available in Ghana but the threats imposed by termites on the production of food and industrial crops are quite evident. Information on population dynamics, outbreaks, damage incidences as well as available management efforts for termite infestation is essential to educate producers in termite-prone zones. In some African countries, information on economic losses is available, for example, in Kenya and Tanzania up to 30% damage has been recorded (Gitonga et al., 1995) while in Ethiopia 60% damage has been recorded (Wood, 1986). Also, interviews held with famers in south western Nigeria revealed that up to 100% damage can occur on maize production (Umeh and Ivbijara, 1997).

Chemical control of termites in plantations and farms is expensive and require skilled labour (Logan et al., 1990) and may not be effective in all cases (Nair, 2007). The excessive application of termiticides causes environmental pollution and may result in the death of non-target organisms as reported by Dennis (1981), which necessitated the ban of some chemical control measures.

Several indigenous methods are used by farmers to prevent and control termites in Ghana. They include wood ash, sand, toads and shell/scallop of tortoise (Akutse et al., 2012). Some of these methods are evaluated and documented for the southern belt of the country only. Information generated on the indigenous knowledge of termite management within the zone will be vital for priority setting and development of pest management strategies that meet local needs (Nyeko et al., 2002). The objective of this study was to identify suitable and sustainable indigenous methods to be adopted by resource poor farmers that best fit the biophysical, economic and socio-cultural conditions of termite control.

MATERIALS AND METHODS

Study area

The study was conducted in Gushegu-Karaga district in the northern region of Ghana located between latitude 9°30' and 10°30' north longitude 0° and 45° west. The district capital Gushegu is about 105 km away from the regional capital, Tamale.

The district is bordered to the north by East Mamprusi district, to the West by Karaga district, to the South by Yendi district and to the East by Saboba-Chereponi district (Figure 1). There are two main seasons in the district, that is, rainy and dry seasons. The district has a total rainfall of 900 mm–1000 mm per annum, the majority falls in the rainy season, which lasts from May to October and

peaks in August and September with the rest of the year being virtually dry. The temperature during the dry season and throughout the year remains high with 38°C or more recorded in March and April.

The vegetation is typically of guinea savannah type characterized by grasses interspersed with drought resistant trees such as Dawadawa, Shea, *Combretum*, Baobab, and Neem amongst others with an overlying shrub layer. Sedimentary rocks are dominant with substantial amounts of concretionary gravel layers near the top horizons and are suitable for road and other constructional works. The soils are mainly savannah ochrosols, groundwater laterites formed over granite and voltaian shale.

The dominant occupation of the people in the area is farming which occurs in mainly the rainy season. The major crops grown include maize, millet, rice, yam, and groundnuts.

Sampling technique

Five communities in the Gushegu district with possible indigenous termite management practices were randomly selected for the study. Snow-ball sampling technique (Lindlof, 1995) was used to identify farmers who practice indigenous methods of termite control in the area. Questionnaires were administered to 100 respondents (20 from each community) on the type of indigenous methods applied in termite control, reasons for the particular indigenous method. The farmers were within the ages of 20 to 65 years and had gained 1 to 30 years of experience in farming and the use of indigenous methods. Thus, years of experience was taken into consideration in order to ascertain the efficacy and problems associated with the use of indigenous methods.

Data collection and analysis

The study made use of semi-structured questionnaires and participatory rural appraisal (PRA) approach (farmers directly involved in identification and assessment of termite in infestation in the field) in achieving the objectives. Field surveys were conducted on the plots where indigenous termite management had been applied. Field observations of possible distance of termite nest/activity from the treatment spots were measured to ascertain the efficacy of indigenous methods.

The data collected using questionnaires were summarized using descriptive statistical packages.

RESULTS AND DISCUSSION

Characteristics of farmers in the study area

Out of a total of 100 farmers/respondents interviewed, one was female and the remaining 99 were males who had all utilized various indigenous treatments on their crop fields against termite infestation. The majority of the farmers were within the 30 to 39 year (42%) followed by the 40 to 49 year (26%) age groups and they were primarily considered the bread winners of their families with farming as their main occupation. They constituted the economically active population age cohort (Ghana Statistical Service, 2000). The farmers belonging to the 60+ year (1%) age group formed the lowest proportion of respondents among those interviewed.

The majority of the farmers (82%) utilized indigenous knowledge for the management of termite infestations on

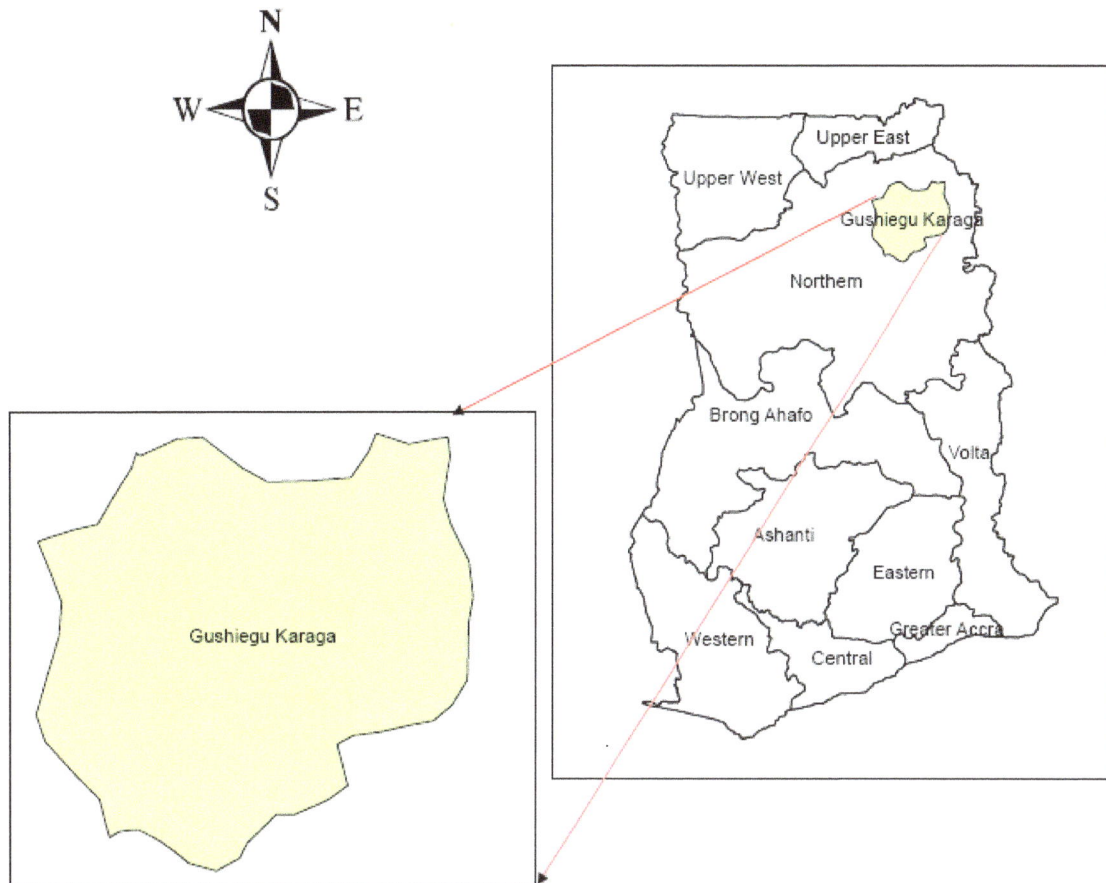

Figure 1. Map of Ghana showing study area, Gushegu-Karaga.

areas between 1 to 3 acres while 18% were working on farm sizes of four acres and above. Banchi is a plant that resembles cassava, *Manihot ultissima* Phol, farmers propagate by cuttings as it is claimed to have the ability to control termites Those farmers that used "banchi" were mostly working on a farm of one acre while those who practiced the burial of plant and animal materials (that is, pounded plant parts together with animals intestines or whole animal buried in the field) mentioned that their method could control termites on a farm size of four acres or more.

Termite species diversity and distribution

Twenty four (24) species of termite were identified in the study area. Out of this number, not all were common in the selected communities (Table 1). The genus *Macrotermes, Odontotermes, and Microtermes* were common throughout the study area comprising 20% of the total. The community of Kanimo had the highest number of 13 different termite species and the community with the least number was Bilsing with nine species (Table 1). The result affirms the findings of Harris (1971) that, the number and species of termites vary within a locality.

Few species were recorded as serious pests of agriculture and other natural resources in the study area as was reported elsewhere by Logan et al. (1990) and Wardell (1990). Though the spatial distribution of the species in the study area varied, the majority of the famers indicated that the pest species present were abundant in number. According to Thacker (2002), most pest species, under favourable conditions, have a high reproductive rate and thus may cause habitat destruction when ensuring their own survival. As such, the availability of suitable food and habitats could have enhanced fecundity for any species of termite pests enabling them to reproduce fast enough to cause enormous destruction. The abundance could as well be attributed to the reduction or absence of natural enemies, such as birds, pangolin, aardvark and amphibians, due to habitat destruction or changes in ecological conditions (Jiru, 2006).

Termite damage and indigenous management methods

Crops that were grown in the study area and more importantly were affected by termites included yam, maize, cassava and pepper (Table 2). The crops were

Table 1. Termite species and spatial distribution in the study area.

Genus	Total number of specie	Specie distribution in the community				
		Kanimo	*Dayoudigli*	*Bilsing*	*Dam*	*Zebihikura*
Termitidae	-	-	-	-	-	-
Macrotermes	4	4	4	1	3	3
Ancistrotermes	3	1	-	2	-	1
Anoplotermes	1	1	1	-	1	-
Acanthotermes	1	-	-	-	1	-
Allondotermes	1	-	-	-	-	1
Odontotermes	3	3	2	2	2	1
Pseudacanthotermes	1	1	-	-	-	-
Pericapritermes	2	1	2	-	2	-
Termes	1	1	-	-	-	-
Basidentitermes	1	-	-	1	-	-
Allognathotermes	1	-	1	-	-	-
Anenteotermes	1	-	-	-	-	1
Microtermes	1	1	1	1	1	1
Rhinotermitidae	-	-	-	-	-	-
Coptotermes	1	-	-	-	-	1
Schedorhinotermes	2	-	-	2	-	1
TOTAL	**24**	**13**	**11**	**9**	**10**	**10**

Table 2. Commonly damaged crops by termites in the study area.

Common name of crop	Scientific name
Maize	*Zea mays*
Yam	*Dioscorea spp.*
Cassava	*Manihot esculenta*
Sweet potato	*Ipomea batata*
Millet	*Pennisetum glaucum*
Groundnuts	*Arachis hypogea*
Rice	*Oryza sativa*
Soya beans	*Glycine max*
Cowpea	*Vigna unguiculata*
Pepper	*Capsicum spp.*
Okra	*Albemulchus esculentus*
Garden eggs	*Solanum melongena*

often observed to be significantly damaged before or after harvest. Though there was no quantitative data on extent of termite damage, the study revealed that the highest crop damage occurred in maize production followed by yam, while other crop damages were minor. High crop damage in maize confirms the findings of Umeh and Ivbijara (1997), obtained by farm interviews held with famers in south western Nigeria, that 100% damage by termites can occur in maize production.

There was no reported incident of sorghum damage on crop fields during the study thus, confirming the assertion that sorghum is protected from termite damage, perhaps due to its role as a reservoir of termite predatory ants (Sekamatte et al., 2003). About 55% of the respondents reported that partial damage occurred in various crops ranging from the seedling to harvest phase but peak damage usually occurred when harvest was delayed with a 100% possibility of damage occurring in storage facilities for every crop. Direct observations revealed that in the study area termite damage was not limited only to crops but to all sorts of resources such as buildings, farm huts, trees, wood and products.

The research outcome as indicated in Table 3 revealed five main methods used by farmers in the study area for the control of termite infestations. Some of the methods are commonly used in the southern belt of the country as reported by Akutse et al., (2012). A single application of any of these methods was said to be enough to protect the field for several seasons, except the wood ash method that required annual application. None of the farmers practiced a combined or an integrated treatment method.

It was realized that all the methods they used did not kill termites but some acted as repellents. This may have

Table 3. Indigenous management methods used in the study area.

Method used	Planting of elephant grass	Planting of "banchi/yoobkarugu"	Burial of plant and animal materials	Wood ash	Salt in shea butter residue
Frequency of application	Once	Once	Once	Any time before storage	Once
Area applied	Farm	Home	Any where	Any where	Termite nest and infested field
Time applied	Rainy season	Rainy season	Any season	Any season	Any season
Method of application	Planting of cuttings in underground tunnel	Planting of cuttings on infested field	Pound plant parts and bury with intestines or whole animal	Spread on floor and keep harvested produce on top	Spray in nest/on field
% of farmer users (100)	46	22	18	10	4

Banchi is a plant that resembles cassava (*Manihot ultissima* Phol) in morphology.

been intentional, as termites are used by farmers as a cheap source of protein feed for chickens during the first four weeks of the chickens' growth in the area. Apart from the provision of chicken feed, termites as reported by Nyavor and Seddoh (1991) provide quite a number of ecosystem services such as soil enrichment through nutrient cycling and the minimization of wildfire hazards through the removal of fuel litter (Lepage et al., 1993).

The method of usage by farmers as outlined in Table 3, planting of elephant grass was the method most commonly used by the people (46%). This could be due to the fact that, planting of elephant grass has a higher efficacy then the rest. It could also be partially due to the readily available and accessible planting materials in the study area. Dissolved salt (sodium chloride) in Shea residue was the method least used by farmers (4%). This could be attributed to cost of the materials for application as salt is primarily used as a cooking ingredient it is fairly expensive for farmers to purchase in the quantities needed.

The results of the questionnaire indicated that the respondents had no clear indication as to the level of control each particular method exerted on their respective fields. However, measurement of termite locations/termitaria from the treatment spots was used to calculate the average distances that the termites were repelled by the various Methods (Figure 2).Fields treated with plant and animal materials recorded the highest mean distance of 374 m, while fields treated with salt dissolved in Shea butter residue recorded the lowest average distance of 33 m. During the entomological survey, it was observed that termites had returned to certain fields where "banchi" was used as a termite management tool. Therefore, this supported the claims made by farmers (21%) during the interview that the method had lost its efficacy in the control of termites. For instance, out of the 22 fields that were treated with "banchi", only one showed good control results while the other 21 were invaded by termites. The efficacy loss of this method could be attributed to a reduction in the number of red ants introduced as a biological control mechanism against termite pests when the "banchi" roots decay. Thus, when the "banchi" roots decay, they attract the red ants to the area which are predators to the termites. The presence of these ants on farmsteads could have been prevented by the massive indoor, farm and farm huts residual spraying intended to control weeds and pests in the study area. This agrees with the findings of Sekamatte and Okwako (2007) that, Ugandan elders linked the increasing termite problem and low abundance of predatory ant species to aerial sprays intended to control tsetse flies (*Glossina sp*) during the 1960s and 1970s.

The control of termites by elephant grass could be due to the presence of antixenose mechanisms in the plant, properties that deter or prevent

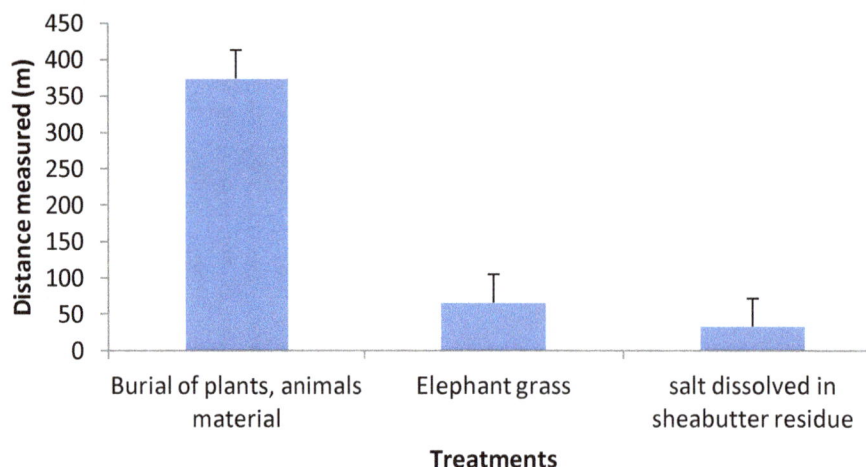

Figure 2. Mean distances of termites/termitaria from fields treated by different indigenous methods.

colonization of plants with termites. Based on the results, the burial of plant and animal materials proved to be the best method for termite control. The reason for this was the method induced the invasion of large numbers of ants on the field to act as biological enemies of termites. This finding supports Logan et al. (1990) that protein-based bait results in greater ants nesting near maize plants and hence reducing termite damage. It also affirms Seka-matte et al. (2001) who reported that reduction in termite damage in plots that received a fish meal treatment was due to the increased number of predatory ants.

Conclusion

The existence of 24 species of termites in five localities in Gushegu-Karaga District has serious implication on natural resources especially the presence of the known pest genera such as *Odontotermes*, *Macrotermes* and *Microtermes* in the area. Farmers' innovation was evident in the diversity of indigenous termite control methods that were employed in the study area. Five methods of termite control identified in the study area were said to protect the fields for several seasons upon a single application. These methods included: planting of elephant grass, "banchi/yoobkarugu", burial of plant and animal materials, wood ash and salt in shea butter residue. Reports, direct observations and field measurements gave evidence towards the efficacy of these methods.

REFERENCES

Akutse KS, Owusu EO, Afreh-Nuamah K (2012). Perception of farmers' management strategies for termites control in Ghana. J. Appl. Biosci., 49: 3394– 3405

Dennis SH (1987). Agricultural insects' pest of temperate regions and their control. Third edition. Press Syndicate of the University of Cambridge. New York. Pp 99, 525.

Dennis SH (1981). Agricultural insects of the tropics and their control. Second edition. Press Syndicate of the University of Cambridge. New York.169-177.

Donald JB, Dweight MD (1970). Introduction to the study of insects. Third edition. Holt, Rinehart and Winston INC, 152-158.

Eggleton P (2000). Global patterns of termite diversity. Kluwer Academic Publishers, Dordrecht, Netherlands. 25-51

Fenemore FG, Prakash A (2006). Applied Entomology. Second edition. New Age International (P) Ltd publishers. 200-203.

Forsyth J (1966). Agricultural Insects of Ghana. Ghana Universities Press, Accra.76-78.

Ghana Statistical Service (2000). Summary of the 2000 Population Census Report. Accra; Ghana: Ghana Statistical Service.

Gitonga W, Kusewa TM, Ochiel GRS (1995). Comparison of chemical and cultural against termites in Western Kenya. In proceedings of second regional Workshop on Termites Research. March 7-9, Nairobi Kenya. 169-216.

Harries WV (1971). Termites: their Recognition and Control. Second edition. Longman Publishers. 15-32.

Jiru D (2006). Trees with insecticidal properties and indigenous knowledge base on copping mechanism against pest. Drylands Coordination Group, Addis Ababa, Ethiopia, 88-91.

Lepage M, Abbadie L, Mariotti A (1993). Food habits of sympatric termite species (Isoptera *Macrotermtinae*) as determined by stable carbon isotope analysis in Guinean savanna. J. Trop. Ecol. Lamto, Cote d'Ivoire. 9:303-311.

Lindlof TR (1995). Qualitative communication research methods. Thousand Oaks: Sage.

Logan JWM, Cowie RH, Wood TG (1990). Termite (Isoptera) control in agriculture and forestry by nonchemical methods: Rev. Bulleting Entomol. Res., 80:309-330.

Nair KSS (2007).Tropical Forest Insects Pest: Ecology, Impact, and Management. The Press Syndicate of Cambridge University. New York, 218-236.

Nyavor CB, Seddoh S (1991). *Biology GAST for Senior Secondary School*. Unimax publishers Ltd in association with Macmillan publishers Ltd.58-60pp.

Nyeko P, Gareth-Jones E, Day RK, Thomas R (2002). Farmers' knowledge and perceptions of pests in agroforestry with specific reference to Alnus species in Kabale District, Uganda. Crop protection. 21(10) 929–41

Nyeko N, Olubayo FM (2005). Participatory assessment of farmers' experience of termite problems in Agroforestry in Tororo district. Agriculture Research and Extension Network paper No 143.Overseas Development Institute, London, UK.

Sekamatte MB, Okwako MJN (2007). The present knowledge on soil pests and pathogens in Uganda. Afric. J. Ecol., 45:9-19.

Sekamatte MB, Latigo OM, Smith AR (2003). Effects of maize- legume intercrops on termites damage to maize, activity of predatory ants and maize yield in Uganda. Ugandan J. Crop Protectn., 22:653:662.

Sekamatte MB, Latigo OM, Smith AR (2001). The potentials of proteins and sugar based baits to enhance predatory ants and reduce termite damage to maize in Uganda. Ugandan J. Crop Protectn., 20:653-662.

Sileshi G, Akinnifesi FK, Ajayi OC, Chakeredza S, Mngomba S, Nyoka BI (2008). Towards sustainable management of soil biodiversity in agriculture and landscape in Africa. J. biodiversity, Zambia. 9:64-67.

Thacker JRM (2002). An Introduction to Arthropods Pest Control. The press Syndicate of the University of Cambridge. 94-97.

UNESCO (1997). Biodiversity Conservation: Traditional Knowledge and Modern Concepts. Proceedings of UNESCO MAB Regional Seminar on Biosphere Reserves For Biodiversity Conservation and Sustainable Development in Anglophone Africa. March, 9-12. Enviro. Protectn. Agency. Accra, Ghana. 140-146.

Umeh VC, Ivbijaro MF (1997). Termite abundance and damage in traditional maize- cassava intercrops in southern Nigeria. J. Insects Sci. Appl., 17: 315-321.

Wardell DA (1990).The African termite: peaceful coexistence or total war? Agroforestry Today. 3: 4-6.

Wood TG (1986). Report on visit to Ethiopia to advice on assessment of termite damage to crops. Report R1347(R).ODNRI, London. 52-80.

9

Evaluation of new molecules against scarlet mite, *Raoiella indica* Hirst in arecanut

B.K. Shivanna[1]*, B. Gangadhara Naik[2], R. Nagaraja[3], S. Gayathridevi[4], R. Krishna Naik[5] and H. Shruthi[6]

[1]Department of Agricultural Entomology, University of Agricultural Sciences (UAS), AINRP(T), ZARS, Shimoga -577204, Karnataka, India.
[2]Department of Plant Pathology, University of Agricultural Sciences (UAS), College of Agriculture, Shimoga -577204, Karnatak, India.
[3]Department of Agricultural Microbiology, University of Agricultural Sciences (UAS), KVK, Shimoga -577204, Karnataka, India.
[4]Department of Agricultural Entomology, College of Agriculture, University of Agricultural Sciences (UAS), Shimoga -577204 Karnataka, India.
[5]Department of Computer Science, College of Agriculture, University of Agricultural Sciences (UAS), Shimoga -577204 Karnataka, India.
[6]Department of Agricultural Microbiology, College of Agriculture, University of Agricultural Sciences (UAS), Shimoga -577204 Karnataka, India.

Scarlet mite *Raoiella indica* Hirst (Acari: Tenuipalpidae) is an important sucking pest on young arecanut palms during dry weather in areca growing tracts. The registered insecticides that provide adequate control of the pests need repeated application in higher doses which result in adverse effects on the environment and health. In order to circumvent the problems, replacement of conventional insecticides with new powerful molecules at lower dose is necessary. Hence, a replicated field experiment was conducted at five different locations for two consecutive years (2008/2009 and 2009/2010). Two sprays each of the new molecules fenazaquin (10EC at 1.5 ml/L), diafenthiuran (50WP at 1.2 g/L) and propargite (57EC at 0.5 ml/L) were compared with wettable sulphur (80% WDG at 2.5 g/L), dicofol (20EC at 2.5 ml/L), azadirachtin1300 ppm (0.03% at 3 ml/L) and untreated control. Pooled results showed that five days after spray, all the treatments recorded significantly less number of mites (per cm^2 leaf) as against control. Propargite and diafenthiuran were on par with each other and were significantly superior over dicofol and wettable sulphur by registering the lowest number of mites. Fenazaquin was on par with dicofol and wettable sulphur with less number of mites. However, the botanical azadirachtin recorded maximum number of mites. Results suggested that the new molecules, propargite (57EC at 0.5 ml/L) or diafenthiuran (50WP at 1.2 g/L) can be used for effective management of mites in arecanut. Further, fenazaquin (10EC at 1.5 ml/L) can also be used as an alternative to existing conventional insecticides.

Key words: *Raoiella indica* Hirst, arecanut mite management.

INTRODUCTION

The arecanut palm, *Areca catechu* L. (Palmae) is the source of arecanut commonly referred to as betelnut or supari in India. Since time immemorial, it is being used in

masticatory (chewing), religious and social ceremonies (Murthy, 1968). Arecanut is largely cultivated in the plains and foothills of Western Ghats and North Eastern regions of India. Karnataka, Kerala and Assam account for over 90% of area and production. Less labour intensive and good price in the last two decades forced the farmers to cultivate the crop with improved varieties in changed agro-climatic conditions. Although arecanut has been an

*Corresponding author. E-mail: bkshivanna@gmail.com.

Table 1. Response of arecanut mites to different insecticidal sprays.

Treatment	Number of mites per cm² leaf in a plant								
	PTC			5 DAT					
				I			II		
	2008	2009	Pooled	2008	2009	Pooled	2008	2009	Pooled
Wettable Sulphur (2.5 g/L)	9.06 (3.06)*	9.13 (3.15)	9.09 (3.09)	4.07 (2.03)	4.2 (2.19)	4.14 (2.15)	1.9 (1.88)	1.9 (1.84)	1.9 (1.55)
Azadaractin (0.03%; 4 ml/L)	9.12 (3.18)	9.10 (3.14)	9.13 (3.17)	4.12 (2.28)	4.21 (2.35)	4.15 (2.16)	4.1 (2.16)	4.6 (2.10)	4.31 (2.22)
Fenazaquin (10 EC; 1.5 ml/L)	9.06 (3.09)	9.13 (3.15)	9.07 (3.17)	1.20 (1.77)	1.2 (1.73)	1.20 (1.30)	1.5 (1.64)	1.6 (1.59)	1.52 (1.43)
Diafenthiuron (50 WP; 1.2 g/L)	9.06 (3.01)	9.10 (3.12)	9.09 (3.09)	1.01 (1.19)	1.43 (1.40)	1.21 (1.33)	1.02 (0.99)	1.3 (1.6)	1.13 (1.30)
Propargite (57 EC; 0.5 ml/L)	9.13 (3.12)	9.13 (3.15)	9.13 (3.10)	1.10 (1.29)	1.43 (1.43)	1.25 (1.34)	1.0 (1.18)	1.4 (1.33)	1.2 (1.32)
Dicofol (20 EC; 2.5 ml/L)	9.2 (3.27)	9.2 (3.18)	9.2 (3.12)	1.4 (1.97)	1.50 (1.89)	1.43 (1.40)	1.6 (1.89)	1.7 (1.93)	1.61 (1.47)
Control	9.30 (3.28)	9.21 (3.17)	9.26 (3.11)	9.06 (3.06)	9.8 (3.0)	9.03 (3.17)	9.40 (3.39)	9.02 (3.01)	9.21 (3.10)
CV (%)	6.5	2.50	0.15	13.74	13.33	1.91	5.81	17.34	2.42
CD at 5%	0.31	0.12	0.01	0.40	0.40	0.06	0.16	0.48	0.08

PTC= Pretreatment count, DAT= Days after treatment; *figures are √x+0.5 transformed values.

important commercial crop, due to lack of scientific knowledge and ignorance by the cultivators on agronomic aspects, pest and diseases, considerable crop losses were encountered in fields. An array of insect and non-insect pests infests all parts of the palm, such as stem, leaves, inflorescence, roots and nuts in one or other stage of the crop growth. As many as 102 insect and non-insect pests have been reported to be associated with arecanut palm (Nair and Daniel, 1982). Among them, mites are the serious pests in young areca plantation on leaves which are active after the onset of hot weather (Patel and Rao, 1958). The two major species of foliage feeding mites are the cholam mite/white mite (Oligonychus indicus Authority) and the palm mite/ red mite (Raoiella indicia Hirst). Both nymphs and adults of R. indica live in colonies on lower surface of leaves by de-sapping, leading to the formation of yellowish speckles on the lamina which later coalesces, become bronze coloured and the leaves wither away. Suggested chemicals against foliage mites, such as wettable sulphur (Bhat et al., 1957; Puttarudriah and Channabasavanna, 1957), dicofol, dimethoate and phosphamidon (Devasahayam and Nair, 1985) that are in vogue, needs to be replaced with safe and efficient molecules.

MATERIALS AND METHODS

A multi location field trial in three districts (five locations) was conducted for two consecutive seasons, during 2008/2009 and 2009/2010 in randomized block design with seven treatments and three replications (Table 1). Two insecticidal sprays were given at an interval of 15 days. The spray fluid was applied to the lower surface of leaves at the rate of 500 L/ha with a knapsack sprayer. Ten plants were randomly selected in each plot by tying with luggage labels. Observations on number of mites/cm^2 on top, middle and bottom leaves of selected plants were recorded a day before spraying (pre-treatmental count, PTC) and 5 days after treatment. The efficacy was computed as reduction in number of mites compared to control. The data on the (average of top, middle and bottom leaf of each plant) mean of three replications were considered for statistical analysis after square root transformation.

RESULTS AND DISCUSSION

The results with respect to mite population were significant, indicating differential efficacy of the treatments imposed. Pooled data of two years in all the locations showed significant treatment differences for number of mites/cm^2 leaf/plant. Least number of mites (1.30 and 1.32 mites/cm^2 leaf/plant) was observed after 2nd spray on the areca palm treated with diafenthiuron and propargite respectively and was significantly superior over rest of the treatments. The level of mite population in standard check dicofol (1.47 mites/cm^2 leaf/plant) was on par with fenazaqin and wettable sulphur (1.43 and 1.55 mites/cm^2 leaf/plant, respectively). However, the plant based azadirachtin displayed moderate level of control (2.22 mites/cm^2 leaf/plant) and was significantly different from the unsprayed control which recorded the highest population of 3.10 mites/cm^2 leaf/plant.

The reduction in mite population was due to the efficacy of newer molecules such as diafenthiuron, propargite and fenazaquin which are target oriented. Literature on these molecules against scarlet mite was meager. However, minimum population of mites observed in present findings in dicofol and wettable sulphur treated plots were in confirmation with the results reported earlier by Bhat et al. (1957), Puttarudriah and Channabasavanna (1957), Kanth et al. (1963), Ponnuswamy (1966), Anonymous (1967) and Devasahayam and Nair (1985).

REFERENCES

Anonymous (1967). *Annual Report of the Central and Regional Arecanut Research Station* 1964-65. Central Arecanut Research Station, Vittal, p. 92.

Bhat KS, Patel GI, Bavappa KVA (1957). Preliminary observations on the yellow leaf disease of arecanut palm. *Arecanut Journal*, 8: 61-62.

Devasanayam S, Nair CPR (1985). Chemical control of palm mite, *Raoiella indica* Hirst on arecanut. In: Arecanut Research and Development. (Shama Bhat K, Nair CPR, eds). Proc. SIJAR, 1982, CRCRI. Kasaragod. Pp. 140-142).

Kantha S, Ray BK, Lal R (1963). Laboratory evaluation of the toxicity of insecticides to palm mite, *Raoiella indica* Hirst. Ind. Coconut J., 16: 63-66.

Loganathan S, Marimuthu S, Ramarthinam S (2000). A survey of phytophagous mites in arecanut plantation in Coimbatore district of Tamilnadu. Insect Environ., 6: 67-68.

Murthy KN (1968). Arecanut growing in north east India. Indian Farming, 18: 21.

Nair CPR, Daniel M (1982). Pests. In: *The Arecanut palm*. (Bavappa KVA, Nair MK, Prem Kumar T, eds). CPCRI. Kasaragod, pp. 151-184.

Patel GI, Rao KSN (1958). Important diseases and pests of arecanut and their control. Arecanut. J., 9: 89-96.

Ponnuswamy MK (1966). Trials with Pesticides against red mite, *Raoiella indica* Hirst in Kerala. Plant Protection Bull., 18: 27-28.

Puttarudriah M, Channabasavanna GP (1957). Preliminary acaricidal tests against the areca mite, *Raoiella indica* Hirst. Arecanut J., 8: 87-88.

10

Histological study on the body wall of *Ascaridia galli* (Nematoda)

Zohair I. F. Rahemo* and Sagida S. Hussain

Department of Biology, College of Science, University of Mosul, Mosul, Iraq

The body wall of Ascaridia galli has been investigated using 1-2 µ thick sections and stained with Toludin Blue. It is found to consist of three layers, cuticle, hypodermis or subcuticle and muscle layer. The cuticle in turn consists of three layers namely cortex or upper layer deeply stained with TB, 1.7 µ in thickness, middle layer or matrix 3.15 µ in thickness with no distinct constitutions, then fiber layer which in turn consists of several layers of dense tissue, its thickness 5.65 µ and settled on basement membrane. On the surface of the cuticle plugs a transverse annuli which divide the body into annuli, the distance between each two 0.22 µ each annulus is divided by three subannuli, large anterior 0.124 µ in length and other two are small, middle and posterior 0.047-0.054, 0.045-0.054 µ respectively. The hypodermis or dermis is thin and not well developed and of syncytial type and its nuclei are constricted to the longitudinal cords that is, dorsal, ventral, and two laterals. Mainly these nuclei are present in the lateral cords and longitudinal excretory canals which pass through these cords. The hypodermis become more thicker as proceed posteriorly to reach 2 - 3 µ in thickness in the inter-caudal region as well as thicker in the four longitudinal cords protruded in the pseudocoel and reach the maximum thickness near the nerve ring. In addition to these cords there also four secondary cords which divide the muscle groups into 8 sectors and. Moreover lateral cells in the middle of the body are larger and more prominent becoming pyriform clusters or cup-shaped forming lateral hypodermal glands. The average depths of these glands are 97 µ in male and 148 µ in female. A canal or duct arises from each gland the two canals united to open together and a prominent cuticular pominance is seen near the opening of these glands.

Key words: *Ascaridia galli*, cuticle, nematode, layer, cortex.

INTRODUCTION

The body wall of nematodes has been subjected to investigation especially those of Ascaris lumbricoides and other ascarids (Bird, 1971; Chitwood and Chitwood, 1974), as they studied cuticle and its layers and its chemical compositions. Roggen et al. (1967) studied the body wall, cuticle, hypodermis, and muscle layer of *Xiphinema ibndex* which is a plant parasitic nematode .A detailed ultrastructural study has been performed on the body wall of potato parasitic nematode, Heterodera rostochiensis (Wisse and Daem, 1968). In addition the body wall of the free living nematodes, *X. xenoplax* has

been investigated especially its cuticle layers, hypodermis, and body muscles (De Grisse, 1972).

Ultrastructure of *Syphacia obvelata*, a roden nematode, observed the cuticular modification especially the middle layer in the lips, papillae, buccal cavity by Wright and Hope (Wright and Hope, 1968), while Batson (1979) demonstrated the ultrastructure of *Gastromermis boophthorae* and its larvae and its relation with nutrition. Lee et al. (1993) made a freeze-fracture study on the cuticle of adult *Nippostrongylus brasiliensis*. Nawab-Al-Deen (1994) studied the cuticle ultrastucture of the fish parasitic nematode, *Rhabdochona tigrae*. DeCraemer et al. (1996) studied the cuticle ultrastructure of the free living nematodes, *Criconema paradoxiger* especially the external annulation.

*Corresponding author, E-mail: zohair_rahemo@yahoo.com

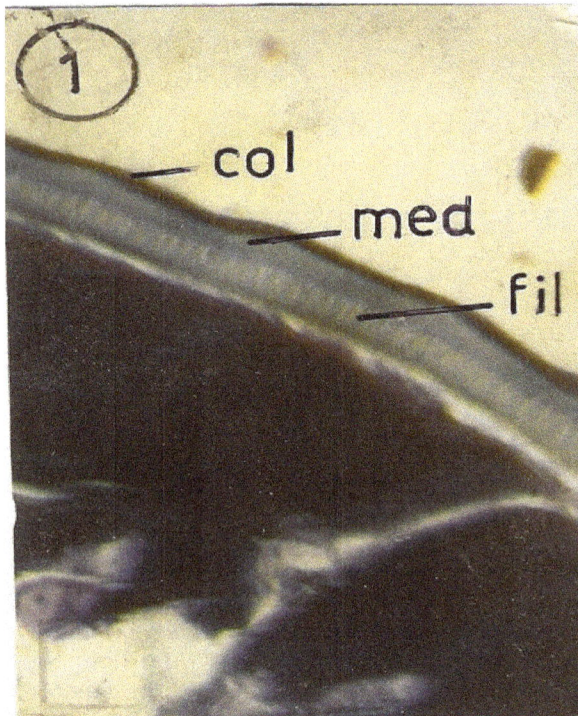

Figure 1. Photomicrograph of A section passing through the cuticle layers of *Ascaridia galii*.
Col; cortical layer, **med**; median layer, **fil**; fibrous layer. Toludin Blue X400.

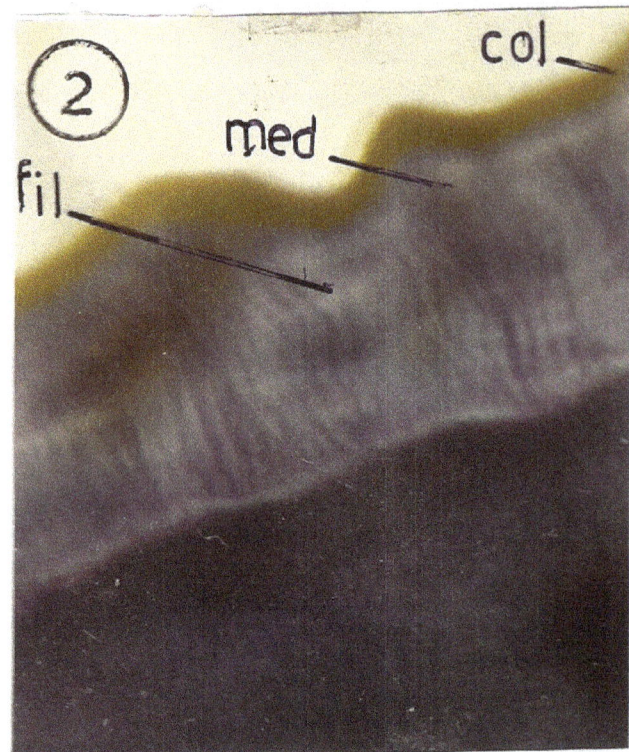

Figure 2. A magnified photomicrograph of a section passing in the body wall of cuticular layers
Col; cortical layer, **med**; median layer, **fil**; fibrous layer. TB ×1000.

MATERIALS AND METHODS

Worms of A. galli recovered from local fowl, Gallus gallus domesticus which have been bought from local markets in Mosul. Worms were removed, washed in Hanks solution, pieces were fixed in gluteraldehyde dissolved in 2% phosphate buffer for 60-90 min in ice bath (4 C), and then specimens were moved to 1% osmium tetra oxide in phosphate buffer for 90-120 min in ice bath. Specimens then washed in distilled water then dehydrated in ascending series of alcohol (50, 70, 90, and 100%) then cleared in proline oxide. Specimens then embedded in Epon-812, then sectioned using ultramicrotome provided with glass knifes with an angel of 55, with thickness ranged between 1-2 µ, put on slides in 60 c for flattening .Then sections were stained in Toludin blue, mounted in DPX and examined and photographed by self-built camera.

RESULTS AND DISCUSSION

The body wall of *A. galli* consists of three layers namely cuticle, subcuticle or dermis and a muscle layer. In this basic layers it resemble most nematodes studied (Chitwood and Chitwood, 1974; Nawab-AlDuin, 1994; Decraemer et al., 1996; Hyman, 1951; Roberts and Janovy, 2005).

Cuticle

Cuticle in the present worm, *A. galli*, which is revealed by light microscopy found to consists of three layers namely: cortex or the upper layer stained deeply with Toludin blue 1.7 µ in thickness ,then middle layer or matrix 3.15 µ in thickness with no distinct constitutions (Figures 1 - 3), middle layer or matrix homogenous layer 5.65 µ in thickness , and fiber layer which in turn consist of several very thin layers , 5.65 µ in thickness arranged around the worm and settle on the basement membrane. Cortex show small round or oval pores connected by small canals (Figure 4).

The cuticle which cover all the body invaginate in different regions as mouth, anus, vulva, rectum, cloaca in addition also to cover the sense organs such as amphids, phasmids, and some time evaginate to cover some other sense organs such as cephalic, labial and caudal papillae. The thickness of cuticle differs in different body regions as the external cuticle which is usually thicker than the internal cuticle. In addition structure of cuticle differ in different regions for example matrix or the middle region in some sense organs and the cortex and basal layers not persist. It is believed that such lost help in accelerating transmission of nerve impulses to nerve endings underneath the thin cuticle as reported by Bird (1971) and as found in *A. lumbricoides*.

On the surface of cuticle of *A. galli* plugs a transverse

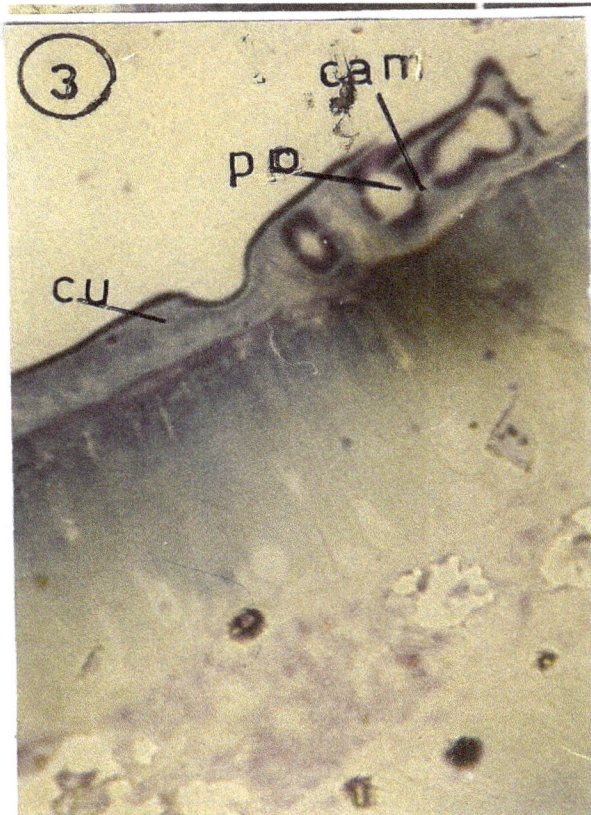

Figure 3. Photomicrograph of a transverse section in the body wall showing pore and canal system.
Can; canal, **po;** pore, **cu:** cuticle. TB ×400.

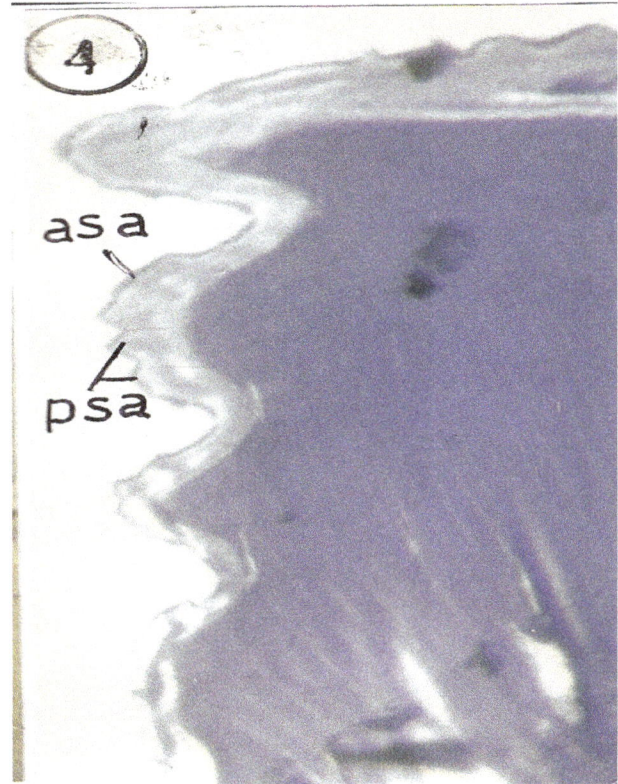

Figure 4. Photomicrograph of tangential section in the body wall showing transverse striations and annuli.
Asa; anterior subannulus, **psa;** posterior annulus. TB × 200.

annuli which divide the body into anunuli, the distance between each two 0.22 µ ,each annulus is subdivided by three subannuli, large anterior 0.124 um in long, and the other two are small, middle 0.047-0.054 µm, and the posterior 0.047-0.054 (Figure 4). Similar typical transverse striation was also observed in *Syphacia caudibandata* sp. by Ghazi et al. (2005). Similar observation was noticed by Decraemer et al. (1996) in free living nematodes in *Criconema parodoxiger*. In the present worm, striations mostly uniform in body while in *Pseudomazzia macrolabiata* are more prominent posteriorly as found by Bilquees et al. (2005). Sometime platelets or scales could be observed in other species (Jairajpuri and Southey, 1984). It is possibly that these rings appear as a result of fast cuticle deposition during ecdysis or during formation of new cuticle. Furthermore extensions from the cuticle of the present worm in the anterior and posterior regions of this worm especially in male forming cervical alae (Figure 5), and caudal alae (Figure 6).

The cuticle of *A*. lumbricoides found to compose of collagen which is secreted by epidermis and this layer is unique in nematodes (Ruppert and Barnes, 1994). Citwood and Chitwood (1974) reported that *Ascaris suum* cuticle is composed of five layers of protein namely albu-

min, glycoprotein(mucoid), fibroid, and collagen and the albumin ratio is 25%, glycoprotein undetermined fibroid 35% while the collagen constitute 29% while keratin 2.2%. Croll (1976) believed that the cuticle is 75% of water and remaining part is protein and little quantity of carbohydrate and fat. In more recent reference it is believed that the outermost covering of nematode body is a non-living proteinaceous cuticle (Shimk, 2008). This protein is collagen which is the major constituent of vertebrate ligaments, and is not elastic and not stretchable. This collagen is basically three layered-structure secreted tightly adjacent to one another giving strength and resiliency to the body wall (Shimk, 2008).

Cuticle thickness differs in different nematode species. In the present worm thickness is ranging from 8.5-11 µm while in plant parasitic nematodes or free living nematode it is much lower, 10-13 nm (Lee 1970; Bonner and Weinstein, 1972; Anderson, 2000).

Bonner and Weinsein (1972).) found that the cuticle thickness in Nippostrongylus brasiliensis reach about 60 nm. Also, thickness differ according to sex as found in the present worm 8.5 and 9.5 µ in male and female respectively. The results coincide to that found in *G. boophthorae* as the cuticle thickness is 3 and 4.5 um in male and female respectively (Batson 1979). Also results are similar to those found in *Rhabdochona tigrae* a para-

Figure 5. Photomicrograph of a hand section passing through oesophagus showing cervical alae (ca). ×200.

site in freshwater fishes as its cuticle measure 1.3 and 3.29 um in male and female respectively 9Nawab-AlDuin, 1994). Such thickness of cuticle in female possibly because it usually posses a double reproductive organ with two uteri filled with eggs as such need protection from external environment.

Cuticle thickness depends mainly on the size of the worm, it is usually thicker as the worm is larger in size, but a ratio is taken between thickness of cuticle and diameter of the worm the condition is reverse, this indicate requirement of thick cuticle in size . The present worm ratio is 1:75 in males and 1:79 in females while in *A. lumbricoides* 1: 100, in *Ancylostoma dudenale* it is 1:51 (Nawab-AlDuin, 1994), accordingly the ratio in *A. lumbricoides* is considered low if compared to the diameter of the worm, similarly R. tigrae.

Curiously, the thickness of cuticle in the present worm, *A. galli*, is thicker than the plant parasitic species, the free living form close to animal parasitic forms as such may be thickness here is a protective tool against digestive enzymes usually present in the alimentary canal of the hosts, and possibly cuticle also secrete some materials to neutralize the alkaloid medium of the intestine.

The thickness of cuticle in the anterior portion of the present worm, *A. galli*, is more than the middle of the body, a phenomenon which is similar to that reported by (1) in *A. duedenale*, it is likely because the anterior and posterior portions have less diameter than the middle. It is similar to the results of Nawab Al-Deen (Nawab-AlDuin, 1994). observed in R. tigrae, she postulated that thickness in anterior and posterior regions is due to thickness of cuticle constitutions. It is possibly also that the thinness of cuticle in the middle region of the worm is due to stretching effect due to the bulk of reproductive organs in the middle of the worm.

In the present study the cortex of *A. galli* appear as one layer with 3.15 μ in thickness and many structures appear as rounded or oval pores or canals connected together by small canals. In ultrastructure study the cortex has two layers, external and internal, the external layer is with more contact with the environment as such this layer is thin and porous as in large ascarid worms (Bird and Deutsch 1957; Watson, 1965). In this situation Tim (1949). suggest through his long experiments on A. lumbricoides that this layer has a permeability to drugs and has very thin fatty layer on the surface of cuticle. Furthermore, Bird (1957) found that the cortex has branched network of canals and pores and has a thin fatty layer measure about 0.1 μ. On the other hand, internal cortex differs in its thickness in different nematodes and is fibrillar in nature (Bird, 1971). Furthermore, this layer in large worms contains transverse structure underneath the transverse furrows which are present on the external surface of the worm which separate the external annuli. Researchers gave many terms to these organelles such as circular lamellae, fibers, strands of condensed materials, pore canals, or thick fibrous masses (Bird, 1971). In the present investigation the best name to be given to these organs are canals or pores as recovered in A. galli. It is likely that these canals and pores make an exoskeleton and pores that materials passing through to the exterior vastly (Bird, 1971). Watson (1965) during his study on A. lumbricoides observed that these canals and pores are connected canals and can reach down to the basal fibriliar layers. On the other hand, Inglis (1964) believed that all connecting canals observed in nematodes depends on system of punctuation canal as a system is evolved to allow the growth by addition of new materials by the epidermal protrusions. Anya (1966a) discovered the presence of

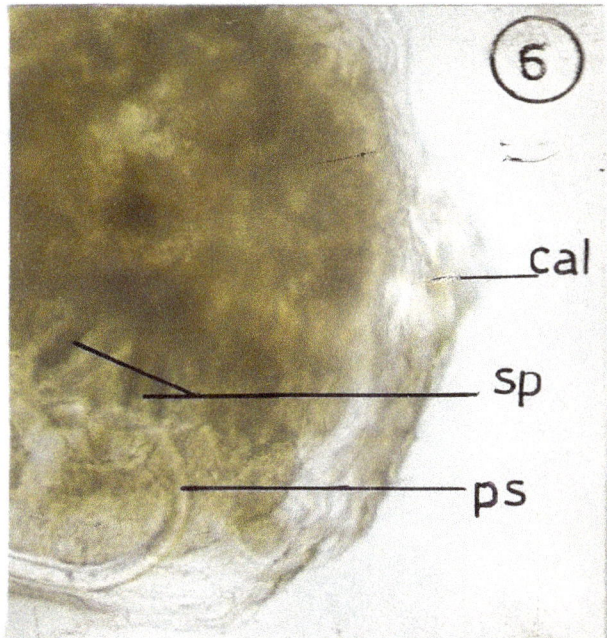

Figure 6. Photomicrograph of a hand section passing in the posterior end of the male showing caudal alae and precloacal sucker.
Cal; caudal alae, **sp**: spicules, **ps;** precloacal sucker. × 40.

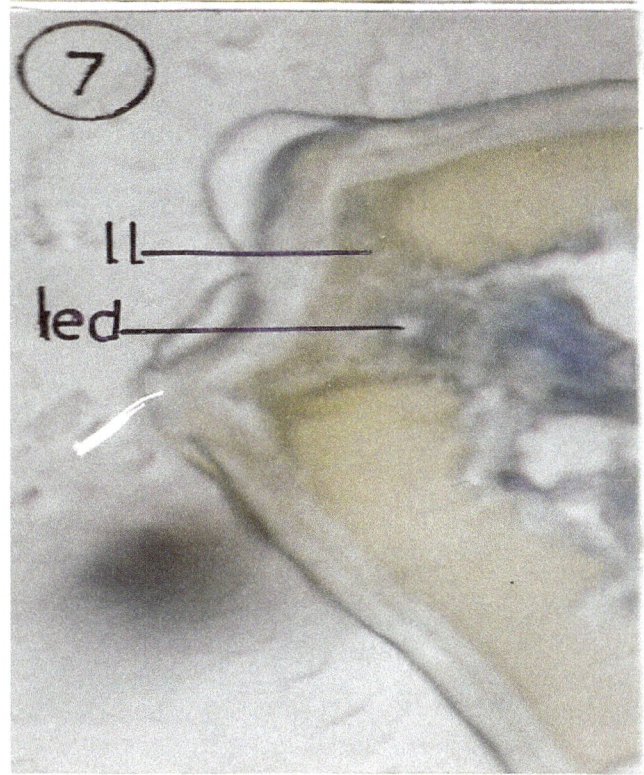

Figure 7. Photomicrograph of a section in the body wall showing excretory canal.
Ll; lateral line, **led;** longitudinal excretory duct. TB ×400.

RNA and ATPase, acid phosphates, and ascorbic acid in the cortex of three nematodes studied including *A. lumbricoides*, while in other of study, Anya (1966b) concluded that the cuticle is able to synthesize protein which it needs.

In the present worm, the middle layer or the homogenous layer is lacking structures and its thickness is 1.7 μ. This lacking is similar to that found by Watson (1965) in large ascarid nematodes.

In other nematodes this layer is not homogenous layer as it is supplied by struts or skeletal rods filled with hameoglobin and are connected by collagen fibers present in this present layer. As concern chemical nature of this layer it consists of proteins which is similar to collagen as well as non-specific esterase and acid mucoplysaccaride and a few lipids but this layer has no metabolic activity such as cortex (Bird, 1971).

The innermost layer of A. galli is the fibril layer which appears to be consists of several thin layers arranged around the worm and its thickness is about 5.65 μ. This layer has been described by many researches and has been given different names such as striped layer with regularly arranged rods or canals regularly spaced striations or regularly arranged crystalloid structures (Bird, 1971). In the present worm, *A. galli,* this layer seems to be close to striped layer. It is obvious from other researches that this fibrilar layer is more prominent in worms suffer from changes in different habitats as found by Lee (1993) that the reasons of the presence of regular spaces between fibers is likely depending on its

chemical nature as it is formed from protein with closely bands in its molecules which became more firm in habitat changing especially in plants parasitic nematodes such as Meloidogyne javanica.

The hypodermis of this worm, A. galli is syncial type, and its nuclei are constricted to the longitudinal cords in the thickening of hypodermis that is, dorsal, ventral, and two lateral cords. Mainly these nuclei are present in the lateral cords and the longitudinal excretory canals pass through these cords (Figure 7) This finding is similar to those found by Hinz (1966) Paraascaris equorum and Davey (1965) in Phocanema decipens and Watson (1965) in *A.* lumbricoides and Rogen et al.(1967) in X. index.

In the present worm the hypodermis is thin and not developed, hypodermis become thicker as proceed posteriorly to reach 2-3 μ in thickness in the inter-caudal region as well it become thicker in the four longitudinal cords protruded in the pseudocoel and reach maximum thickening (growth) near the nerve ring. Moreover, in addition to the main four longitudinal cords which divide the muscle bundles in four groups there are another four secondary cords which divide the muscle groups into 8 sectors (Figure 8).

This is similar to that found by Roggen et al. (1967) in X. index as they found that the hypodermis is very thin

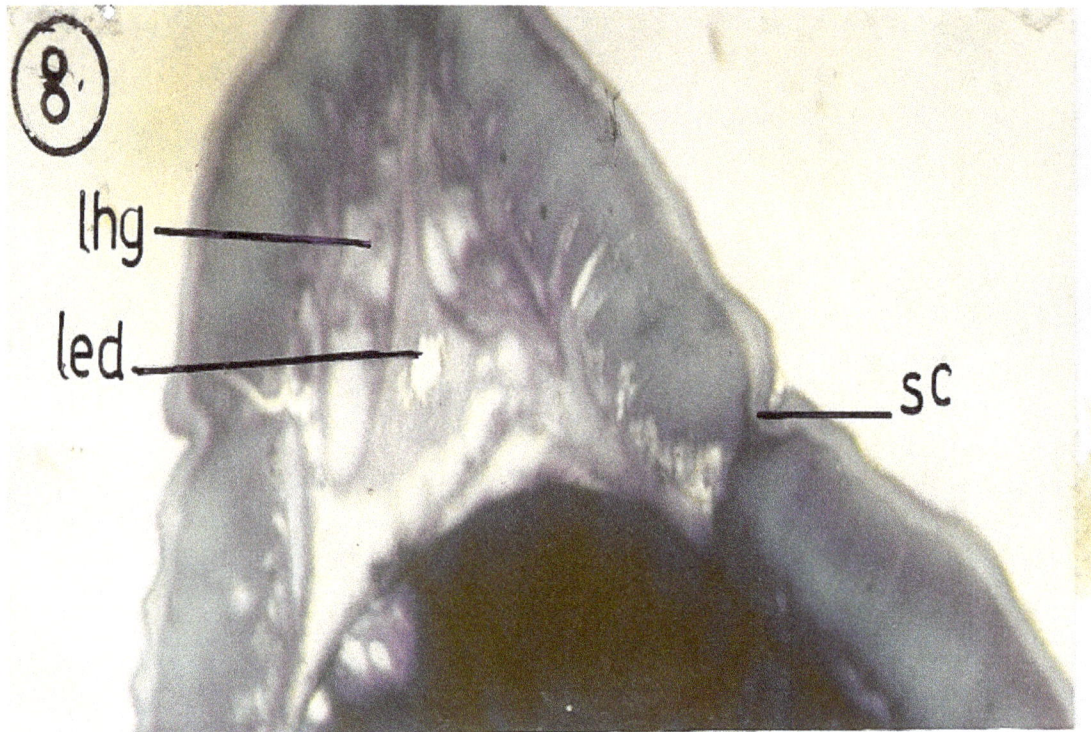

Figure 8. Photomicrograph of a section passing through oesophageal region showing secondary striations (constrictions) in the body wall.
Lhg; lateral hypodermal gland, **led;** longitudinal excretory duct, **sc;** secondary cord. TB × 400.

lie underneath the cuticle and also has four small longitudinal cords devoid of nuclei in the anterior region, and the best growth of ventral, dorsal, two lateral cords in just posterior to the nerve cord. Batson (1979) found that hypodermis in *G. boophthorae* is very thin and its thickness is 0.5 um in the inter-cordal region and became thickened in the four longitudinal cords.

Nuclei are present in epidermis of the four main cords of the present worm, *A. galli* especially in lateral cords. This phenomenon is similar to those observed in *A. lumbricoides* and *P. equonum, P. decipens* (Bird, 1971).

As concern lateral cells in *A. galli* present in lateral cords especially in the region posterior to the esophagus in the middle of the body are larger and more prominent and are continuous just before the posterior end of the body (Figure 9). These cells are present in the form of pyriform clusters, lateral cells, clup-shaped. These cells enlarges towards the pseudocoel and each posses large nucleus forming lateral hypodermal glands .The depth of these glands are 97(78-101) μ in male and 148(112-126) μ in female (Figure 10). Furthermore, three spines protruded from the surface of the cuticle situated near the opening of the hypodermal glands (Figure 11). From each canal arises two canals are united to open with small lateral pore underneath the cuticle to protrude in space situated underneath the cuticle in which open the lateral glands. These protrusions might help in the move-

ment of materials thrown in the space or transport some materials from cuticle. These cells and their ducts were noticed in some marine nematodes (Hyman 1951; Chitwood and Chitwood 1974). Birds (1971) reported the presence of proteins, nucleic acids in hypodermis and histochemical changes happen and structural change before moulting. On the other hand McKerrow and Huima (1999) indicated that the hypodermis is main protein synthesis site which supply the hypodermis itself and the cuticle.

In the present study the hypodermal cells present in the four main cords are bulged into the pseudocoel through out the body very close to the alimentary canal (Figure 10) which mean a possible transport of nutrients to this canal. Furthermore, hypodermis is syncytial which gave one path with resistance of cell membranes for the transport of protein in the bodies of hypodermal cells. This conclusion is similar to that given by McKerrow and Huima (1999). Furthermore, in the present, *A. galli*, lateral glands were larger in female than male, it may be because the female deposit large number of eggs and require large amount of proteins.

Further study especially at molecular level, as performed by Kampfer et al. (1998) and Litvaitis et al. (2000), such as phylogenetic analysis of rDNA sequences, nucleotide sequences of D3 expansion segments (26/28s rDNA) can reveal more about the

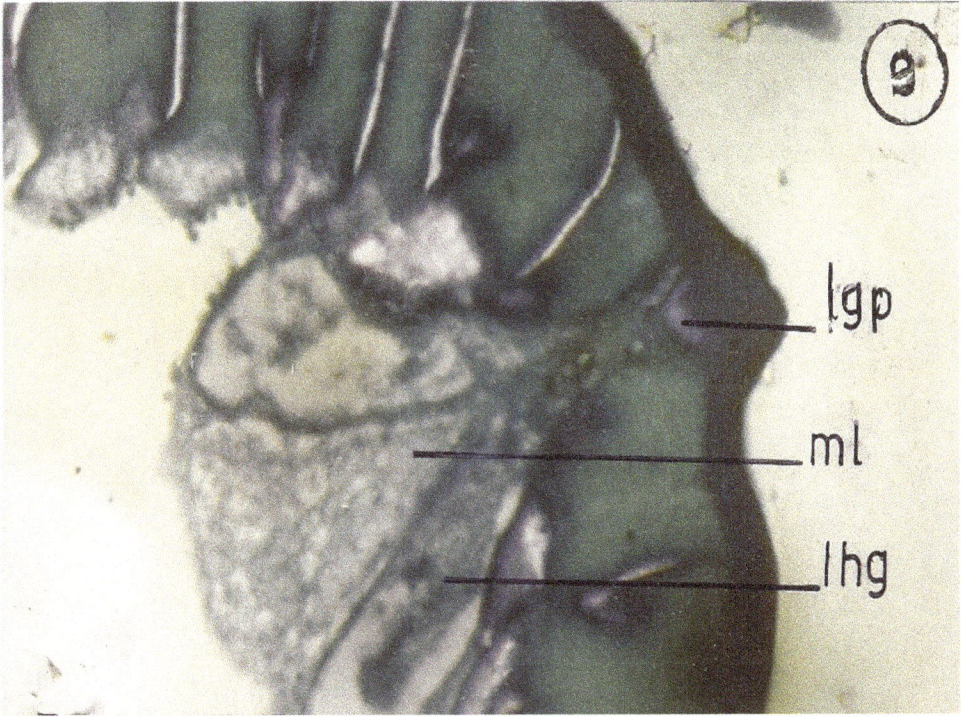

Figure 9. Photomicrograph of a section passing in the oesophagea region showing lateral glands. **Lgp;** lateral gland pore, **ml;** median line, **lhg;** lateral hypodermal gland. TB × 1000.

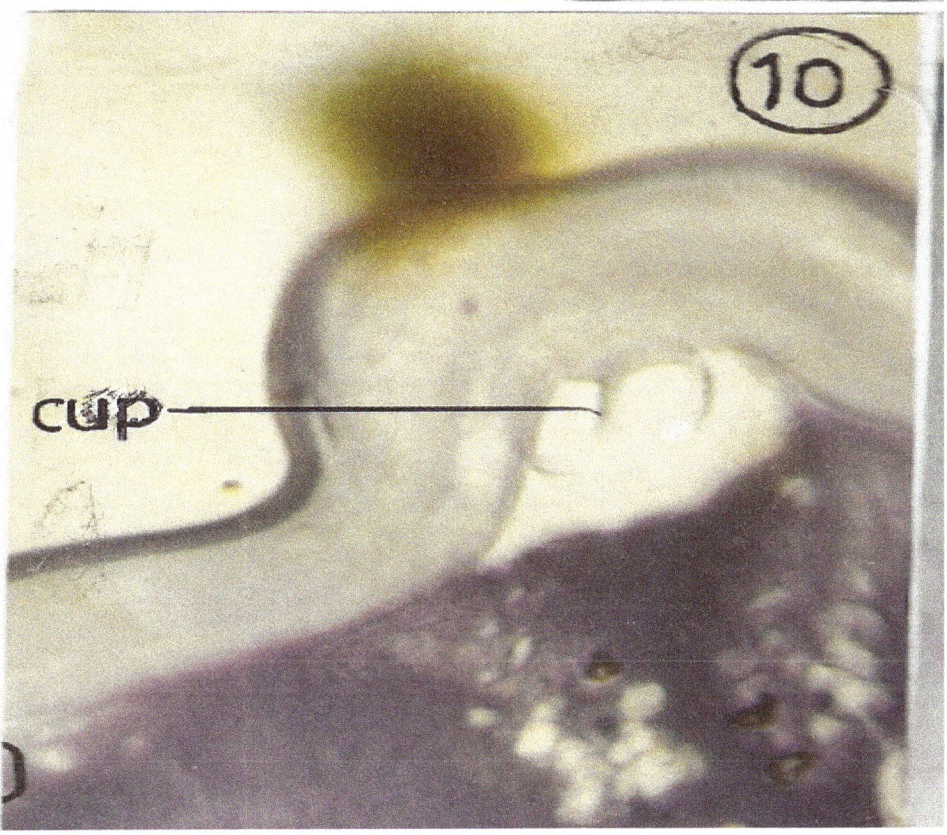

Figure 10. Photomicrograph passing in the body wall showing cuticular processes near the open Cup; cuticular process .TB × 1000.

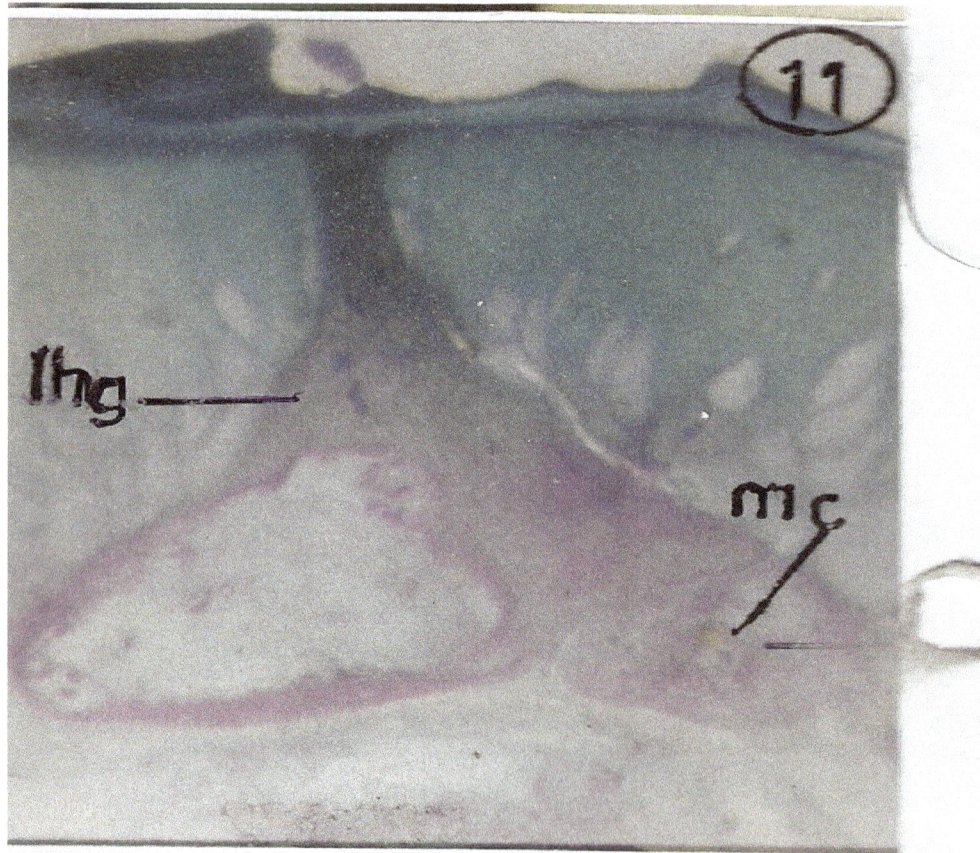

Figure 11. Photomicrograph of a section passing through the body wall showing lateral gland and its nervous supply.
lhg; lateral hypodermal gland. TB × 400.

different body structures of the present nematode in addition to tracing its affinities among other ascarids nematode to explore its diversity.

REFERENCES

Anderson RC (2000). Nematode Parasites of Vertebrates, 628 pp., CABI publishing.

Anya AO (1966a).Studies on the chemical composition of the nematode cuticle. Observations on some oxyurids and Ascaris. Parasitology 56: 179-1

Anya AO (1966b). Studies on the structure and Histochemistry of the male reproductive tract of *Aspiculuris tetraptera* (Nematoda: Oxyuridea). Parasitology 56:347-358.

Batson BS (1979). Body wall of juvenile and adult *Gastromermis boophorae* (Nematoad: Mermithidae). Ultrastructure and nutritional role. Int. J. Parasitiol. 9: 495-503

Bilqeese FM, Rehana Ghazi R, Hasseb MF (2005). *Pseudomazzia macrolabiata* n. gen., n. sp. (Nematoda: Spiruridae; Mazziinae) from the fish *Pomadasys olivaceus* of Karachi coast. Turkiye Parazitologi Dergisi 29(4); 295-297.

Bird AF (1971)). The Structure of Nematodes. Academic Press. Inc. (Lond.), Ltd p.318.

Bird AF (1957). The adult female cuticle and egg sac of the genus Meloidogyne Goeldi (1887) . Nematologica 3: 205-212.

Bird AF, Deutsch K (1957). The structure of *Ascaris lubricoides* var suis.

Parasitology 47: 319-328.

Bonner TP, Weinstein PP (1972). Ultrastructure of cuticle formation in nematodes *Nippostrongyles brasilienses* and *Nematospiroides dubius* J. Ultrastructure Res. 40; 261-271.

Bonner TP, Weinstein PP (1972). Ultrastructure of cuticle formation in nematodes Nippostrongyles brasilienses and *Nematospiroides dubius* J. Ultrastructure Res. 40; 261-271

Chitwood BC, Chitwood MB (1974). Introduction to Nematology. University Park press, Baltimore, London, Tokyo, p.334.

Croll NA (1976). The organization of Nematodes. Academic Press. Inc. 434p.

Davey KG (1965). Neurosectretory cells in the nematode, Ascaris lumbricoides. Can. J. Zool. 42: 731-735.

De Grisse AT (1972). Body wall ultrastructure of *Macroposthonia xenoplax* (Nematoda). Nematologia 18: 25-30.

Decraemer W, Baldin JG, Eddleman C, Geraert E (1996). Criconema paradoxiger (Orton Williams, 1982) raski and Luc, 1985: Cuticle ultrastructure and revalidation of the genus, Amphisbaena. Nematologica 42: 408-416.

Ghazi RR, Khatoon NB, Bilquees FM, Rathore SM (2005*). Syphacia caudibandata* sp. (nematode: Oxyridae) from a lagomorphan host *Lepus capensis* Linn in Karachi, Sindha, Pakistan. Turkiye Parazitologi Dergisi 29(2): 131-134.

Hinz E (1966). Einfach-und Mehr-fachhelfall mot Darm-helminthen in der Bovol-Kerungder. Westafribkanischen Regantropen Z. Tropenned. Parasit. pp.427-442.

Hyman LH (1951). The Invertebrates: Acanthocephala, Aschelminthes and Entoprocta. McGraw-Hill Book Company, Inc, 3: 192-455.

Ingles LG (1964). The functional and developmental significance of the cephalic septum in the Ascaridoidea (Nematoda). Proc. Linn. Soc. London 176: 23-36.

Jairajpuri MS, Southey JF (1984). *Notaocriconema sherheradae* n.sp. Nematode; Criconematidae with observations on extracuticular layer formations. Revuel de Nematologie 7: 73-79.

Kampfer S, Sturmbauer C, Ott J (1998). Phyllogenetic analysis of rDNA sequences from adenophorean nematodes and implications for the adenophorea-secernenta Controversy. Invert. Biol. 117: 29-36.

Lee DL (1970). Moulting in nematodes, the formation of adult cuticle. Tissue Cell. 2: 225-231.

Lee DL, Wright KA, Shivers RR (1993). A freeze-fracture study of the cuticle of adult *Nippostrongylus brasiliensis* (Nematoda). Parasitology 107: 545-552

Litvaitis MK, Bates JW, Hope WD, Moens V (2000). Inferring a classification of the adenophorea (Nematoda) from nucleotide sequences of the D3 expansion segment (26/ 28s rDNA). Can. J. Zool. 78: 911-922

McKerrow JH, Huima T (1999). Do filarid Nematodes have a vascular system ? Parasitol. Today 15: 123.

Nawab-AlDuin FM (1994). Studies on the nematode parasites in many species of freshwater fishes in Iraq. M.Sc. thesis, Mosul University.

Roberts LS, Janovy J (2005). Foundations of Parasitology, 702 pp. McGraw Hill International Editions, Biology series.

Rogenn DR, Raski DL, Jones ND (1967). Further electron microscopic observations of *Xiphinema index*. Nematologica 13: 1-16.

Ruppert E, Barnes R (1994). Invertebrate Zoology. 6th ed. Saunders College Publishing.

Shimk RL (2008). A spineless Column. Reefkeeping an online magazine for marine aquarist. Reef Central, L LC.

Tim AP (1949). The kinetics of the penetration of some representative anthelmintic and related compound into *A.* lumbricoides. Parasitology 3-4 .Cited by Chitwood and Chitwood (1974). ng of lateral gland.

Watson BC (1965). The fine structure of the body wall and the growth of the cuticle in the adult nematode, *Ascaris lumbricoides*. Q.H. Microsc. Sci. 106: 83-91

Wisse E, Daem WT (1968). Electron Microscopic observations on second stage larvae of Potato toot eelworm *Heterodera rostochinensis*. J. Ultra. Res. 210-231.

Wright KA, Hope WD (1968). Elaboration of the cuticle of Acanthonchus duplications Wieses, 1959 (Nematoda: Cyatholaimidae) as revealed by light and electron microscopy. Can. J. Zool. 46: 1005-1911.

11

Nematode diversity in a soybean-sugarcane production system in a semi-arid region of Zimbabwe

M. D. Shoko[1] and M. Zhou[2]

[1]Department of Agronomy, Stellenbosch University, Box X1, Matieland, RSA.
[2]School of Plant, Environmental and Soil Sciences, Louisiana State University, Baton Rouge, LA 70803, USA.

This study was done to investigate the nematode diversity in a soybean-sugarcane production system in a semi- arid region of Zimbabwe .Results indicated that *Pratylenchus, Helicotylenchus* and *Scutellonema* had the highest populations per 100 cubic centimeters of soil before the planting of soybeans. *Pratylenchus* and *Criconemella* were not present after the harvesting of soybeans. *Xiphinema* and *Scutellonema* were absent in the subsequent cane roots. Basically there was a considerable reduction in nematode population in soils after harvesting soybeans

Key words: Nematodes, soybeans, sugarcane.

INTRODUCTION

There is a long history of association between nematodes and sugarcane. Root knot nematode (*Meloidogyne*), cyst nematode (*Heterodera*) and the lesion nematode *(Pratylenchus)* are common in sugarcane production systems (Mills and Elephantine, 2000; Shoko, 2005; Shoko and Tagwira, 2005). Research done in Australia on Decline in yield venture in sugarcane indicated that nematode population increased with years of continuous sugarcane production (Stirling and Blair, 2001). According to Magarey (1994) annual sugarcane yield losses attributed to nematodes was 0.2% (Australia), 3% in Peru, >5% in South Africa, 6% in USA, 11% in Cote d'Ivoire and 14% in Burkina Faso. Research done in Australia has shown that the use of soybean as a break crop reduced the population of the root knot, lesion and cyst nematodes compared to monoculture sugarcane production (Bell, 2001). The use of soybeans in Cote D'ivore has also decreased nematode populations as compared to the use of artificial nitrogen (Coyne et al., 2003). Berry et al. (2009) found out the use of fallow legume crops like *Mucuna deeringiana* and *Dolichos lablab* in sugarcane rotations can reduce the infestation levels of some sugarcane nematodes. Work done by Rhodes et al. (2009) in Kwazulu Natal showed that monoculture cane increases the nematode populations. She also noted that longer periods of legume fallow cane also lead to an increase in nematode infestation levels.

MATERIALS AND METHODS

The overall objective of this research is to assess the dynamics of nematodes in Zimbabwean soils when soybean has been used during fallow periods in sugarcane production systems.

The experiment was done in Block Z4 at the Zimbabwe Sugar Association Experiment Station (ZSAES) in the South Eastern Lowveld of Zimbabwe during 2003 - 2005. ZSAES is 430 m above sea level, at 21°01'S latitude and 31° 38'E Longitude. The experimental plots were arranged in a Completely Randomized Block Design (RCBD) with three treatments in the first experiment, namely vegetable soybeans *cv. S114*, grain soybeans, *cv storm* and fallow. They were replicated four times. The second experiment had two sugarcane varieties CP72-2086 (nematode resistant) and N14 (susceptible to nematodes) planted after soybeans. Cane planted on fallow plots was used as control crop (monoculture cane. The subsequent cane experiment was replicated four times in a CRBD.

Soil samples were taken for extraction of nematodes before the planting of soybeans and after harvesting soybeans in December, 2003 and April, 2004 respectively. Soil samples were collected using a 50 mm augur. Soil samples were collected at depth of 0 - 30 cm from each plot. Samples were thoroughly mixed to one composite sample for each sampling depth and plot. From each composite sample, a sub-sample of 500 g was stored in a cold room at 4°C. The soils were taken for analysis and extraction of nematodes at the Plant Protection and Research Institute, in Harare. The nematodes were extracted using a combination of sieving and modified flotation method.

On 19 February, 2004 soybeans were planted with an interow spacing of 0.75 m and inrow spacing of 0.05 m and a seed rate of 80 kg ha⁻¹. The seed was inoculated with soybean innoculant at the

---

*Corresponding author. E-mail: shokom@africau.ac.zw

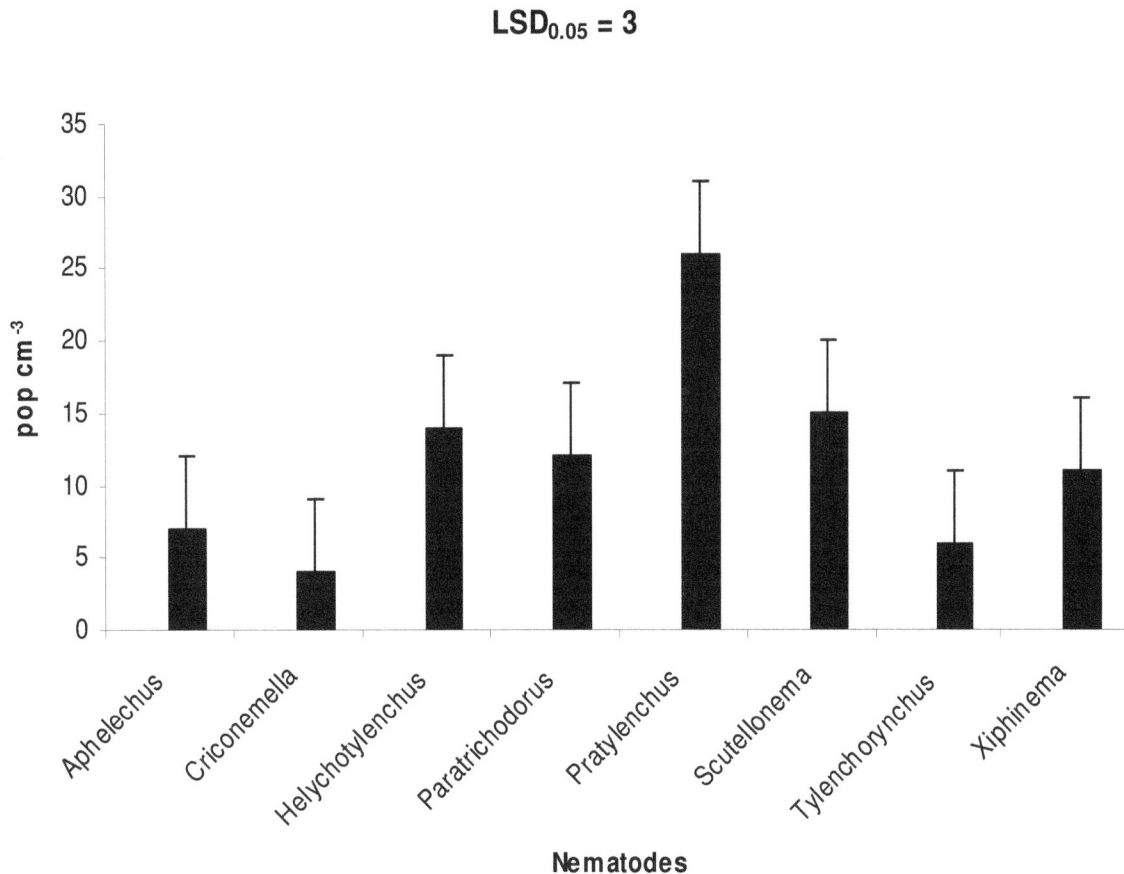

Figure 1. Nematode types and populations extracted from the soils sampled before planting soybeans.

rate of 200 g innoculant to 50 kg seed. Phosphorus was applied to the furrow at 100 kg P_2O_5 ha^{-1} before planting soybeans. Roots of six plants of soybeans per plot, randomly chosen were sampled and came with a composite sample of 50 g. Roots and soil from the rhizosphere were removed using a trowel from a depth of 300 cm. Nematodes were extracted from 100 cm^3 and 5 g fresh root sub samples using the Baermann filter technique. Nematode densities were extracted after 48 h using a stereomicroscope. The nematode suspension was reduced to 10 cm^3 and the population densities estimated from 2 x 1 cm^3 aliquots. Motile nematodes were only assessed.

On 9 and 10 July 2004 sugarcane after soybeans was planted at an interow spacing of 1.5 m, using two, three-eyed-cane setts .Varieties N14 and CP72-2086. Soil sampling for nematodes was done when the cane was still in the field. The soil sampling was done before planting, at 6 months after planting (8 January, 2005) and after harvesting of cane (10 September, 2005). Root samples were also taken during these dates following the protocol on soybean production. The data was subjected to ANOVA using MSTAT version 4 and means were compared at probability P < 0.05.

Nematode types and populations extracted from the soils sampled before planting soybeans are shown on Figure 1. This data represents the dynamics of nematodes after sugarcane monoculture production system. The *Pratylenchus*, *Helicotylenchus* and *Scutellonema* had the highest populations per 100 cm of soil. These species of nematodes are of economic importance to sugarcane production world over (Stirling and Blair, 2001). This paragraph should go in result and discussion section.

RESULTS AND DISCUSSIONS

Nematode types and populations in soil

Table 1 shows the nematode types and populations in soil after harvesting soybeans [vegetable and grain] and on monoculture cane plots [fallow/control] or before planting of the subsequent cane crop. There were significant differences (P < 0.05) between treatments with soybean plots showing a decline in all the nematodes of sugarcane. *Tylenchorynchus and Criconemella* were not found in the soil where soybeans were grown.

Nematode types and populations in soybean roots

Nematode types and populations in soybean roots are shown in Table 2. There were significant differences (P < 0.05) between the soybean treatments. The *Pratylenchus* was the highest in both cases.

Nematode dynamics in roots

The nematode dynamics in roots of subsequent sugarcane cv, CP72 - 2086 and N14 as well monoculture cane

Table 1. Nematode types and populations in soil after harvesting soybeans and on monoculture cane plots 100 cm^{-3} of soil. Figures followed by the same letter in a row are not significant at P = 0.05. This was before the planting of the subsequent cane crop.

Nematodes	Monoculture cane	Vegetable soybean	Grain soybean
Aphelechus	9c	4b	2a
Helychotylenchus	15b	9a	9a
Paratrichodorus	16c	4b	1a
Pratylenchus	32c	14b	10a
Scutellonema	14c	6a	8b
Xiphinema	15c	5b	3a

Table 2. Nematode types and populations in soybean roots. Figures followed by the same letter in a row are not significant at P = 0.05.

Nematodes	Vegetable soybean	Grain soybean
Aphelechus	5b	3a
Helychotylenchus	6b	3a
Paratrichodorus	7b	3a
Pratylenchus	10a	9a
Scutellonema	5b	2a

Table 3. Nematode types and populations in subsequent sugarcane roots and monoculture cane. Figures followed by the same letter in a row are not significant at P = 0.05.

Nematodes	Monoculture cane	CP 72- 2086	N14
Aphelechus	10c	5b	3a
Helychotylenchus	16c	6b	3a
Paratrichodorus	12c	7b	3a
Pratylenchus	28c	10a	9a
Tylenchorynchus	13c	5b	2a

Table 4. Nematode types and populations in soil after harvesting sugarcane 100cm^{-3} of soil. Figures followed by the same letter in a row are not significant at P = 0.05. This was before the planting of the subsequent cane crop.

Nematodes	Monoculture cane	CP 72-2086 plots	N14 plots
Aphelechus	8c	5b	3a
Helychotylenchus	13c	6b	4a
Paratrichodorus	14c	7b	2a
Pratylenchus	26c	9a	5a
Tylenchorynchus	21c	5b	3a

are shown in Table 3. *Xiphinema* and *Scutellonema* were absent in the sampled roots. This may imply that they these are soil nematodes and immobile unlike others like *Pratylenchus* (Mills and Elephantine 2000; Shoko, 2005; Shoko and Tagwira, 2005).

Nematode dynamics after the harvesting

The nematode dynamics after the harvesting of subse-
quent cane and monoculture cane in shown in Table 4. There were significant differences (P < 0.05) between treatments. Soil from the subsequent cane crop had low number of nematodes. CP72-2086 had the least numbers thus confirming it resistance to infestation by nematodes, unlike N 14 which is susceptible (Shoko, 2005).

This study showed that soybeans as a soil fertility ameliorant can equally reduce the populations of some of the economic nematodes of sugarcane.

ACKNOWLEDGEMENTS

The authors would like to that the Zimbabwe Sugar Industry for funding and providing a site for this research

REFERENCES

Bell MJ, Berthelsen JE, Garside AL, Halpin NV (2001). Yield Response to Breaking The Sugarcane Monoculture. Proceedings of Australian Society of Sugarcane Technologists. 22: 68-*76*.

Berry SD, Rhodes R, Rutherford RS (2009). Green Manure Crops: Their Growth and Effect on Nematodes of Sugarcane. Combined Congress Proceedings Abstracts, p. 24.

Coyne DI., Sahrawat KI,.Plowright RA (2003). The influence of mineral fertilizer application and plant nutrition on plant-parasitic nematodes in upland and lowland rice in Cote d'Ivore and its implications in long term Agricultural research trials. Exp. Agric. 40(2),233-245.

Magarey RC (1994). Microbial aspects of sugarcane yield decline. Bureau of Sugarcane Experiment Station, Queensland.

Mills G, Elphinstone G (2000). Costal Soybean Cropping Guidelines. Proceeding of Australia society of sugarcane Technologists. 21: 57-69.

Stirling GR, Blair B (2001). Nematodes are involved in the yield decline syndrome of sugarcane in Austarlia. International Society of sugarcane Technologists Proceedings XXIV Congress. 17-21 (2) 23-26.

Shoko M (2005). Soybean [*glycine max (l) merr.*] in sugarcane [*saccharum officinarum (l)*] breakcrop systems: An assessment of potential nutrient and economic benefits. Unpublished, MSc thesis, Africa University, Zimbabwe

Shoko MD, Tagwira F (2005). Assessment of the potential of vegetable and grain soybeans as breakcrops in sugarcane production systems in Zimbabwe. Proceedings of African Crop Science Society 7: 59- 65

12

Assessment of *Drosophila* diversity during monsoon season

Guruprasad B. R.*, Pankaj Patak and Hegde S. N.

Kannada Bharthi College, Kushalnagar, Madikari. Atreya Ayruvedic Medical College, Bangalore
Department of Zoology and Genetics, University of Mysore, Mysore, India.

Two months survey was conducted to analyze the altitudinal variation in diversity of *Drosophila* in Chamundi hill of Mysore, Karnataka state, India. *Drosophila* flies belonging to 15 species were collected from 680, 780, 880 and 980 m altitudes. The species diversity according to the biodiversity indices was very high in 680 m compare to other higher altitudes.

Key words: *Drosophila,* Simpson, Berger-Parker indices.

INTRODUCTION

Drosophila is being extensively used in biological research, particularly for genetical, cellular, molecular, developmental and population studies. It has been used as model organism for research for almost a century. It has richly contributed to our understanding of pattern of inheritance, variation, mutation and speciation. Studies have also been made on the population genetics of different species of this genus. However, most of these studies have been carried out in the laboratory by many workers. Though early studies on *Drosophila* in India were mainly concerned with taxonomy, 1970 onwards studies on other field have also been initiated. Significant progress has been made in the field of cytogenetics, developmental genetics and molecular biology of *Drosophila*. The taxonomical and population genetical studies have progressed little due to lack of interest of people in it. Although many workers feel that the taxonomical work shall not be neglected, people show little interest because of the hardship during work and lack of opportunity in the field. To fill up this gap at least partially, we took this work for the study of *Drosophila* population and their species diversity in a given locality.

MATERIALS AND METHODS

To study the altitudinal variation of *Drosophila* and their distribution,

*Corresponding author. E-mail: gurup2006@yahoo.co.in.

the collection was done in the Chamundi hill during 2008-2009. Chamundi hill is a famous tourist spot with altitude 1100 m, 6 km from the Mysore city. Karnataka, India, The altitude of the hill from the foot (base) is 580 m, the temperature ranges from 17 to 35°C and relative humidity varies from 19 to 75%. The collections of flies were made during monsoon season (June and July once in 15 days of the months). For this method, flies were collected by using sweeping and bottle trapping method from the all altitude, such as 680, 780, 880 and 980m (base of the hill) lower altitude of Chamundi hill: 1) Bottle trapping method 2) Net sweeping method. In bottle trapping method, regular banana baits in quarter pint 250 ml milk bottles sprayed with yeast were tied to the twigs of tree at two and half feet above the ground in cool shaded areas that is covered by scrubs. Next day flies were attracted by the bait and thus the bottles were collected during early morning by plugging with cotton to the mouth of the bottles.

In net sweeping methods, rotting fruits are spread usually beneath shaded areas of the bushes of plantation, various fruits, such as *Musca paradisca* (banana), *Ananas comuses* (pineapple), *Vitas vanifera* (grape), *Artcarpus hetrophylles* (jack fruit), *Pyrus malus* (Apple) , *Carica papaya* (papaya), *Arthras* (guava) and *Citrous auranthium* (lime),are mixed and used for spreading. After one day of spreading, the flies are swept using fine net, this is done in all the altitude (680, 780, 880, 980 and 1100 m) height of the hill. The flies are transferred to the bottles containing wheat cream–agar medium and then brought to the laboratory isolated, sexed and identified according to the texas publication 1975 records, and then they were examined under the microscopy.

Vegatation at 680 m: The foot of the hill is surrounded by mango orchards along with trees such as *Acacia concinna, Acacia catechu, Anacardium occidentale, Bombax ceiba, Breynea restusa, Cassia spectabilis, Celastrus paniculata, Cipadessa baccifera, Clematis trifolia, Dalbergia paniculata, Dioscorea pentaphylla, Ficus religiosa, Ficus bengalensis, Glyrecidia* species, *Gymnima sylvestres, Hibiscus malva, Ichnocarpus frutescens, Lantana camera, Pongamia glabra, Phyllanthus* species, *Tamarindus* indica, Thunbergia species, Tectona grandis, Sida retusa, and many shrubs including cactus.

The vegetation both at 780 and 880 m was the same. Major plants found in these localities were Albizzia amara, Andrographis serpellifolia, Argyria species, Bignonia species, Breynea restusa, Bridalia species, Cassia fistula, Cassine glauca, Eucalyptus grandis, Garcinia species, Lantana camera, Phyllanthus microphylla, Sida rhombifolia, Terminalia paniculata, Terminalia tomentosa, Vitex negundo, Zizipus oenoplea and Zizipus jujuba. The vegetation at the top of the hill (980 m) includes, Acacia catechu, Anacardium occidentale, Autocarpus integrifolia, Jasminum species, Jatropa curcus, Lantana camera, Leus aspera, Mallotus philippensis, Murraya paniculata, T. indica and Zizipus jujuba.

Analysis of species diversity of flies collected in monsoon was assessed by Simpson (D) and Berger-Parker (1/d) indices (Mateus et al., 2006). Shannon-Weiner index was also calculated, but the result was the same as Berger-Parkar index, hence not included here. Among these, Simpson index (D), which measures the probability that two individuals randomly selected from a sample that belong to the same species, was calculated using the formula:

$$D = \frac{\sum n(n-1)}{N(N-1)}$$

Where, n = the total number of organisms of a particular species, N = the total number of organisms of all population

Berger- Parker index (1/d) which shows the relative abundance was calculated using the formula:

$$\frac{1}{d} = \frac{N}{N_{Max}}$$

Where, N= Number of individuals of all species; N_{max} = Number of individuals in the most common species

RESULTS AND DISCUSSION

Results of our experiments shows that as altitude increase, there was a decrease in the biodiversity (using biodiversity indices, such as Berger- parker and Simpson Index) of Drosophila (Guruprasad et al., 2009), Monsoon is the best season building large community, and in terms of population (number of flies), it is significant in different altitude. The community and biodiversity was big in lower altitude compare to higher. These results were due to micro and macro climatic conditions (Guruprasad and Hegde, 2007). Only 15 species were collected from all the altitude, which decreased compared to 20 species that were listed in our published data. About five species, such as D. takahashii, D. suzukii, D. repleta, D. immigrans and D. buskii, were not found in the collections that were found in earlier data during 2005-2006. A totally of 956 species were collected and belonged to 4 subgenera namely Sophophora, Drosophila, Dorsilopha and Scaptodrosophila. According to the biodiversity index, lower altitude showed higher biodiversity. D. nasuta and D. malerkotilana species are the common species found in the hill and it is regard as the common

and abundant species in the hill. Another most important finding was that all species were not found in all altitude and D. nasuta, D. neonasuta, D. malerkotliana, D. rajasekari, D. jambulina and D. bipectinata were common abundant species found in all altitudes. The highest number and species of flies were found in the 680 m altitude. Further, our intention is not only to study the taxonomy of Drosophila, but also the relationship of ecology and phenotypic traits that is longevity (life span). From the aforementioned study, we realize the importance of Drosophila in two resources: its powerful genetic tools as a model system, and a natural ecology that provides substantial genetic variation across significant environmental heterogeneity. To know this heterogeneity, isofemale lines derived from low altitude have longer longevity compare to higher. This confirms the work of Trotta et al. (2006), where derived lines from temperate European populations, tropical Central American and African populations also show differences in mean life span, and mean life span under different thermal environments.

Cakir and Bozcuk (2000) showed that the differences in longevity have also been observed between inbred lines recently derived from natural populations near Ankara, Turkey. Longevity also varies significantly within populations (Schmidt and Paaby, 2008). This shows that the ecological factors play a role in the determination of longevity. Mueller et al. (2008) demonstrate a method for determining age specific survival and mortality in natural populations by marking individuals sampled from the wild at unknown age and subsequently constructing the life tables from recorded times -of-death. This technique has been used to describe the survival and death schedule of the medfly, Ceratitis capitata, and could be used for all the wild type of species found around us.

Drosophila populations have been surveyed in order to study the mechanisms of maintaining genetic variability of quantitative characters particularly morphological traits (Das et al., 1994; Garcia-vazquez et al., 1989; Sheldon and Milton, 1972). Morphological differences among natural populations are frequently attributed to natural selection, but the role of non-gentic modification by the environment has been neglected (Coyne and Beecham, 1987). According to the Carson and Stalker (1949), a population of one locality might adapt itself to the cyclic climatic changes associated with season, and undergo morphological change by a rapid type of natural selection, while Anderson (1973) is of the opinion that morphological change by a rapid type of natural variations may be simply a phenotypic response to environment, reflecting developmental plasticity or it may be partly or wholly genetic. In Drosophila, evidence on the adaptive nature of body size come from the observation of latitudinal clines and cyclic seasonal changes in several species (David and Bocquet, 1975) and from experiments with population cages (Anderson, 1966; Yadav and Singh, 2006), from all these aforementioned evidence my second objective were to analyse the variation in morphometric traits of *D. malerkotliana* at different localities. Altitudes of Chamundi hill is one of them (Guruprasad and Hegde, 2006). From this objective, we confirmed that as altitude increase, there was increase in morphometric traits, for example wing length, which is the index of body size (Hegde and Krishna, 1997). This is highly significant in case of male compare to female. This also shows that the male are more heterogeneous compare to female, female is too expose to more selection pressure than male. Thus, the present study implies that morphological variation of the species is inevitable consequence of the effects of environmental on it.

Earlier investigations have shown that these traits are expected to play important role in adaptations of flies to different environmental conditions (Griffith, et al., 2005). Anderson (1973) has shown that populations kept at different temperatures show divergence in wing length. There was clear association of body size with environmental temperature, with lower temperature favoring relatively larger size and higher temperature favoring smaller size. Tantawy (1964) has also shown that in addition to temperature, humidity also plays an important role in maintaining morphological differences. Tantawy (1964) compared 12 strains of *D. melanogaster* collected in Cameron at different altitudes. Their studies suggest that environmental variations correlated with altitude and play a direct role to bring about genetic variations. All these studies suggest that morphological variation in a given population is as a result of interplay of genotype and environment. Thus, the present study shows that morphometric variation is an inevitable consequence of the effects of environment. The length of wing and other parts of the body serve as indices of the body size, and in the present study, the author noticed the variability of these traits. These results, thus, suggest that the variability of body size is an inheritant property of the natural populations of *Drosophila* and in particular *D. ananassae*. Furthermore, the present studies contradict with the findings of Kitagawa et al. (1982) who have demonstrated lack of genetic divergence of morphometric traits of different populations of *D. nusuta* disturbed in a given area. On the other hand, Takanashi and Kitagawa (1977) have observed significant differences in the populations of the same species collected from different countries. Therefore, the author is of the opinion that the morphometric variability in the natural population does not only depend on the species or environment, but also on the response of the species in question to that environment.

From the point of future studies, one has to evaluate courtship behavior of *Drosophila* itself by aspirating the mating pair directly from nature or wild localities and evaluate the body size and morphometric traits. This can be continued frequently for not less than five year. Moreover, we can come across some question, such as, is there any increase in body size over years? From the

results of the aforementioned question, one can predict it for many years and study the body size of *Drosophila*, which seems to be molded by the action of natural selection. In the present studies, it is noticed that the density of *Drosophila* at different altitudes of Chamundi hill decreased with increasing altitude. Thus the presence or absence of a species in an ecological niche, its richness or abundance in that area is an indicator of both biological and ecological diversity of that ecosystem. In addition to physical and biotic factors, the topography and season also affect the animal distribution. The list of plant species available at the collection sites indicates that the plant diversity also decreases with increasing altitude. Thus, the present study shows that the *Drosophila* community does not only depend on vegetation, but also on altitude.

REFERENCES

Anderson WW (1966). Genetic divergence in M. Vetukhiv's experimental populations of *Drosophila pseudoobscura*. Genet. Res., 7: 255-266.

Anderson WW (1973). Genetic divergence in body size among experimental populations of *Drosophila pseudoobscura* kept at different temperatures. Evolution, 27: 278-274.

Cakir S, Bozcuk AN (2000). Longevity in some wild–type and hybrid strains of *Drosophila melanogaster*. Turk. J. Biol. 24: 321-326.

Coyne JA, Beecham E (1987). Heritability of two morphological characters with in and among natural populations of *Drosophila melanogaster*. Genetics, 177: 727-737.

Das A, Mohanty S, Parida BB (1994). Inversion polymorphism and extra bristles in Indian natural populations of *Drosophila ananassae*. Heredity, 73: 405-409.

David JR, Bocquet C (1975). Similarities and difference in latitudinal adaptation of two *Drosophila* sibling species. Nature, 257: 590.

Garcia-Vazquez E, Sanchez-Refusta F, Rubio J (1989). Chromosomal inversions and frequency of extra bristles in natural populations of *Drosophila melanogaster*. Heredity, 67: 183-187.

Griffiths JA, Schiffer M, Hoffmann AA (2005). Clinal variation and laboratory adaptation in the rain forest species *D. birichii* for stress resistance, wing size, wing shape and developmental time. J. Evol. Biol., 18: 213-222.

Guruprasad BR, Hegde SN (2006). Altitudinal and seasonal flucatuation of *Drosophila* fauna of Chamundi hill. Drosophila. inform. Serv., 89: 10-11.

Guruprasad BR, Hegde SN (2006). Altitudinal variation of Morphometric traits of *Drosophila malerakotliana* of Chamundi hill. Drosophila Info Serv.. 89: 29-31.

Guruprasad BR, Hegde SN (2008). Clinal variation of Morphometric traits of *Drosophila malerakotliana* of Chamundi hill. J. Ecol. Fisheries., 1: 1-5.

Guruprasad BR, Hegde SN, Krishna MS (2009). Seasonal and altitudinal changes in population density of 20 species of *Drosophila* in Chamundi hill. J. Insect. Sci., 10: 129-139.

Hegde SN, Krishna MS (1997). Size-assortative mating in *Drosophila malerkotliana*. Anim. Behav., 54: 419-426.

Kitagawa O, Wakahama KI, Fuyama Y, Shiada Y, Takanashi E, Hatsumi M, Umabop M, Mitay Y (1982). Genetic studies of the *D.nasuta* subgroup with notes on disitrubtion and morphology. Jap. J Genet., 57: 141-149.

Mueller HG, Wang JL, Carey JR, Caswell-Chen C, Papadopoulos N, Yao F (2004). Demographic window to aging in the wild constructing life tables and estimating survival functions from marked individuals of unknown age. Aging Cell, 3: 125-31.

Mateus RP, Buschini MLT, Sene FM (2006). The *Drosophila* community in xerophytic vegetations of the upper Parana- Paraguay

River Basin. Brazlian. J. Biol., 66(2): 719-729.

Schmidt PS, Pabby AB, Heschel MS (2005). Genetic variance for diapause expression and associated life histories in Drosophila melanogaster. Evolution. 59: 26166-25.

Sheldon BL, Milton MK (1972). Studies on the Scutellar bristles of *Drosophila melanogaster*. II Long term selelction for high bristle number in Oregon RC strain and correlated response in obdominal chaetae. Genetics., 71: 567-595.

Takanashi E, Kitagawa O (1977). Quantitative analysis of Genetic differentiation among geographical strains of *Drosophila melanogaster*. Drosophila info. Serv., 52: 28-37

Tantawy AO (1964). Studies on natural populations of *Drosophila*. III Morphological and genetic differences of wing length in *D. melanogaster* and *D. simulans* in relation to season. Evolution, 18: 560-570.

Trotta V, Calboli FCF, Ziosi M, Gerra D, Pezzoli MC, David JR, Cavicchi S (2006). Thermal plasticity in *Drosophila melanogaster*. a comparison of geographic populations. BMC Evol. Biol., 6: 67-69.

Yadav JP Singh BN (2006). Evolutionary genetics of *D. ananassae* I. Effect of selection on body size and inversion frequencies. J. Zool. Syst. Evol. Res., 44: 323-329.

13

Purification and characterization of an endo-beta-D-xylanase from major soldier salivary glands of the termite *Macrotermes subhyalinus* with dual activity against carboxymethylcellulose

H. S. Blei[1], S. Dabonné[1], Y. R. Soro[2] and L. P. Kouamé[1]*

[1]Laboratory of Biocatalysis and Bioprocessing, University of Abobo-Adjame, Abidjan, Côte d'Ivoire.
[2]Laboratory of Biotechnology, University of Cocody, Abidjan, Cote d'Ivoire.

An enzyme with apparent dual functions as a xylanase and carboxymethylcellulase was purified from the major soldier salivary glands of the termite *Macrotermes subhyalinus*. The preparation was found homogeneous by gel electrophoresis after successive chromatography on anion-exchange, cation-exchange and hydrophobic interaction columns. The specific activities towards carboxymethylcellulose and xylan were respectively 3.59 and 5.68 U/mg of protein. The molecular weight was measured to be 57.12 kDa by gel filtration and 14.47 kDa by sodium dodecyl sulfate-polyacrylamide gel electrophoresis. The purified enzyme showed a pH optimum of 5.6 for carboxymethylcellulase activity and 5.0 for xylanase activity. The optimum temperature for carboxymethylcellulase and xylanase activities was respectively 60 and 65 °C. The enzyme was capable of hydrolyzing both beta-1,4-glycosidic and beta-1,4-xylosidic bonds in carboxymethylcellulose and xylans, respectively. Based on thin-layer chromatographic analysis of the degradation products, the carboxymethylcellulase activity produced glucose, cellobiose and cellodextrins from carboxymethylcellulose as the substrate. When xylan from Birchwood was used, end products were xylobiose and xylodextrins. The salivary glands of *M. subhyalinus* soldier apparently produce an endo-beta-xylanase with dual activity against carboxymethylcellulose. The apparent role of this enzyme in the digestive tract is the hydrolysis of xylan and potentially cellulose.

Key words: Endo-beta-D-glucanase, endo-beta-D-xylanase, major soldier, physiological role, salivary glands, termite *Macrotermes subhyalinus*.

INTRODUCTION

Lignocellulose is the most abundant organic material on earth, representing 50% of the total plant biomass and presenting an estimated annual production of 5×10^{10} tons (Rajarathnam and Bano, 1989). It consists of three types of polymers, cellulose, hemicellulose and lignin that are strongly intermeshed and chemically bonded by non-covalent forces and by covalent cross linkages (Pérez et al., 2002).

The lignin is rather difficult to biodegrade and reduces the availability of the other polymers by means of a physical restriction (Ladisch et al., 1983). Hemicelluloses are biodegraded to monomeric sugars and acetic acid. Xylan is the main carbohydrate found in hemicellulose. Its complete degradation requires orchestrated actions of various enzymes including endo-xylanase, beta-D-xylosidase, alpha-glucuronidase, acetyl esterase and alpha-L-arabinofuranosidase (Beg et al., 2001). Endo-xylanase randomly hydrolyzes the xylan backbone, while beta-D-xylosidase converts xylobiose and xylo-oligosaccharide into xylose (Polizeli et al., 2005). In addition, alpha-L-arabinofuranosidase, alpha-D-glucuronidase, acetyl

---

*Corresponding author. E-mail: kouame_patrice@yahoo.fr.

xylan esterases, ferulic-acid esterases, and *p*-coumaric-acid esterases are required for the removal of the side chains (Subramaniyan and Prema, 2002). Enzymatic hydrolysis of cellulose is carried out by cellulase enzymes which are highly specific (Béguin and Aubert, 1994). These include several endo-1-4-beta-glucanase, exo-1-4-beta-glucanase and beta-glucosidase. The former two enzymes can degrade native cellulose synergistically to generate cellobiose which is a product inhibitor for these enzymes (Bhat and Bhat, 1997). Beta-Glucosidase plays an important role of scavenging the end product cellobiose by cleaving the beta (1 - 4) linkage to generate D-glucose and also in the regulation of exo and endo-cellulases synthesis. Furthermore, when a beta-glucosidase preparation is added to lignocellulosic materials, it plays a major role in release of phenolic compounds, suggesting that cellulose degrading enzymes may also be involved to facilitate the breakdown of polymeric phenolic matrices (Zheng and Shetty, 2000). These enzymes are widely spread in nature, predominantly being produced by micro organisms such as molds, fungi, bacteria (Bayer et al., 1998) and insects (Martin and Martin, 1978, 1979). Cellulases and xylanases from termite workers have been characterized extensively (Rouland et al., 1988a, b; Veivers et al., 1991; Kouamé et al., 2005a, b; Faulet et al., 2006a, b). However, little attention has been paid to enzymes from termite soldiers. Therefore, no data have been published on purification and characterization of termite major soldier enzymes. Here we report for the first time, the purification and characterization of an endo-beta-D-xylanase from major soldier salivary glands of the termite *Macrotermes subhyalinus* with dual activity against carboxy-methylcellulose. This was done in order to elucidate it role in the digestive tract and increase our knowledge on trophallaxis phenomenon.

MATERIALS AND METHODS

Chemicals

Polysaccharides, oligosaccharides and *p*-nitrophenyl-glyco-pyranosides were purchased from Sigma Aldrich. ANX-Sepharose 4 Fast-Flow, CM-Sepharose CL-6B and Phenyl Sepharose CL-4B gels were obtained from Pharmacia-LKB Biotech. The chemicals used for polyacrylamide gel electrophoresis (PAGE) were from Bio-Rad. All other chemicals and reagents were of analytical grade.

Enzymatic source and preparation of crude extract

The major soldier of the termite *M. subhyalinus* originated from the savanna of Lamto (Abidjan, Côte d'Ivoire). They were collected directly from their nests and then stored frozen at -20°C. Salivary glands (9 g) were dissected and homogenized with 20 ml 0.9 % NaCl (w/v) solution in an Ultra-Turrax and then sonicated as previously described by Rouland et al. (1988a). The homogenate was centrifuged at 20000 x g for 15 min. The collected supernatant constituted the crude extract. After freezing at -180°C in liquid nitrogen, the crude extract was stored at -20°C (Kouamé et al., 2005a).

Enzyme and protein assays

Under the standard test conditions, xylanase or cellulose activity was assayed spectrophotometrically by measuring the release of reducing sugars from Birchwood xylan or carboxymethylcellulose (CMC). The reaction mixture (0.38 ml) contained 0.2 ml of 0.5% xylan or CMC (w/v) dissolved in 20 mM acetate buffer (pH 5.0) and 0.1 ml enzyme solution. Determination of other polysaccharidase activities was carried out under the same experimental conditions. The reference cell contained all reactants except the enzyme. After 30 min of incubation at 45°C, the reaction was terminated by adding 0.3 ml of dinitrosalicylic acid solution (Bernfeld, 1955) followed by 5 min incubation in a boiling water bath. The tubes were cooled to room temperature for 10 min and 2 ml of distilled water was added. The product was analysed by measuring the optical density at 540 nm.

The disaccharidase activity was determined by measuring the amount of glucose or xylose liberated from disaccharide by incubation at 45°C for 30 min in a 20 mM acetate (pH 5.0), containing 10 mM disaccharide. The reference cell contained all reactants except the enzyme. The amount of glucose was determined by the glucose oxidase-peroxidase method (Kunst et al., 1984) after heating the reaction mixture at 100°C for 5 min. The hydrolysis of xylobiose was assayed by withdrawing aliquots (100 µl) which were heated at 100°C for 5 min. After filtration through a 0.45 µm hydrophilic Durapore membrane (millipore), the reaction mixture (20 µl) was analysed quantitatively by HPLC at room temperature. Chromatographic separation of sugars (xylobiose and xylose) were performed on a Supelcosyl LC-NH$_2$ (5 µm) column (0.46 x 25 cm) from supelco using acetonitrile/water (75: 25; v/v) as the eluent, and monitored by refractometric detection. The flow rate was maintained at 0.75 ml min^{-1} (Kouamé et al., 2001).

Enzymatic activity against *p*-nitrophenyl-glycopyranoside was measured by the release of *p*-nitrophenol. An assay mixture (0.25 ml) consisting of a 20 mM acetate buffer (pH 5.0), 1.5 mM *p*-nitrophenyl-glycopyranoside and enzyme solution was incubated at 45°C for 10 min. The reference cell contained all reactants except the enzyme. The reaction was stopped by the addition of sodium carbonate (2 ml) at a concentration of 2% (w/v) and absorbance of the reaction mixture was measured at 410 nm (Kouamé et al., 2005a; Yapi et al., 2007).

One unit (U) of enzyme activity was defined as the amount of enzyme capable of releasing one µmol of reducing sugar per min under the defined reaction conditions. Specific activity was expressed as units per mg of protein (U/mg of protein).

Protein concentrations were determined spectrophotometrically at 660 nm by method of Lowry et al. (1951) using bovine serum albumin as a standard.

Purification procedures

Fifteen (15) ml of crude extract was loaded onto an ANX-Sepharose 4 Fast-Flow (2.2 x 7.3 cm) that had been equilibrated previously with 20 mM acetate buffer pH 5.0. The unbound proteins were removed from the column by washing with two column volumes of the same buffer pH 5.0. The retained proteins were eluted with a gradient of NaCl (0 - 2 M). Fractions (2 ml each) were collected at a flow rate of 90 ml/h and assayed for enzyme activity. The fractions containing the highest xylanase and cellulase activities were pooled and submitted to cation-exchange chromatography in a CM-Sepharose CL-6B column (1.6 x 4.0 cm) equilibrated with 20 mM acetate buffer pH 5.0 at a flow rate of 72 ml/h. The column was washed with the same buffer and eluted with a gradient of NaCl (0 - 2 M). Fractions of 0.5 ml were collected and active fractions were pooled together. The pooled fraction from the previous step was saturated to a final concentration of 1.7 M sodium thiosulfate and applied on a Phenyl-Sepharose CL-4B column (1.6 x 2.7 cm) previously equilibrated with 20 mM acetate buffer pH 5.0 containing 1.7 M sodium thiosulfate. The column was washed with equilibration buffer and the retained proteins were then eluted using a gradient with sodium thiosulfate (1.7 M). Fractions of 0.5 ml were collected at a flow rate of 78 ml/h and active fractions (cellulase and xylanase activities) were pooled. The pooled fraction was dialysed against 20 mM acetate buffer pH 5.0 overnight in a cold room.

Electrophoretic methods

To check purity and determine molecular weight, the purified enzyme was analysed using polyacrylamide gel electrophoresis on a 10% separating gel and a 4% stacking gel (Hoefer mini-gel system; Hoefer Pharmacia Biotech, San Francisco, USA), according to the procedure of Laemmli (1970) at 10 °C and constant current 20 mM. Proteins were stained with silver nitrate according to Blum et al. (1987). In denaturing conditions, the sample was denatured by a 5 min treatment at 100 °C. Electrophoretic buffers were contained sodium dodecyl sulfate (SDS) and beta-mercaptoethanol. The molecular weight (M_w) of the purified enzyme was determined using the plot of log M_w of standard protein markers versus their relative mobility. In native conditions, the sodium dodecyl sulfate and beta-mercaptoethanol were not used. The sample was not heated.

Native molecular weight determination

The native molecular weight of the enzyme was determined using gel filtration on Sephacryl S200 HR. The column Sephacryl S200 HR (1.2 × 48 cm) equilibrated and eluted in 20 mM acetate buffer (pH 5.0) was calibrated with beta-amylase (206 kDa), cellulase (26 kDa), bovine serum albumin (66.2 kDa), ovalbumine (45 kDa) and amyloglucosidase (63 kDa). Fractions of 0.5 ml were collected at a flow rate of 10 ml/h. The M_w of the purified enzyme was determined using the plot of log M_w of standard protein markers versus their elution volume.

Temperature and pH optima

Optimum pH was estimated using the cellulase (CMC) or xylanase activity (Birchwood xylan) assay over a pH range between 3.0 and 8.0: acetate (20 mM) buffer for pH range 3.6 to 5.6; citrate-phosphate (20 mM) buffer for pH range 3.0 to 7.0; phosphate (20 mM) buffer for pH range 5.6 - 8.0. Optimum temperature was estimated using the cellulase or xylanase activity assay at temperatures between 30 and 80 °C.

pH and temperature stabilities

The stability of the enzyme was followed over the pH range of 3.0 to 8.0 in 20 mM buffers. The buffers were the same as in the study of the pH and temperature optima (above). After 2 h incubation at 25 °C, aliquots were taken and immediately assayed for residual xylanase or cellulase activity.

The thermal stability of the enzyme was determined at 45, 60 and 65 °C after exposure to each temperature for a period from 30 to 360 min. The enzyme was incubated in 20 mM acetate buffer pH 5.0. Aliquots were drawn at intervals and immediately cooled in ice-cold water. Residual activities, determined in both cases at 45 °C under the standard test conditions, are expressed as percentage activity of zero-time control of untreated enzyme.

Determination of kinetic parameters

The kinetic parameters (K_M, V_{max} and k_{cat}/K_M) were determined in 20 mM acetate buffer (pH 5.0) at 45 °C. Hydrolysis of xylans (Birchwood and Beechwood) or carboxymethylcellulose was quantified on the basis of released reducing sugars similarly as in the standard enzyme assay. K_M and V_{max} were determined from Lineweaver-Burk plot using different concentrations of xylan (2.0 - 10.50 mg/ml, w/v) and carboxymethylcellulose (2.0 to 10.50 mg/ml, w/v).

Thin-layer chromatography analysis of hydrolysate

The reaction mixture consisting of 0.1 ml of carboxymethylcellulose or xylan from Birchwood (0.5%, w/v) in 20 mM acetate buffer (pH 5.0) and 0.1 ml of enzyme was incubated at 45 °C. At definite intervals (30 min, 3 h, 6 h, 12 h and 24 h), 0.05 ml aliquots were taken and the reaction was terminated by heating at 100 °C for 5 min. Mono and oligosaccharides were analyzed by thin-layer chromatography on silica Gel G-60, using butanol/ethanol/water (3,5,2 v) as the mobile phase system. The bands were visualised with 3% (w/v) phenol in sulphuric acid/ ethanol (5 - 95, v).

Effect of chemical agents

The enzyme was incubated with 1 mM or 1 % (w/v) of different chemical agents for 2 h at 25 °C (various cations in the form of chlorides). After incubation, the residual activity was determined by the standard enzyme assay using xylan from Birchwood or CMC as a substrate. The activity of enzyme assayed in the absence of the chemical agents was taken as 100%.

RESULTS

Purification of enzyme

A single enzyme was purified from crude salivary gland extracts prepared from major soldier of the termite *M. subhyalinus*. Xylan from Birchwood and CMC were used as the major substrates to monitor enzymatic activity. The purification protocol involves three steps of column chromatography: anion-exchange, cation-exchange and hydrophobic interaction. A single peak of activity was eluted from ANX-Sepharose 4 Fast-Flow (Figure 1A). Pooled fractions (13 to 18) showing xylanase and cellulase activities after this first step were subjected to a cation-exchange chromatography on a CM-Sepharose CL-6B. One peak showing xylanase and cellulase activities were resolved in this step (Figure 1B). The enzyme (pooled fractions 16 to 22) was further purified in a final step using hydrophobic interaction on Phenyl Sepharose CL-4B (Figure 1C). After purification, the

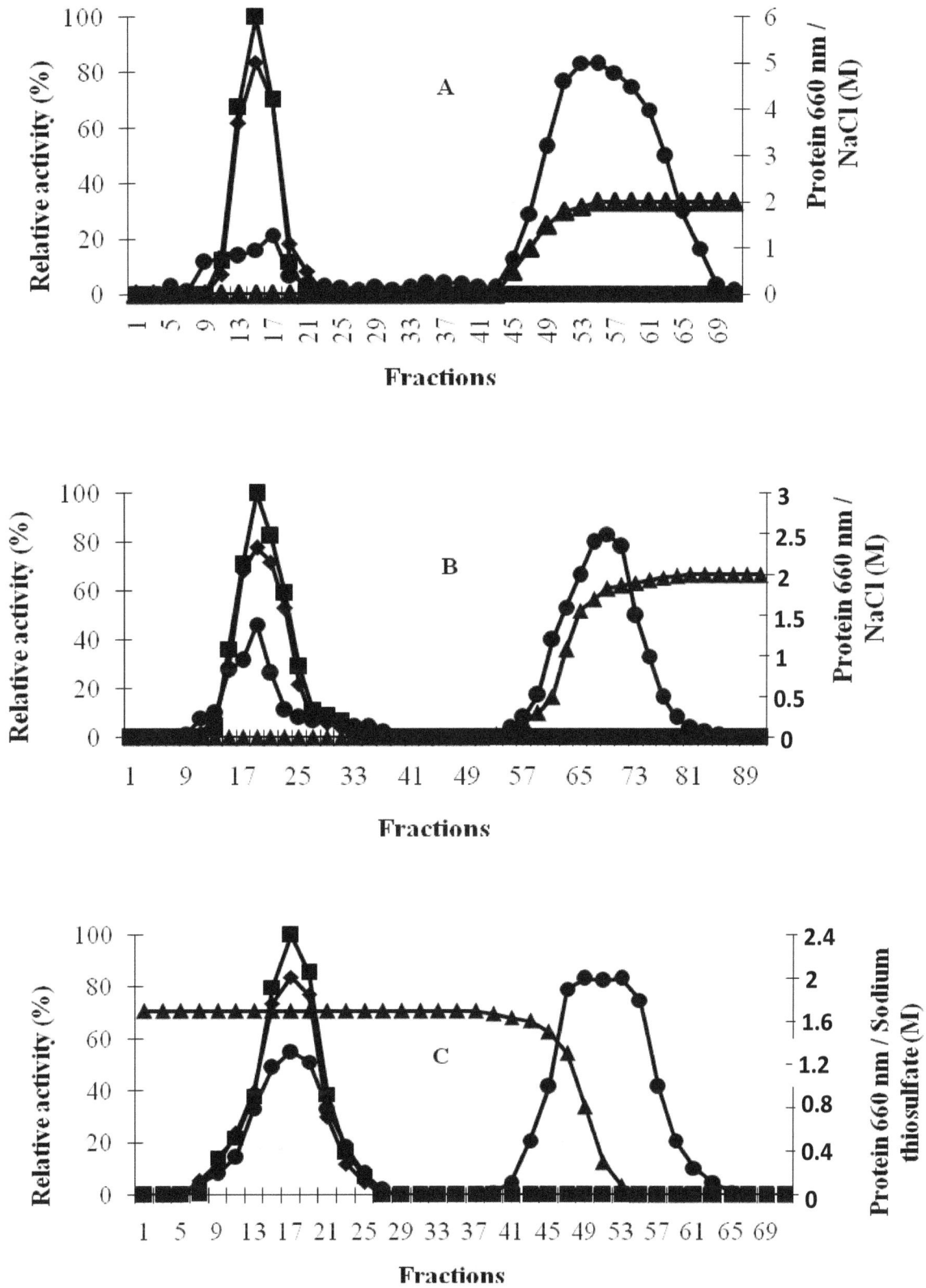

Figure 1. Purification profile of an endo-beta-D-xylanase from major soldier salivary glands of the termite *M. subhyalinus* with dual activity against carboxymethylcellulose. (A) Ion exchange chromatography (ANX-Sepharose 4 Fast Flow); (B) Ion exchange chromatography (CM-Sepharose CL 6B); (C) Gel hydrophobic chromatography (Phenyl-Sepharose CL-4B). Xylanase activity (■), cellulase activity (♦), chloride sodium or sodium thiosulfate (▲) and protein contents (●).

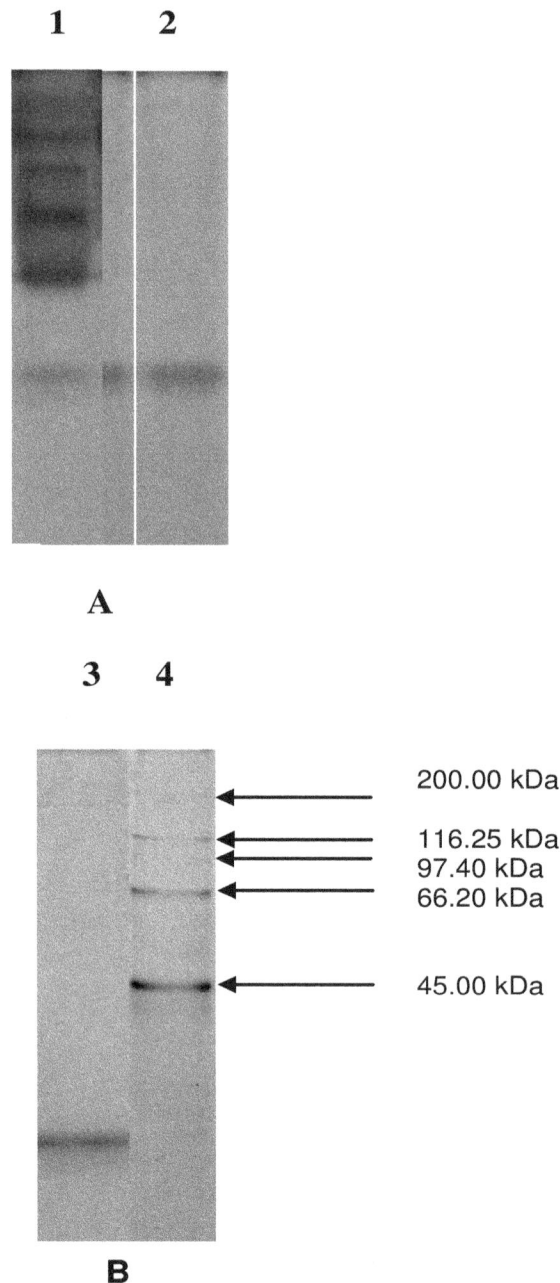

Figure 2. Polyacrylamide gel electrophoresis in native (A) and denaturing (B) conditions of an endo-beta-D-xylanase from major soldier salivary glands of the termite *M. subhyalinus* with dual activity against carboxymethyl-cellulose. Lanes 1 and 3, purified enzyme; lane 2; crude extract; lane 4, molecular weight markers.

specific activities towards carboxymethylcellulose and xylan from Birchwood (pooled fractions 14 to 18) were respectively 3.50 and 5.35 U/mg of protein (Table 1). The enzyme showed a single protein band by polyacrylamide gel electrophoresis in native (Figure 2A) and denaturing conditions (Figure 2B).

Molecular weight

From the migration pattern of the standard, the molecular weight was calculated to be 14.47 kDa. The molecular weight determined by gel filtration chromatography was57.12 kDa (Table 2).

Table 1. Purification of an endo-beta-D-xylanase from major soldier salivary glands of the termite *Macrotermes subhyalinus* with dual activity against carboxymethylcellulose.

Purification steps	Total protein (mg)	Total activity (U)	Specific activity (U/mg)	Yield (%)	Purification factor
Crude extract					
Carboxymethylcellulase	109.41	2.73	0.03	100	1
Xylanase	109.41	4.70	0.04	100	1
ANX-Sepharose 4 Fast Flow					
Carboxymethylcellulase	6.29	1.77	0.28	64.84	9.38
Xylanase	6.29	2.67	0.43	56.86	10.33
CM-Sepharose CL-6B					
Carboxymethylcellulase	0.67	0.59	0.80	21.61	29.33
Xylanase	0.67	0.89	1.33	32.60	33.25
Phenyl-Sepharose CL-4B					
Carboxymethylcellulase	0.08	0.28	3.50	10.25	116.67
Xylanase	0.08	0.43	5.35	9.15	133.75

Table 2. Some physicochemical characteristics of an endo-beta-D-xylanase from major soldier salivary glands of the termite *M. subhyalinus* with dual activity against carboxymethylcellulose.

Physicochemical properties	Carboxymethylcellulase activity	Xylanase activity
Optimum pH	5.6	5.0
pH stability range	4.6-6.0	4.6-5.6
Optimum temperature	60	65°C
Activation energy (KJ/mol)	48.18	26.87
Temperature coefficient (Q10)	1.74	1.35
Half life		
at 60°C (min)	100	120
at 65°C (min)	30	90
Michaelis Menten equation	Obeyed	Obeyed
Presence of transglycosylation activity	Yes	Yes
Mode of action	Endo	Endo
Molecular weight		
Mobility in SDS-PAGE[a]	14.47	
Gel filtration	57.12	

a = sodium dodecyl sulfate-polyacrylamide gel electrophoresis.

pH and temperature optima

The enzyme showed an optimum pH of 5.6 for cellulase activity and 5.0 for xylanase activity in acetate buffer (Table 2). It retained more than 60% of its cellulase activity in the range pH 5.0 to 6.0 (Figure 3A). Concerning xylanase activity, the enzyme retained more than 60% of its activity in the range pH 4.6 to 6.0 (Figure 3B). The optimum témperature of the enzyme with xylan from Birchwood and CMC hydrolysis were found to be respectively 65 and 60°C (Table 2). The enzyme retained more than 70% of its cellulase activity in the range 55 to65°C (Figure 4). The value of the temperature coefficient (Q_{10}) calculated between 45 and 55°C was

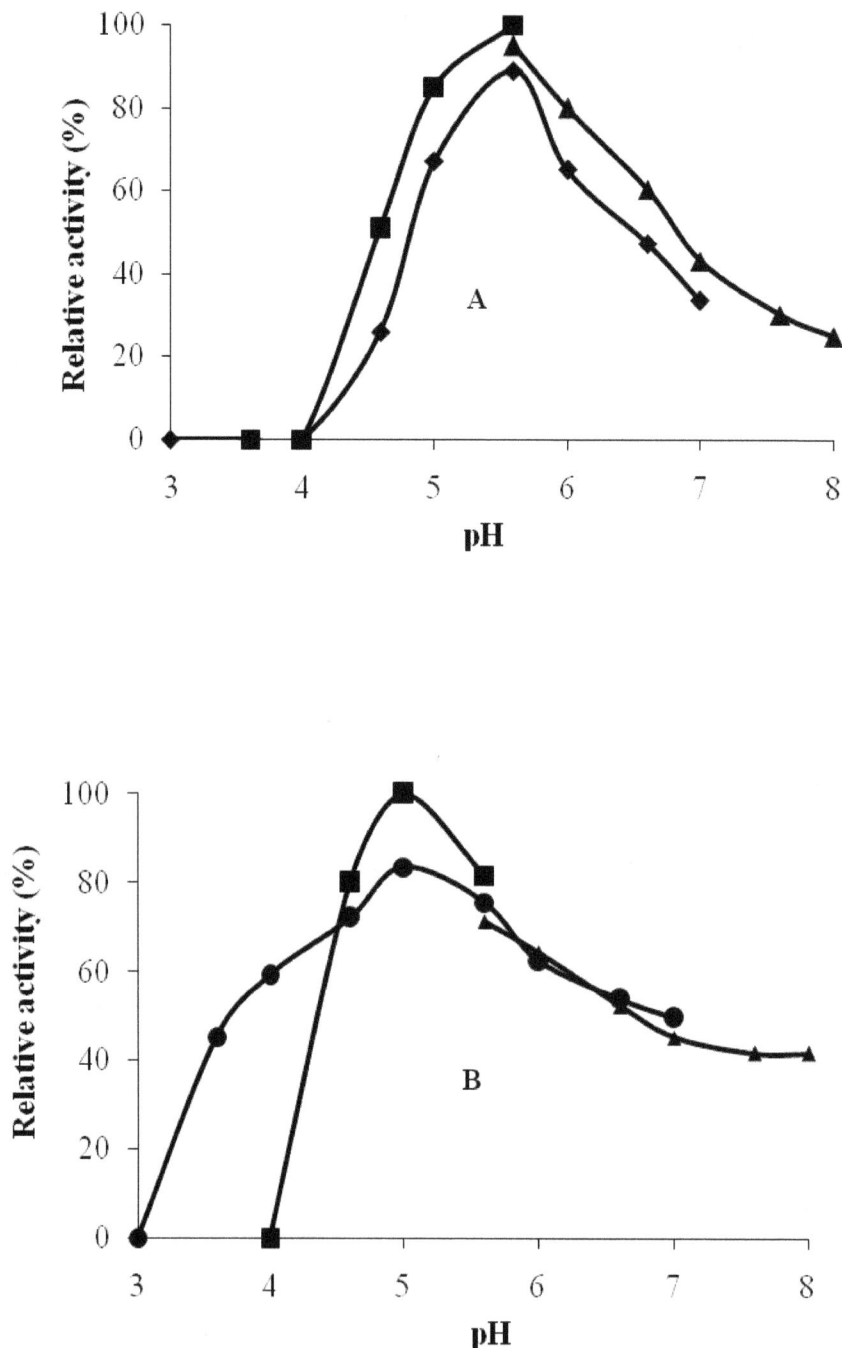

Figure 3. Effect of pH on an endo-beta-D-xylanase from major soldier salivary glands of the termite *M. subhyalinus* with dual activity against carboxymethylcellulose. Cellulase activity (A), xylanase activity (B), Acetate (■); citrate-phosphate (●); phosphate.

1.74 (Table 2). From the Arrhenius plot, the activation energy was found to be 22.58 KJ/mol (Table 2). Concerning xylanaseactivity, the enzyme retained more than 70% of its activity in the range 55 to 70°C (Figure 4). The value of the temperature coefficient (Q_{10}) calculated between 50 and 60°C was 1.35 and the activation energy

was found to be 23.52 KJ/mol (Table 2).

pH and temperature stabilities

At 25°C, the cellulase activity of the purified enzyme was

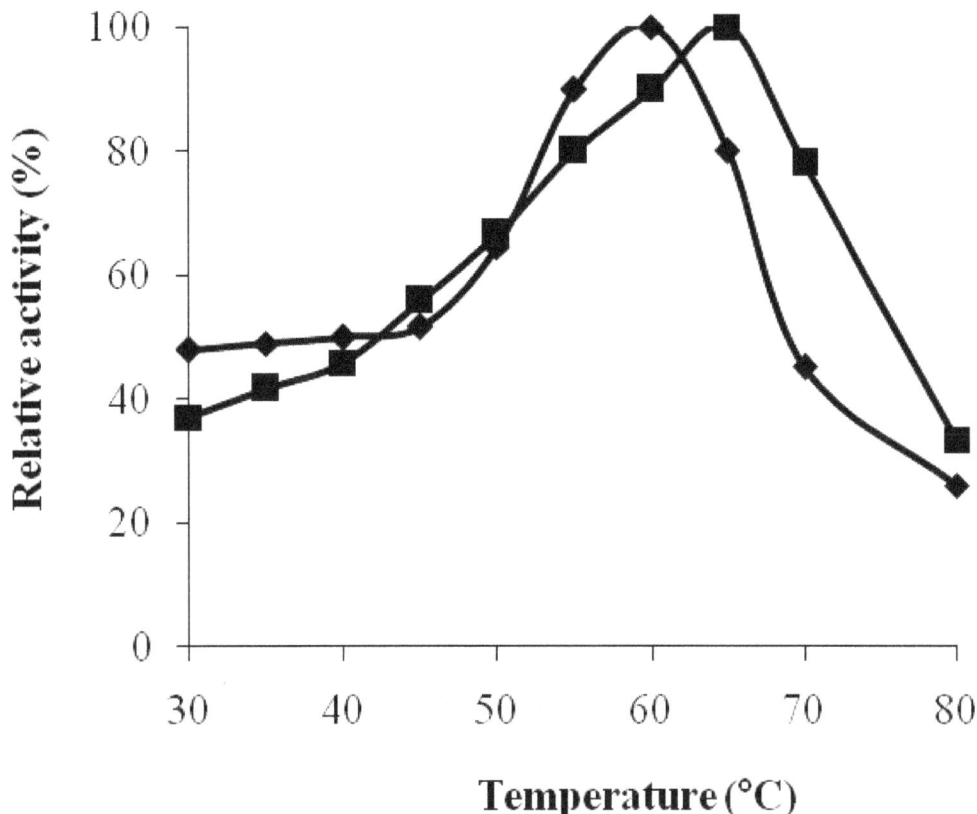

Figure 4. Effect of temperature on an endo-beta-D-xylanase from major soldier salivary glands of the termite *M. subhyalinus* with dual activity against carboxymethylcellulose. Xylanase activity (■), cellulase activity (♦).

stable over a wide pH range of 4.6 to 6.0 for 120 min (Table 2). The same activity was fully stable for 30 min at 60°C, but at 45°C, it was stable for 360 min in 20 mM acetate buffer pH 5.0. At 60°C, the half life of the cellulase activity was 100 min (Figure 5). Concerning xylanase activity, the purified enzyme was stable over a wide pH range of 4.6 TO 5.6 for 120 min (Table 2). At 60 and 65°C, this activity was unstable in 20 mM acetate buffer pH 5.0, but at 45°C, it was stable for 360 min (Figure 5). At 65°C, the half life of the xylanase activity was 90 min (Table 2).

Substrate specificity and kinetic parameters

The purified enzyme did not attack *p*-nitrophenyl glycopyranosides, cellobiose, sucrose, lactose, xylobiose, maltose, inulin, avicel, *sigmacel* 50-cellulose and starch (Table 3). Although, the enzyme degraded carboxymethylcellulose and xylans (Beechwood and Birchwood) (Table 3). The effect of substrate concentration on enzymatic activity was studied with carboxymethylcellulose and xylans. With these substrates, the enzyme obeyed the Michaelis-Menten equation (Table 2). The K_M, V_{max} and V_{max}/K_M values are reported in Table 4. The catalytic efficiency of the enzyme, given by the V_{max}/K_M ratio is much higher for the carboxymethylcellulose than the xylans (Table 4).

Thin-layer chromatography analysis of hydrolysate

Within the first 12 h of the reaction, the hydrolysis of carboxymethylcellulose produced the expected oligosaccharides, disaccharides and monosaccharides. From xylan (Birchwood), the hydrolysate contained oligosaccharides and disaccharides. After 12 h of incubation, the hydrolysis products such as monosaccharides and disaccharides disappeared in the reaction (Figure 6).

Effect of chemical agents on enzyme activity

Chemical agents KCl, $FeCl_2$, SDS, DTNB and *p*CMB showed an inhibitory effect on cellulase activity of the

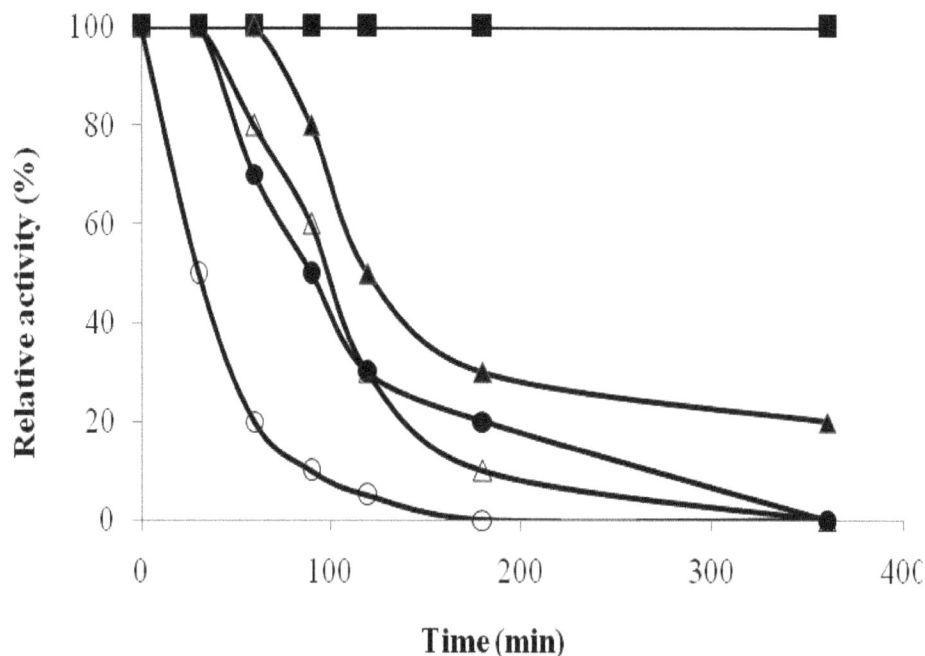

Figure 5. Thermal stability of an endo-beta-D-xylanase from major soldier salivary glands of the termite *M. subhyalinus* with dual activity against carboxymethylcellulose. Xylanase activity 45°C (■), cellulase activity 45°C (□), xylanase activity 60°C (▲), cellulase activity 60°C (△), xylanase activity 65°C (●), cellulase activity 65°C (○).

Table 3. Activities of an endo-beta-D-xylanase from major soldier salivary glands of the termite *M. subhyalinus* on synthetic chromogenic, disaccharide and polysaccharide substrates.

Substrate	Concentration in assay	Relative rate of hydrolysis (%)
Carboxymethylcellulose	2.6 mg/ml	100.00
Xylan (Birchwood xylan)	2.6 mg/ml	160.80
Xylan (Beechwood xylan)	2.6 mg/ml	116.80
Avicel	2.6 mg/ml	0.00
Sigmacel 50-cellulose	2.6 mg/ml	0.00
Inulin	2.6 mg/ml	0.00
Starch	2.6 mg/ml	0.00
Maltose	10.0 mM	0.00
Sucrose	10.0 mM	0.00
Lactose	10.0 mM	0.00
Cellobiose	10.0 mM	0.00
Xylobiose	10.0 mM	
p-Nitrophenyl-glycopyranoside	1.5 mM	0.00

purified enzyme. However, BaCl$_2$, CuCl$_2$, EDTA and CaCl$_2$ had no effect on the same enzyme activity (Table 2). Concerning xylanase activity, the enzyme was inhibited by ZnCl$_2$ and SDS. However, it was activated by MgCl$_2$, BaCl$_2$ and CaCl$_2$. EDTA, DTNB, *p*CMB, CuCl$_2$, FeCl$_2$, KCl and NaCl had no effect on the same enzyme activity (Table 5).

DISCUSSION

A single enzyme with dual activity (carboxymethyl cellulase and xylanase) was purified from the major soldier salivary glands of the termite *Macrotermes subhyalinus*. The preparation showed only one band in sodium dodecyl sulfate polyacrylamide gel electrophoresis

Table 4. Kinetic parameters of an endo-beta-D-xylanase from major soldier salivary glands of the termite *M. subhyalinus* towards carboxymethylcellulose, xylan from Birchwood and xylan from Beechwood.

Substrate	K_M (mg/ml)	V_{max} (U/mg)	V_{max}/K_M (Uxml/mg^2)
Carboxymethylcellulose	0.46	9.92	21.56
Xylan from Birchwood	1.10	15.61	14.19
Xylan from Beechwood	1.21	11.74	9.70

A

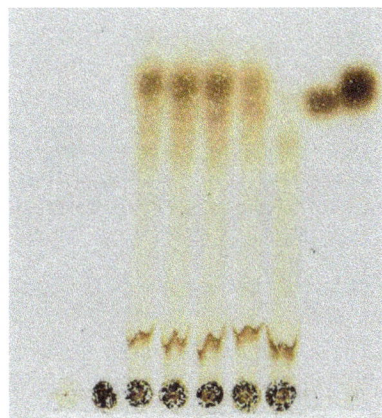

0_1 0_2 1/2 3 6 12 24 G_2 G_1

Time (h)

B

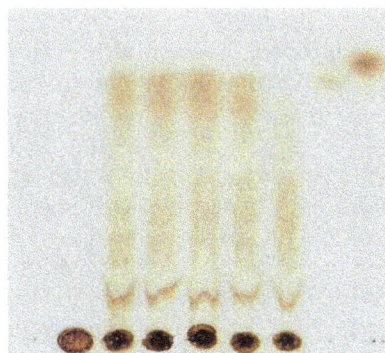

Time (h)

0_1 0_2 1/2 3 6 12 24 X_2 X_1

Figure 6. Time course of end products from (A) carboxymethylcellulose or (B) xylan from Birchwood hydrolysis for an endo-beta-D-xylanase from major soldier salivary glands of the termite *M. subhyalinus* with dual activity against carboxymethylcellulose. G_1= glucose, G_2= cellobiose, X_1= xylose, X_2 = xylobiose, 0_1= enzyme, 0_2= substrate (xylan Birchwood or carboxymethylcellulose).

Table 5. Effect of chemical agents on an endo-beta-D-xylanase from major soldier salivary glands of the termite *M. subhyalinus* with dual activity against carboxymethylcellulose.

Chemical agents	Concentration in assay	Relative activity (%)	
		Carboxymethylcellulase activity	Xylanase activity
Control	0	100	100
$ZnCl_2$	1	127.20	88.13
$MgCl_2$	1	135.96	156.15
$BaCl_2$	1	97.36	163.04
$CaCl_2$	1	100	169.56
$CuCl_2$	1	100	100
$FeCl_2$	1	61.27	100
KCl	1	80.70	95.24
NaCl	1	100	100
EDTA[b] (%, w/v)	1	100	100
DTNB[c] (%, w/v)	1	47.14	100
pCMB[d] (%, w/v)	1	79.11	100
SDS[e] (%, w/v)	1	0	0

b = sodium ethylendiamintetraacetate, **c** = 5,5-dithio-bis(2-nitrobenzoate), d = *p*-chloromercuribenzoate, **e** = sodium dodecyl sulfate.

and in polyacrylamide gel electrophoresis. The purified enzyme had no contaminating glycosidase activities such as glucosidase, fucosidase, mannosidase, arabinosidase and xylosidase. The only substrates that were hydrolyzed by the enzyme were xylan and carboxymethylcellulose. These results are in agreement with those for the bifunctional polysaccharidase from the symbiotic fungus *Termitomyces* sp of the termite *M. subhyalinus* (Faulet et al., 2006b). But, they differ to cellulolytic and xylanolytic enzymes from the termite *Trinervitermes trinervoides* (Potts and Hewitt, 1974), *Reticulitermes sperutus* (Watanabe et al., 1997), *Coptotermes fornosunus* (Nakashima et al., 2002), *Odontotermes formosanus* (Yang et al., 2004) and *M. subhyalinus* workers (Faulet et al., 2006b).

Within the first 12 h of the reaction, the hydrolysis of the two substrates carboxymethylcellulose and xylan from Birchwood produced the expected oligosaccharides, disaccharides and monosaccharides from them. These results indicate that the purified enzyme randomly cleaved internal beta-1,4-glucosidic and beta-1,4-xylosidic bonds in these substrates respectively as a polysaccharidase possessing endo-glucanase and endo-xylanase activities. Carboxymethylcellulose, which measures endo-beta-1,4-glucanase activity, is one of the most popular artificial substrates for measuring cellulase activity because of its high solubility in water. Thus, carboxymethylcellulose has been preferentially used in most studies of cellulose digestion in termites and other insects. However, it has long been recognized that evaluation of cellulose digestibility by carboxymethyl-cellulose degradation assays is somewhat insufficient.

This observation is supported by the incapacity of the major soldier salivary glands to degrade *in vitro* crystalline cellulose, a pure constituent of native cellulose, into which penetration by water-soluble enzymes is difficult due to tightly packed cellulose fibres joined to each other by hydrogen bonds and van der Waals forces (Gardner and Blackwell, 1974). The digestibility of native cellulose is highly dependent upon its crystallinity and its association with other structural polymers, especially lignin (Wood and Saddler, 1988). In insects, the grinding action of the mandibles and the highly alkaline conditions that prevail in the midgets of some species might also serve to reduce the crystallinity of ingested cellulose (Martin, 1991). It seems that the apparent role of this enzyme in the digestive tract is the hydrolysis of xylan (hemicellulose) and potentially cellulose.

The specific activities towards xylans from Birchwood and Beechwood are considerably lower than those obtained for the three xylanases (Faulet et al., 2006a, b) and the two cellulases (Séa et al., 2006) purified previously from the worker of the same termite. These observations suggest how trophallaxis might serve socio-nutritional needs in termite colonies. The worker of the termite *M. subhyalinus* could regurgitate its own salivary glands secretions to supplement the digestive needs of the soldier of the same insect that live in close association with the worker, receiving these liquids by trophallaxis. It is possible that this natural phenomenon is done essentially to complete the soldier enzymatic activities in its digestive tract.

The different temperature and pH activity profiles determined for the purified enzyme with the substrates carboxymethylcellulose and xylan from Birchwood suggest that this protein has two active sites: one for each activity. This pattern seems to reflect the activity of the bifunctional polysaccharidases from *Ruminococcus flavefaciens* (Flint et al., 1993) and *Cellulomonas flavigena* (Pe´rez-Avalos et al., 2008). The apparent bifunctional protein is significantly different from the termite *M. subhyalinus* worker xylanases (Faulet et al., 2006b) and cellulases (Séa et al., 2006). This is a strong indication that the catalytic domains of these enzymes (from termite *M. subhyalinus*) can be grouped into well defined families, indicating that their genes evolved divergently from relatively few ancestral sequences. This observation is in close agreement to reports of Han (1987), who reported that in the course of *M. subhyalinus* caste differentiation, the most dynamic morphogenesis occurs in the stage of moulting from minor worker to presoldier. Miura et al. (1999) recently identified a gene expressed specifically in the mandibular glands of soldiers, but not workers, and subsequent studies have found numerous transcription factor, structural and enzyme-coding genes that differ in expression between soldiers and workers (Scharf et al., 2003).

After 12 h of incubation, the carboxymethylcellulose or xylan (Beechwood) hydrolysis products such as monosaccharides and disaccharides disappeared in the reaction. It seems that the transglycosylation reaction occurred with the hydrolysis of xylan or carboxymethylcellulose, judging from the fact that the polysaccharidase can transfer part of the monosaccharide-based polysaccharide to monosaccha-ride-based oligosaccharides. The rate of transglyco-sylation product formation was largely favored relative to the rate of hydrolysis. These enzymatic activities are analogous to the dual activities of XTH enzyme, which cuts the glucan backbone of xyloglucan and either attaches the newly created reducing end to xyloglucan derived oligosaccharides (xyloglucan endo-transglycosylase activity; XET), or to water (xyloglucan endohydrolase; XEH) (Rose et al., 2002).

The relative molecular weight of the purified enzyme was estimated to be 14.47 and 57.12 kDa by sodium dodecyl sulfate polyacrylamide gel electrophoresis and gel filtration on Sephacryl S 200 HR, respectively. This result would suggest that, in contrast to the cellulases (Séa et al., 2006) and xylanases (Matoub and Rouland, 1995; Faulet et al., 2006b) obtained from workers of the termites *M. subhyalinus,* and *Macrotermes bellicosus*, the purified enzyme is a homotetrameric protein.

Conclusion

The salivary glands of *M. subhyalinus* soldier apparently produce an endo-beta-xylanase with dual activity against carboxymethylcellulose. The role of this enzyme in the digestive tract is the hydrolysis of xylan (hemicellulose) and potentially cellulose. We hypothesize that the low activity is one of the causes of the trophallaxis phenomenon. This protein is significantly different from the termite *M. subhyalinus* worker xylanases and cellulases.

ACKNOWLEDGEMENTS

This work was supported by Laboratory of Biocatalysis and Bioprocessing at the University of Abobo-Adjame, Abidjan, Côte d'Ivoire. The authors thank Dr. Kablan Tano for technical assistance.

REFERENCES

Bayer EA, Chanzy H, Lamed R, Shoham Y (1998). Cellulose, cellulases and cellulosomes. Curr. Opin. Struct. Biol., 8: 548-557.

Bhat MK, Bhat S (1997). Cellulose degrading enzymes and their potential industrial applications. Biotechnol. Adv., 15: 583-630.

Beg QK, Kapoor M, Mahajan L, Hoondal GS (2001). Microbial xylanases and their industrial applications: A review. Appl. Microbiol. Biotechnol., 56: 326-338.

Béguin P, Aubert JP (1994). The biological degradation of cellulose. FEMS Microbiol. Rev., 13: 25-58.

Bernfeld P (1955). Amylase α and β. Methods, in Colswick, S.P. & N.O.K. Enzymology. Ed. Academic Press Inc, New-York, pp. 149-154

Blum H, Beier H, Gross B (1987). Improved silver staining of plant proteins, RNA and DNA in polyacrylamide gels. Electrophoresis, 8: 93-99.

Faulet MB, Niamké S, Gonnety TJ, Kouamé LP (2006a). Purification and biochemical properties of a new thermostable xylanase from symbiotic fungus Termitomyces sp. Afr. J. Biotechnol., 5: 273-283.

Faulet MB, Niamké S, Gonnety TJ, Kouamé LP (2006). Purification and biochemical characteristics of a new strictly specific endoxylanase from termite Macrotermes subhyalinus workers. Bull. Insectology, 59: 17-26.

Flint HJ, Martin J, McPherson CA, Daniel AS, Zhang JX (1993). A bifunctional enzyme, with separated xylanase and beta (1,3-1,4)-glucanase domains, encoded by the xynD gene of Ruminococcus flavefaciens. J. Bacteriol., 175: 2943-2951.

Gardner KH, Blackwell J (1974). The hydrogen bonding in native cellulose. Biochim. Biophys. Acta 343: 232-237.

Han SH (1987). Fondation et croissance des colonies de termites supérieurs. Thèse Université de Dijon, p.305.

Kouamé LP, Niamke S, Diopoh J, Colas B (2001). Transglycosylation reactions by exoglycosidases from the termite Macrotermes subhyalinus. Biotechnol. Lett., 23: 1575-1581.

Kouamé LP, Kouamé AF, Niamke SL, Faulet BM, Kamenan A (2005). Biochemical and catalytic properties of two beta-glycosidases purified from workers of the termite Macrotermes subhyalinus (Isoptera: Termitidae). Int. J. Trop. Insect Sci., 25: 103-113.

Kouamé LP, Due AE, Niamke SL, Kouamé AF, Kamenan A (2005b). Synthèses enzymatiques de néoglucoconjugués catalysées par l'alpha-glucosidase purifiée de la blatte Periplaneta americana (Linnaeus). Biotechnol. Agron. Soc. Environ., 9: 35-42.

Kunst A, Draeger B, Ziegenhorn J (1984). Colorimetric methods with glucose oxydase and peroxydase. In: Bergmeyer, H.U. (Ed). Methods of enzymatic analysis. Verlag chemie, weinheim, 6: pp. 178-185.

Laemmli UK (1970). Cleavage of structural proteins during the assembly of the head of bacteriophage T4. Nature, 227: 680-685.

Ladisch MR, Lin KW, Voloch M, Tsao GT (1983). Process considerations in the enzymatic hydrolysis of biomass. Enzym. Microb. Tech., 5: 82-102.

Lowry OH, Rosebrough NJ, Farr AL, Randall RJ (1951). Protein measurement with the Folin phenol reagent. J. Biol. Chem., 193: 265-275.

Martin MM, Martin JS (1978). Cellulose digestion in the midgut of the fungus-growing termite *Macrotermes natalensis*: The role of acquired digestive enzymes. Science, 199: 1453-1455.

Martin MM, Martin JS (1979). The distribution and origins of the cellulolytic enzymes of the higher termite *Macrotermes natalensis*. Physiol. Zool. 52: 11-21.

Martin MM (1991). The evolution of cellulose digestion in insects. Phil. Tram. R. Soc. Lond. B: 333: 281-288.

Matoub M, Rouland C (1995). Purification and properties of the xylanases from the termite *Macrotermes bellicosus* and its symbiotic fungus *Termitomyces* sp. Comp. Biochem. Physiol., 112B: 629-635.

Miura T, Kamikouchi A, Sawata M, Takeuchi H, Natori S, Kubo T, Matsumoto T (1999). Soldier caste-specific gene expression in the mandibular glands of *Hodotermopsis japonica* (Isoptera: Termopsidae). Proc. Nat. Acad. Sci. U S A, 96: 13874-3879.

Nakashima K, Watanabe H, Saitoh H, Tokuda G, Azuma J-I (2002). Dual cellulose digesting system of the wood-feeding termite, *Coptotermes formosanus* Shiraki. Insect Biochem. Mol. Biol., 32: 777-784.

Pe´rez-Avalos O, Sa´nchez-Herrera LM, Salgado LM, Ponce-Noyola T (2008). A bifunctional endoglucanase/endoxylanase from *Cellulomonas flavigena* with potential use in industrial processes at different pH. Curr. Microbiol., 57: 39-44.

Pérez J, Muñoz-Dorado J, De-la-Rubia T, Martínez J (2002). Biodegradation and biological treatments of cellulose, hemicellulose and lignin: an overview. Inter. Microbiol., 5: 53-63.

Polizeli MLTM, Rizzatti ACS, Monti R, Terenzi HF, Jorge JA, Amorim DS (2005). Xylanases from fungi: Properties and industrial application. Appl. Microbiol. Biotechnol., 67: 577-591.

Potts RC, Hewitt PH (1974). Some properties and reaction characteristics of the partially purified cellulase from the termite *Ihneruitermes trinervoides* (Nasutitermitinae). Comp. Biochem. Physiol., 47B: 327-337.

Rajarathnam S, Bano Z (1989). Pleurotus mushrooms. Part III. Biotransformations of natural lignocellulosic wastes: commercial applications and implications. Crit. Rev. Food Sci. Nut., 28: 31-113.

Rose JKC, Braam J, Fry SC, Nishitani K (2002). The XTH family of enzymes involved in xyloglucan endotransglucosylation and endohydrolysis: current perspectives and a unifying nomenclature. Plant Cell Physiol., 43: 1421-1435.

Rouland C, Civas A, Renoux J, Petek F (1988). Purification and properties of cellulases from the termite *Macrotermes mulleri* (Termitidae, Macrotermitinae) and its symbiotic fungus. *Termitomyces* sp. Comp. Biochem. Physiol., 91B: 449-458.

Rouland C, Civas A, Renoux J, Petek F (1988). Synergistic activities of the enzymes involved in cellulose degradation, purified from *Macrotermes mulleri* fungus *Termitomyces* sp. Comp. Biochem. Physiol., 91B: 459-465.

Scharf ME, Wu-Scharf D, Pittendrigh BR, Bennett GW (2003). Caste and development-associated gene expression in a lower termite. Genome Biol., 4: R62.

Séa TB, Saki SJ, Coulibaly AF, Yeboua AF, Diopoh KJ (2006). Extraction, purification et caractérisation de deux cellulases du termite, *Macrotermes subhyalinus* (Termitidae). Agron. Afr., 18: 57-65.

Subramaniyan S, Prema P (2002). Biotechnology of microbial xylanases: enzymology, molecular biology, and application. Crit. Rev. Biotechnol., 22: 33-64.

Veivers PC, Muhlemann R, Slaytor M, Leuthold RH, Bignell DE (1991). Digestion, diet and polyethism in two fungus-growing termites: *Macrotermes subhyalinus* Rambur and *M. michaelseni* Sjostedt. J. Insect Physiol. 37: 675-682.

Watanabe H, Nakamura M, Tokuda G, Yamaoka I, Scrivener AV, Noda H (1997). Site of secretion and properties of endogenous endo-beta l,4-glucanase components from *Reticulitermes speratus* (Kolbe), a

Japanese subterranean termite. Insect Biochem. Mol. Biol., 27: 305-313.

Wood TM, Saddler JN (1988). Increasing the availability of cellulose in biomass. Methods Enzymol., 160: 3-10.

Yang TC, Mo JC, Cheng JA (2004). Purification and some properties of cellulase from *Odontotermes formosanus* (Isoptera: Termitidae). Entomol. Sin., 11: l-10.

Yapi DYA, Niamke SL, Kouamé LP (2007). Biochemical characterization of a strictly specific beta-galactosidase from the digestive juice of the palm weevil *Rhynchophorus palmarum* larvae. Entomol. Sci., 10: 343-352.

Zheng Z, Shetty K (2000). Solid-state bioconversion of phenolics from cranberry pomace and role of *Lentinus edodes* beta-glucosidase. J. Agric. Food Chem., 48: 895-900.

14

Soil arthropods recovery rates from 5 – 10 cm depth within 5 months period following dichlorov (an organophosphate) pesticide treatment in designated plots in Benin City, Nigeria

B. N. Iloba[1] and T. Ekrakene[2]*

[1]Department of Animal and Environmental Biology, Faculty of Life Sciences, University of Benin, Benin City, Nigeria.
[2]Department of Basic Sciences, Faculty of Basic and Applied Sciences, Benson Idahosa University, P. M. B. 1100 Benin City, Nigeria.

Soil arthropods recovery rate was monitored for five months (April - August) 2007 to ascertain whether the application of dichlorov (an organophosphate) pesticide, in varying concentration levels of 0 L(control), 0.25 L (low) and 0.75 L (high) per 25 m² would adversely affect the rate of sampling soil arthropods within a 5 - 10cm depth. Berlese Tullgren Extraction method, sorting and identification of sampled species were adopted and soil physiochemical properties were measured. Insects from eight different groups were consistently sampled. They are members Collembola, Coleoptera, Acarina and Isoptera. Others include Hymenoptera, Myriapoda, Crustaeca and Arachnida. There was an initial decrease in the monthly number of sampled soil arthropods in the treated plots from April to May but increased from June to August. Members of Acarina, Coleoptera and Myriapoda showed the highest fauna abundance while species from Hymenoptera, Crustaeca and Arachnida showed least fauna abundance. Members of Acarina (mites) exhibited the highest recovery rate while Arachnidan species were least. The result revealed that, the mean number of sampled soil arthropods was significantly different ($p < 0.05$) on the basis of the amount of dichlorov pesticide concentration used compared with the control with high concentration region being the most toxic to the arthropods, hence recording the least number of sampled soil arthropods. On the basis of concentration of applied organophosphate, the soil hydrocarbon content (0.03 - 3.95), soil pH (6.3 - 6.9), soil temperature (25.0 - 29.7°C) and soil moisture (3.2 - 6.9) were not significant ($p > 0.05$). However, increase in soil moisture from April to August was observed to result in the increase in mean numbers of soil arthropod groups sampled. The implication of this study is that, the depth of 0 - 5 cm mark into the soil litter is not the only arthropod bound zone and soil micro arthropod abundance in the soil is dependent among others on the concentration of pesticide applied. Where application is not indiscriminate, soil micro arthropods have high recovery rate which could enhance high productivity from the soil in the long run.

Key words: Dichlorov, organophosphate pesticide, soil, microarthropod, recovery rate, Benin City.

INTRODUCTION

The soil can be referred to as a world of its own life and biodiversity, consisting of various forms of life in an end-less series of interlinked caves with lots of food and sta-ble environmental conditions like a rainforest (Williams, 1999). It is a natural body, comprised of solids, liquids and gases that occur on the land surface, occupies space, and is characterized by one or both of the following; horizons, or layers that are distinguishable from the initial materials as a result of additions, losses, trans-fer and transformations of energy and matter or the ability to support rooted plant in a natural environment (Coleman, 2000). Soil is not a solid indivisible block but

---

*Corresponding author. E-mail: taidi_e@yahoo.com.

consist of innumerable number of pores. Life in the soil is lived on a micro scale and these small pores are largely habitable spaces to organisms that use them including micro arthropods. Soil pore size and number play important role in the population structure and ecology of the soil. Soil dwelling organisms include bacteria, fungi, nematodes, protozoa, molluscs, arthropods and even some vertebrates.

Soil arthropods are a vital link in the food chain as decomposer and without these organisms, nature would have no way of recycling organic material on its own (Trombetti and Williams, 1999). The process of decomposition are controlled largely by soil arthropods in conjunction with some soil invertebrates like protozoa and worms which also contribute to the soil community by mixing, loosening and aerating the soil (Evans, 1984). There is therefore an increasing need to ascertain the ease with which these valuable soil dwellers that contribute immensely to soil fertility and general nutrient recycling processes in nature to be studied on the premise of their ability to re-colonise pesticide treated farmlands.

Many studies have found that community structure, abundance and diversity of soil micro arthropods are influenced by the availability of organic matter, substrate quality, concentrations of macro and micro nutrients, and age and biodiversity of the rehabilitating habitat (Loranger et al., 1998). Environmental fate and behaviour of source component (e.g. mobility, volatility and biodegradability) is affected by time and edaphic factors (e.g. soil organic matter content, moisture, temperature and pH), and biological activities and management such as tillage, nutrient addition, moisture or thermal manipulations all interact to make possible predictions of toxic concentration from gross parameters (Mehlman, 1992).

A number of workers have studied the effect of insecticides or pesticides on the ecosystem. Among them are Brown and Gange (1989); Frampton (1994); Janseen et al. (2006); Trombetti and Williams (1999); Jones and Hopkin (1996); Reed (1997) and Frouz (1999). In their various work, they examined the toxic effects of the pesticides as well as the responses of the individual groups and the consequences of other environmental factors. Badejo (1982); Badejo and Akinwole (2006) and Badejo et al. (2002) emphasised the relationship between soil moisture content and the density of micro arthropods within the 0-5 cm soil litter. This present work became imperative In view of the numerous benefits accruing from the continual presence of soil micro arthropods to the field of Agriculture and ecosystem balance. Its strength is also hinged in the fact that, the use of pesticides (organophosphates) on soil in farmland management has become a general routine by farmers and Agriculturists and research data focussed on the deeper earth are few. These have prompted this investigation on the rate of recovery of soil micro arthropods within 5 - 10 cm depth, following treatment with organophosphate (dichloroy)

pesticide.

MATERIALS AND METHODS

Study area

This study was carried out at the Research Field of Animal and Environmental Biology Department of the University of Benin, Ugbowo main Campus, Benin City. It is situated on the Southern part of Nigeria (6° 19'N°, 6° 36' E), located in the rain forest zone of humid tropic. Benin City is characterized by both rainy and dry seasons, with rainy season and dry season lasting March to October and November to March respectively.

Sampling sites

The investigated area is an expanse of land measuring about 10 × 10 m of the study area. The study area was delineated into four stations numbered 1, 2, 3 and 4. Each station was further divided into three sub-stations marked as A, B and C thus giving a total sampling units of twelve (12). Sub-stations A, B and C represented field areas treated with 0.75 L of wide spectrum organophosphate pesticide in 20 L of water, 0.25 L of wide spectrum organophosphate pesticide in 20 L of water and 20 L of water respectively. By these formulations, Sub-stations A, B and C represented field areas with high concentration of organophosphate, low concentration of organophosphate and Control respectively. The sub-stations were well delineated and marked out as presented in Figure 1 to avoid any form of interference.

Collection and extraction

Samples from the stations were collected with a split core sampler (5 × 5.7 cm). Collection of soil samples was done on fortnight basis from April - August (months). The split core sampler was first pushed into the soil by the vertical application of pressure which was used to turn the split core sampler until it reached the 5 cm mark. This exercise was used to remove the top 0 – 5 cm before repeating the same exercise to sample the 5 - 10 cm depth. The obtained soil samples from the different stations and sub-stations were placed in separate black cellophanes and labelled accordingly. This was followed by their movement to the Laboratory where the multifaceted extractor (Berlese Tullgren Funnel) was adopted. Extraction methods were designed to suit, behaviours and body structures of the organisms (Wallwork, 1970). The Berlese Tullgren funnel extractor is best for extracting soil micro arthropods with efficiency of about 90% (Hopkins, 1997). A volume of 128cc of soil sample was placed on the sieve mesh size of (1 mm) at the top of each funnel and the organisms collected in containers with 70% alcohol within 3 days.

Sampling was done fortnightly between the hours of 10 - 11 am and 12 samples were collected at each sampling period from all stations.

Sorting and preservation

After the organisms were extracted and collected, they were immediately sorted under a binocular dissecting microscope where individuals were removed from the lot by using a sucking pipette. Individual species were then placed in separate specimen bottles with 70% alcohol for preservation and were later mounted and used for identification.

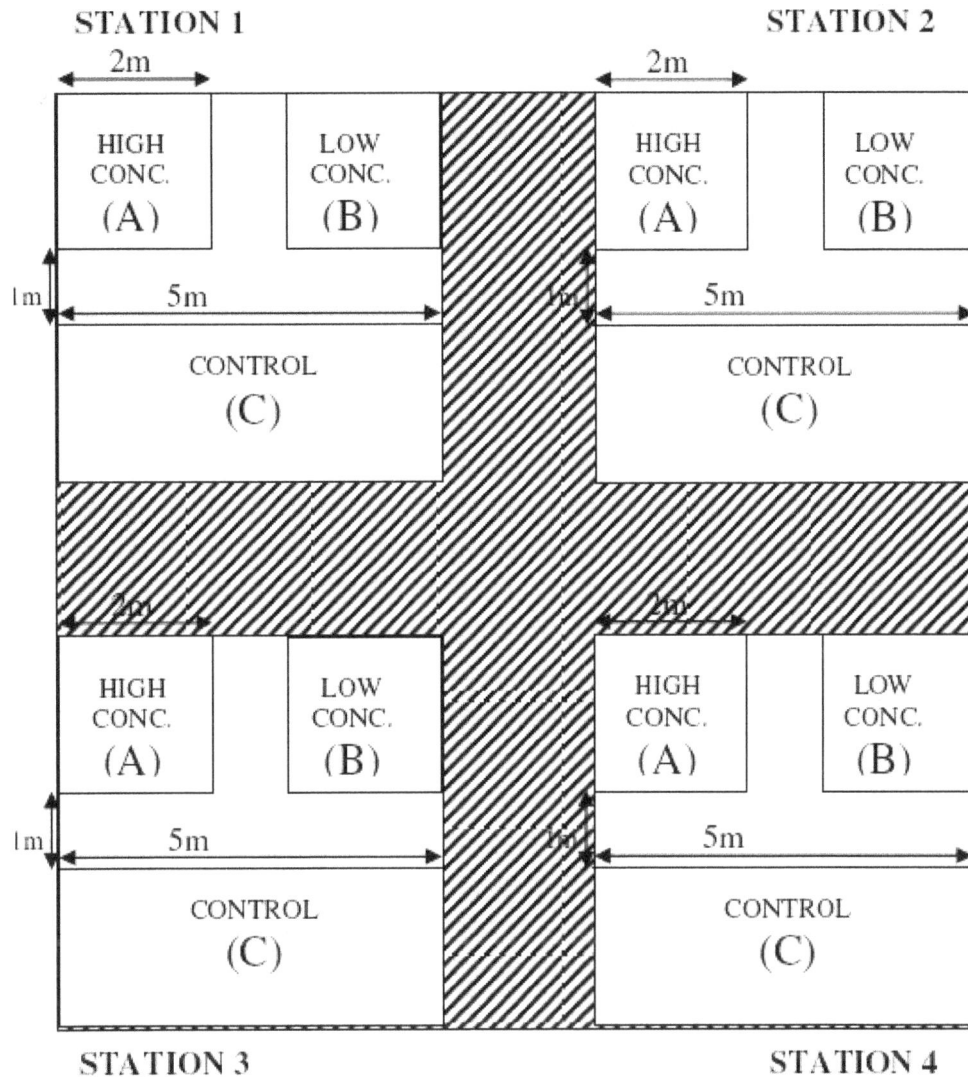

Figure 1. Description of the experimental layout adopted.

Preparation of slides

As result of the small sizes of organisms involved, it was necessary to mount them on slides for examination. The method of making permanent slide described by Hopkin (2000) was adopted to mount the organisms in Canada balsam.

Identification of collected soil micro arthropods species

Species identification was carried out at the International Institute for Tropical Agriculture, Entomology Unit, Ibadan, Nigeria.

Measurement of physiochemical parameters

Soil pH, soil temperature and soil moisture content were the parameters monitored and measured.

Soil pH: The method described by Bate (1954) was adopted. 20 g of air dried soil from each station collected from 5 - 10 cm below the soil surface was put in a 50 ml beaker and 20 ml of distilled water

was added and allowed to stand for 30 min. The mixture was stirred occasionally with a glass rode. The electrode of each pH meter was then inserted into partly settled suspension from each station and reading recorded. The pH meter was calibrated to 7.0, pH 4.0 before use with soil pH readings taken fortnightly.

Soil moisture content: 50 g of soil sample each from the stations were collected from the 5 - 10 cm below the soil surface and weighed and were thereafter placed in the oven for 24 h till constant weights were obtained.

Final weight of sample recorded
Loss in weight = initial weight – final weight

$$\text{Soil moisture content in \%} = \frac{\text{loss in weight} \times 100}{\text{Oven dried}}$$

The soil moisture content was also taken fortnightly along with the sampling time of other parameters.

Soil temperature: Temperature readings were collected between

Table 1. Monthly Mean number of soil micro arthropods sampled at the different concentrations (±S.D).

Sampling months	Conc. (pesticide vol. in litres per 20 L of water)	Mean numbers of arthropod sampled (±S.D)
	.00	7.25 ± 4.37[a]
April	.25	4.00 ± 2.45[b]
	.75	3.63 ± 2.50[c]
	.00	6.75 ± 3.15[a]
May	.25	0.50 ± 1.07[b]
	.75	0.00 ± 0.00[b]
	.00	10.88 ± 5.87[a]
June	.25	3.25 ± 2.38[b]
	.75	2.25 ± 1.91[b]
	.00	13.88 ± 6.42[a]
July	.25	8.63 ± 5.60[b]
	.75	6.63 ± 3.38[c]
	.00	16.25 ± 8.21[a]
August	.25	11.25 ± 4.46[b]
	.75	9.38 ± 4.17[b]

Each value is the mean of four replicates. Means followed by the same letter are not significantly different (P > 0.05) from each other, using New Duncan's Multiple Range Test.

8 - 9 am in the morning and 5 - 6 pm during the evening hours. Temperature reading was achieved by digging first a small 5 cm deep and 5 cm wide hole followed by the insertion of a thermometer so as to enable a depth of the lower 5 - 10 cm to be covered. Reading on the thermometer was obtained after 2 min. This was repeated thrice and average value taken for both the morning and evening sampling periods.

Soil total hydrocarbon: The soil total hydrocarbon was determined using a spectrophotometer, pipette, and 250 ml separating glass funnel, mechanical shaker and n-hexane. A 5 g weight of soil from each site collected from 5 - 10 cm deep was dried and kept in bottle containers. To each bottle container was added 25 ml of n-hexane to extract the soil total hydrocarbon from the soil. These were placed on the mechanical shaker and shaken for 10 min to ensure thorough mixing and thereafter left to stand. A standard of n-hexane was prepared and used to standardize the spectrophotometer before introducing the THC from the soil into the spectrophotometer for the absorbance reading. The soil total hydrocarbon content (THC) concentration in part per million for each was then calculated as follows;

Soil total hydrocarbon content (ppm) = Instrument Reading × Reciprocal of slope × 25 ml/5 g

Where; Instrument reading (IR) was from the spectrophotometer.

The reciprocal of slope was calculated for each based on spectrophotometer reading,

Volume of extraction reagent was 25 ml,
Weight of each soil sample used was 5 g.

RESULTS

The monthly mean number of soil micro arthropods from the different plots (stations) treated with varying concentrations is presented in Table 1. Monthly mean number of sampled arthropods from the treated stations significantly decreased from April to May (high conc. 3.63 - 0.00; low conc. 4.00 - 0.5) and increased steadily from June to August (high conc. 2.25 - 9.38; low conc. 3.25 - 11.25). Though a monthly mean decrease of arthropod sampled was observed from April to May (7.25 - 6.75) in the control station, this was insignificant compared to decreases recorded in the treated stations for the same period. It is also pertinent to note that, the decrease recorded in the treated stations was more pronounced in the plot treated with more of the pesticide as presented in Table 1.

The soil micro arthropod groups showed varying mean values within the 5 months period of investigation. The mean number of soil micro arthropod group sampled is presented in Table 2 and Figure 2. For all soil arthropod groups implicated in this investigation, the level of concentration of organophosphorous pesticide used significantly affected the mean number of sampled soil arthropod group. The least number of soil arthropods was sampled from the plot treated with higher concentration of pesticide compared with the plot treated with low concentration while the control plot recorded the highest number of soil arthropods sampled. On group abundance, members of Acarina, Coleoptera, and Myriapoda were most abundant while Arachnida and Crustacea were least in abundance as presented in Table 2.

The mean value of investigated physiochemical parameters (soil temperature, soil moisture, soil pH and Soil hydrocarbon) is present in Figure 3. The period, April to

Table 2. Mean number of soil micro arthropods sampled at the different concentrations (±S.D).

Soil microarthropod groups	Conc. (pesticide vol. per 20 L of water)	Mean numbers of arthropod sampled (±S.D)
Collembola	.00	9.4 ± 2.88^a
	.25	5.4 ± 4.34^b
	.75	3.8 ± 2.59^b
Coleoptera	.00	$16.8 \pm 6..30^a$
	.25	7.4 ± 5.32^b
	.75	6.6 ± 5.64^b
Isoptera	.00	9.0 ± 4.47^a
	.25	3.6 ± 3.05^b
	.75	3.2 ± 2.77^b
Hymenoptera	.00	7.6 ± 2.51^a
	.25	4.0 ± 3.08^b
	.75	4.0 ± 2.74^b
Acarina	.00	17.6 ± 6.69^a
	.25	9.6 ± 6.50^b
	.75	7.4 ± 5.55^b
Myriapoda	.00	13.6 ± 6.69^a
	.25	9.4 ± 7.80^b
	.75	6.0 ± 6.48^c
Crustacea	.00	5.2 ± 2.28^a
	.25	2.8 ± 2.77^b
	.75	2.6 ± 2.61^b
Arachnida	.00	4.8 ± 2.05^a
	.25	2.0 ± 3.08^b
	.75	1.4 ± 2.19^c

Each value is the mean of four replicates. Means followed by the same letter are not significantly different ($p > 0.05$) from each other, using New Duncan's Multiple Range Test.

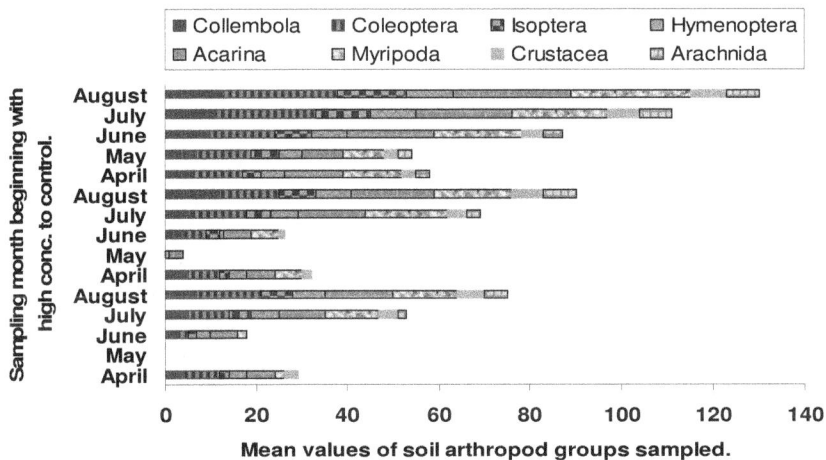

Figure 2. Monthly mean number of soil arthropod groups sampled progressively from High concentrated stations to controlled ones.

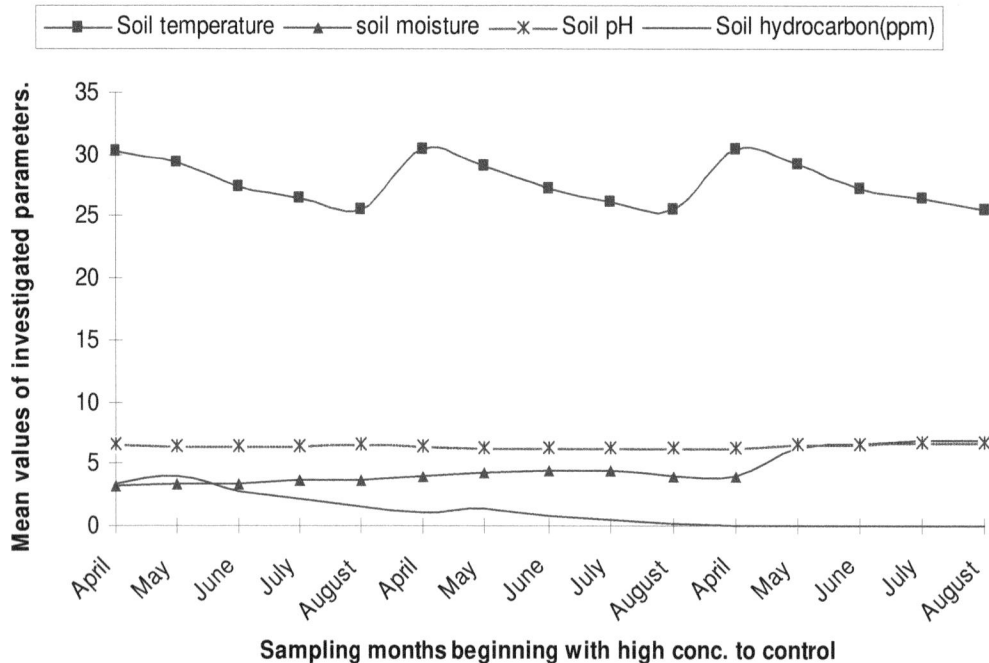

Figure 3. Mean values of investigated parameters.

August recorded decrease in mean value of soil temperature (29.6 - 25.0°C) while there was increase in soil moisture (3.2 - 6.9) irrespective of plots. However, Soil hydrocarbon content decreased from April to August in treated stations with no net change in values in the control station (High Conc. 3.36 -1.59; Low Conc. 1.12 - 0.22; Control 0.03 - 0.03) for the period. Soil pH mean values did not significantly change within the period.

DISCUSSION

Many workers including Badejo (1982) and Badejo and Akintola (2006) have emphasised that, most soil fauna especially the orbited mites enjoy better conducive micro environment within the top 5 cm of soil. The high number of soil arthropods sampled from the 5 - 10 cm depth in this investigation is an indication that, the soil ecosystem is not just a world with array of different species of life but the number that could be found per time at given depths is dependent among other factors, on the physiochemical nature of the soil at that time. This is an affirmation of the richness of the soil ecosystem when viewed from a broad perspective as soil arthropods are one group of living organisms found in the soil. Similar observation was made by Williams (1999). The monthly mean number of arthropod and the fluctuation (increase or decrease) shown in Table 1 and Figure 2 during the period could be attributable to two factors. One possible reason for the initial decrease in the number of soil arthropods as observed in the plots treated with the pesticide from April to May was perhaps the toxic effect of the dichloroy (an organophosphate pesticide) that was applied to the affected plots. The toxic effect of this pesticide has the ability to create harsh environment that could cause death of the soil fauna, thereby preventing them from responding to the extraction method of light rays. This would lead to low number of arthropod that could be sampled. This observation agreed with the ones earlier made by Frouz (1999); Jones and Hopkins (1998); Reed (1997) and Frampton (1994). Though they were not particular on the monthly decrease or increase, they observed that, the application of pesticide affects the environmental condition, thus affecting the number of micro arthropods present in such treated areas. Also significant in the reason for the decrease in the treated plots within April to May period is the amount of decrease observed in both the high and low pesticide concentrated treated plots. The reduction in number of soil arthropod was significantly more in plot treated with high concentration of the pesticide which recorded a zero value in the month of May. This could imply that, more concentration of pesticide would either have caused more of the soil arthropod to die or due to more toxic harsh environment created; it would lead to more downward migration of soil arthropods dwelling there. Hence, the justification for the differences in the arthropod reduction between high and low concentrated plots compared to the control. Significantly, the observed arthropod increase (June - August) after the period of decrease (April to May) may probably be due to either the effect of dilution of water on the pesticide (as its coincided with wet season), temporary absence of parasites or due to the low persistence nature of the pesticide (dichlorov) which ranges between 4 - 8 weeks.

Arthropod species richness in soil is not contestable as 8 groups were consistently sampled. The arthropod groups and how the concentration of the pesticide affected the number sampled are presented in Table 2. The survival and re-colonisation ability of individual insect groups differs. Members of the Acarina were more abundant followed by Coleoptera and Myriapoda while members of Arachnida and Crustacea groups were least in abundance. All arthropod groups implicated were affected by the level of pesticide concentration used as none was sampled in the month of May. Members of Collembola, Isoptera and Hymenoptera groups exhibited the weakest ability in being able to withstand the application of the pesticide as they were observed to have drastically reduced between May and June but resurfaced strongly in July. This drastic reduction in these groups of soil fauna could be as a result of their soft body which possibly offered least protection against the toxicity of the pesticide. The soft-bodied morphology contrast those of Coleoptera, Myriapoda and Crustacea which enjoy protection based on morphological toughness and fast movement away from areas of contamination. Though the Hymenopteran, Collembolan and Isopteran groups showed least ability among others to withstand the application, they exhibited a great tendency to re-colonise with Acarina group showing the greatest tendency of re-colonisation of the treated areas while Crustacean was least as shown in Figure 2. This may have been facilitated by two factors. It might be that, either the pesticide affected the parasites or predators that parasitise or prey on these groups of soil fauna or the toxicity of the pesticide reduced considerably, thus leading to an initial rapid increase in their numbers from July to August as shown in Figure 2.

The soil pH, soil temperature and soil moisture content did not significantly varied among the different areas treated with the different concentrations of dichlorov pesticide and the control. Though these parameters did not vary significantly, field observation revealed a steady increase in soil moisture as the period coincided with rainy season. This increase in soil moisture shown in Figure 3 was observed to lead to increase in soil fauna sampled. This observation is similar to that made by Badejo (1982), when he asserted that there was an increase in the density of soil arthropods with increase in soil moisture. Increased soil water has the ability to dilute the pesticide thereby reducing its toxic effect on both the soil fauna and the environment. However, the concentration of pesticide used resulted in varying amount of THC in (ppm) in the studied stations. The station treated with high concentration of pesticide recorded highest mean of 0.41 ppm while the control station obtained least mean value of 0.03 ppm. The increase in total hydrocarbon content in the stations treated with the pesticide may be due to the chemistry of the applied pesticide, thus leading to the THC values of treated stations as compared to the control in Table 2. How this increase in THC value affected the sampled soil micro arthropods was not clearly understood in this investigation.

The findings of this investigation have significant implications. Whereas, one would have thought that, pesticide application could have a persistently reducing effect on the soil fauna, it revealed that all things being equal, arthropod reduction is with time and re-colonization after a period is imminent. This notion tends to alleviate the fears of soil ecosystem imbalances as a temporary phenomenon with no much adverse effect on the productivity ability of the soil in the long run when the pesticide is not indiscriminately applied. Also, the depth below the immediate zone of litter fermentation is significantly important due the number of soil arthropods it houses. It is expected that more research would be focussed on even deeper depth for the purpose of comparism of arthropod groups on the basis of depth.

REFERENCES

Bate RG (1954). Electrometric pH Determination. John Willeys and Sons, Inc. New York.

Badejo MA 1982. The distribution and abundance of soil microarthropods in three habitats at the University of Ife, M.Sc. Thesis, University of Ife, Nigeria.

Badejo MA, Akintola PO (2006). Micro environmental preference of orbatid mite species on the floor of a tropical rainforest in Nig Exp. Appl. Acarol. (4): 145-156.

Badejo MA, Jose AZE, Adriana M, Elizabeth FC (2002). Soil orbatid mite communities under three species of legumes in an ultisol in Brazil, J. Econ.Entomol. 94(1): 55-59.

Coleman DC (2000). Soil Biota, soil systems and processes. Encyclopedia Biodiver. 5: 305-314.

Evans FR (1984). Soil maintenance by soil dwelling invertebrates. Austr. J. Ecol. 34: 713-720.

Frampton DE (1994). Effect of silivicultural practices upon collembolan population in coniferous forest soil. Acta Zool. Fen. (4): 87-145.

Frouz J (1999). Use of soil dwelling Diptera as bioindicator: A review of ecological requirement and response to disturbance. Agric. Ecosyst. Environ. 74: 107-186.

Hopkins SP (1997). The biology of springtails (insects: collembolan) oxford university of press Inc. New York.

Janssen MA, Schoon ML, Weimao Ke, Borner K (2006). Scharly networks on resilience, Vulnerability and adaptation within the human dimensions of global environmental change. Global Environ. Change 16(3): 240-252.

Jones DT, Hopkins SP (1998). Reduced survival and bodysize in the terrestrial isopod Porcellio and Scaber from a metal-Polluted environment. Environ. Pollut. 99: 215-223.

Loranger G, Ponge JF, lavelle P (1998). Influence of agricultural practices on arthropod communities in a vertisol (Martinique). Eur. J. Soil Biol. 34: 157-165.

Mehlman DW (1992). Effect of fire on plant community composition of North Florida growth pineland. Bull. Torrey Bot. Club 119: 376-383.

Reed CC (1997). Responses of soil microarthropods to pesticides. Nat. Areas J. 17: 59-66.

Trombetti S, Williams C (1999). Investigation of soil dwelling invertebrates. Ecol. 70:220-260.

Williams C (1999). Biodiversity of the soil ecological communities: in dwelling fauna of the soil environment. Ecol. 50: 456-460.

Wallwork JA (1970). Ecol. soil animals. McGraw Hill Publisher, London.

15

Managing nematode pests and improving yield of pineapple with *Mucuna pruriens* in Ghana

K. Osei [1]*, R. Moss[2], A. Nafeo[2], R. Addico, A. Agyemang[1] and J.S. Asante[1]

[1]CSIR- Crops Research Institute, P. O. Box 3785, Kumasi, Ghana.
[2]West Africa Fair Fruit, P. M. B. KD11, Kanda, Accra, Ghana.

We investigated the potentials of a non-traditional legume crop, *Mucuna pruriens* and fallow treatments in managing plant parasitic nematodes population and improving the yield of pineapple in 2008 to 2009 at Asamankese in the Eastern region of Ghana. Nematode population density/200 cm^3 soil at 6 months of application of treatments and at harvest, population/g^{-1} of pineapple root and yield of pineapple were analysed. *M. pruriens* reduced nematodes population and improved the yield of pineapple considerably. At harvest, rhizosphere soil from pineapple cultivated on bush fallow treated plots recorded significantly P = 0.05) higher populations: 76, 211, 100 and 68 for *Helicotylenchus multicintus, Meloidogyne* (juveniles) spp., *Pratylenchus brachyurus* and *Rotylenchulus reniformis,* respectively while soil from pineapple on weed free-*M. pruriens* treated plots recorded lower populations: 39, 50, 64 and 31 representing population reduction of 95, 322, 56 and 119%, respectively. Three nematode species: *Meloidogyne* spp., *P. brachyurus* and *H. multicintus* were recovered from pineapple roots at harvest in which the nematicidal potential of *M. pruriens* was further demonstrated. While significantly (P < 0.01) higher populations (165, 89 and 61) of *Meloidogyne* spp., *P. brachyurus* and *H. multicintus* were recovered from the bush fallow treated pineapple roots, lower populations (39, 35 and 24) of the pests were recovered from weed free-*M. pruriens* treated pineapple roots, respectively. Thus, *M. pruriens* reduced populations by 323, 154 and 154%, respectively. The yield improvement potential of *M. pruriens* was manifested when the main pineapple crop from weed free-*Mucuna* plots recorded 51 t/ha, whilst a bush fallow treatment recorded 34 t/ha out-yielding the bush fallow treatment by 50%. The use of higher plants in managing nematodes and improving yield of crops is environmentally acceptable and sustainable.

Key words: *Ananas comosus*, horticultural crop, plant parasitic nematodes, sustainable agriculture.

INTRODUCTION

Pineapple (*Ananas comosus* L. (Merr.) is an important horticultural crop cultivated for export in the southern part of Ghana. The smooth cayenne cultivar is the most largely cultivated in the Eastern and Greater Accra regions, while the sugar loaf dominates the Central region (Trienekens, 2003). Major production of the crop, especially for the export market is done in the Eastern, Central, Greater Accra and Volta regions. The role of fruits and fruit juices in nutrition and health cannot be overemphasised (Wardy et al., 2009). The pineapple fruit supplies significant amount of vitamins and minerals necessary for healthy growth. In Ghana, local industries process pineapple into fruit juice, oven-dried, salad and jam. The pineapple industry is a source of income and offers employment to a majority of people in southern Ghana (HEII, 2006). It is among the top-five non-traditional export commodities in terms of foreign earnings in Ghana (Wardy et al., 2009). Current Ghana's export earnings of pineapple hovers around 32 million dollars a year. Two of the major constraints to the production of pineapple are infertile soils and the menace of plant parasitic nematodes. Three species of nematodes: the root–knot nematodes, *Meloidogyne* spp., the reniform nematode, *Rotylenchulus reniformis* and the root lesion nematode, *Pratylenchus brachyurus* are of the greatest economic importance (Spies et al., 2005). These pests delay leaf emergence and reduce leaf weight (Sarah, 1986), cause stunted growth and generally reduce pineapple yield significantly (Spies et al., 2005). In Ghana, management of nematode pests has relied extensively on the use of agro-chemicals (nematicides). Similarly, the improvement of soil fertility has been largely through chemical fertilizer usage. Degreening of pineapple is done with chemicals such as ethrel which is the most commonly used ethephone product in Ghana. Ethephon is an organophosphate and placed in the US EPA's top toxicity category. Indiscriminate use of agro-chemicals is an abuse of the environment and the price associated with this system of farming is really high. More importantly, the future of the pineapple industry is bleak because the new legislation on Minimum Residue Levels (MRLs) in the European Union (EU) will reject imports of produce with residue of pesticides not approved under its regulations (Gogoe et al., 2001). The purpose of this paper was to find a sustainable system of pineapple production by divorcing the over reliance on synthetic chemicals method of production and marrying the organic system of production. The improper use of synthetic products constitutes an assault on the environment and the practice should be discontinued. An experiment was therefore conducted at Asamankese in the Eastern region of Ghana, to evaluate *Mucuna pruriens* and fallow treatments for their nematode suppression and yield increasing potentials in pineapple production.

MATERIALS AND METHODS

Treatments and experimental design

A two factor experiment mounted on randomized complete block design was conducted to assess the potential of treatments at reducing plant parasitic nematodes population and their effect on yield of pineapple. The factors were four land preparation methods which were the main plot treatments and seed bed preparation methods (the sub plots). The main plot treatments were: weed free-*M. pruriens* (*M. pruriens* was cultivated and weeds controlled), clean fallow (plots were weeded whenever the weed free-*Mucuna* plots were weeded), unweeded-*M. pruriens* (*Mucuna* was cultivated but weeds were not controlled) and bush fallow (plots were not weeded at all). The sub plot treatments were, planting on ridges with plastic mulch, planting on ridges without plastic mulch and planting on the flat. The land preparation methods and the main plots, lasted for five months (April to September, 2008) before the seed bed preparation methods, the sub plots were superimposed. The treatments were replicated three times. A plot measured 5 x 4 m.

Experimental site

The experiment was conducted at Asamankese in the Eastern region of Ghana. The prominent weed on the experimental site was *Setaria* spp. The experimental plot had previously been cultivated with pineapple cv. smooth cayenne in 2005 and the current study started in April 2008. The cultivar used in the current study was smooth cayenne which was planted at a double row spacing of 30 x 30 cm. Hand weeding was done six times before harvesting the main crop.

Sampling and extraction

Soil samples were collected at three time periods; at the start of the trial before the planting of *M. pruriens*, six months of application of treatments and at harvest of the main crop with a 5 cm soil auger to a dept of 20 cm. The samples (200 cm^3) per treatment were extracted using the modified Baermann funnel method. Nematodes were also extracted from one gram pineapple roots per treatment at harvest. After 24 h of extraction, samples were fixed with TAF and nematodes were identified under a stereo microscope at magnification 100x. Yield data was not transformed but nematode based data was log transformed (ln (x+1)) to conform to the assumption of normal distribution before analysis using Genstat 8.1. Means were separated using Duncan's multiple range test (DMRT).

RESULTS

From the initial soil samples assayed, nematodes were found on all the plots. In all, eight species were encountered. The nematodes in order of abundance were; *Meloidogyne* (juveniles) spp. > *H. multicintus* > *P. brachyurus* > *R. reniformis* > *Tylenchulus semipenetrans* > *Hoplolaimus* spp. > *Xiphinema* spp. > *Paratrichodorus minor*. The last two nematodes transmit plant viruses. Of the eight nematodes extracted from the soil samples, *Meloidogyne* spp., *P. brachyurus* and *R. reniformis* are the most pathogenic on pineapple.

Five plant parasitic nematodes namely; *H. multicintus*, *Meloidogyne* spp., *P. brachyurus*, *R. reniformis* and *T. semipenetrans* reacted differently to the treatments six months after application (Table 1). There was no interaction between weed free-*Mucuna pruriens* and planting on ridges with plastic mulch regarding the management of the five plant parasitic nematodes. However, weed free-*M. pruriens* treatment significantly (P < 0.01) reduced population of *H. multicintus* and *T. semipenetrans* than the clean and bush fallow treatments. Weed free-*M. pruriens* treatment was also different from the unweeded-*Mucuna pruriens* treatment. Similarly, unweeded-*Mucuna pruriens* treatment was not different from the fallow treatments in reducing the population of *H. multicintus* and *T. semipenetrans*.

Weed free-*M. pruriens* treatment was significantly effective in reducing the population of *Meloidogyne* spp., *P. brachyurus* and *R. reniformis* than the other treatments, except unweeded-*M. pruriens* treatment. Unweeded-*M. pruriens* treatment was also different from the two fallow treatments.

No interactions were observed between the main plots and the sub plots treatments at harvest. *T. semipenetrans* encountered six months of application of treatments was never found in soil samples. In root samples, three species *Meloidogyne* spp., *P. brachyurus* and *H. multicintus* were recovered. In both soil and root

Table 1. Nematode population density/200 cm^3 of soil six months after application of treatments.

Treatments	Nematodes				
	Heli	Meloi ^{z}	Praty	Roty	Tyl
Weed-free Mucuna	52(3.3)^{y}a	34(3.0)a	49(4.2)a	51(4.2)a	57(4.0)a
Unweeded Mucuna	66(4.5)b	63(4.4)a	64(4.4)a	67(4.2)a	80(4.5)b
Clean fallow	108(4.4)b	143(4.7)b	129(4.6)a	162(4.4)b	100(4.6)b
Bush fallow	222(4.6)b	283(4.7)b	216(4.9)b	365(4.7)b	168(4.6)b
Grand mean	4.1	4.2	4.4	4.6	4.4
Lsd at P <0.01	0.4	0.4	0.5	0.3	0.2

Values following different letters in a column are significant at P < 0.01.z, Juveniles; ^{y}ln (x + 1) transformed data used in ANOVA in parenthesis. *Heli: H. multicintus.; Meloi: Meloidogyne* spp.; *Praty: P. brachyurus; Roty: R. reniformis; Tyl: T. semipenetrans.*

Table 2. Nematode population density/200 cm^3 of soil at harvest.

Treatments	Nematodes			
	Heli	Meloi ^{z}	Praty	Roty
Weed-free Mucuna	39(4.7)^{y}a	50(5.5)a	64(7.0)a	31(4.2)a
Unweeded Mucuna	53(6.2)ab	91(8.7)ab	74(7.7)a	42(5.0)ab
Clean fallow	59(7.1)ab	92(9.1)b	84(8.4)a	37(5.0)ab
Bush fallow	76(8.1)c	211(12.8)c	100(9.4)a	68(7.4)bc
Grand mean	6.5	9.0	8.1	5.4
Lsd at P = 0.05	0.8	1.7	2.0	0.9

Values following different letters in a column are significant at P = 0.05. ^{z}, Juveniles; ^{y}, ln (x + 1) transformed data used in ANOVA in parenthesis; *Heli: H. multicintus; Meloi: Meloidogyne* spp.; *Praty: P. brachyurus; Roty: R. reniformis.*

samples, results followed the same trend observed during six months of application of treatments (Tables 2 and 3). Weed free-*Mucuna* treatment significantly (P = 0.05 and P < 0.01) reduced the population density of *H. multicintus* and *Meloidogyne* spp., respectively than did in the two fallow treatments. However, the weed free-*Mucuna* was not different from the unweeded *Mucuna* treatment in soil samples at harvest. The clean fallow and the bush fallow treatments were different with many more *H. multicintus, Meloidogyne* spp. and *R. reniformis* found on bush fallow plots than the clean fallow plots (Table 2). In root samples, weed free-*Mucuna* treatment significantly (P < 0.01) reduced the population of *Meloidogyne* spp., *H. multicintus* and *P. brachyurus* than the fallow treatments (Table 3). Pineapple yield from weed free-*Mucuna* treated plots was a high of 51 t/ha, while the yield from bush fallow plots was a low of 34 t/ha (Table 4).

DISCUSSION

To prevent an assault on the environment, a sustainable agriculture should be promoted. The components of sustainable agriculture include integrated pest management (IPM) and integrated crop management (ICM). IPM emphasizes the growth of a healthy crop with the least possible disruption to the agro-ecosystems and encourages natural pest control mechanism. ICM on the other hand, involves managing crops profitably with respect for the environment, in ways which suit local soil, climatic and economic conditions. The conventional crop production system with the attendant indiscriminate use of agro chemicals is not sustainable. Not only would the environment be destroyed to the disadvantage of posterity, crops produced through this system would not be competitive on the world market. Nematicidal properties have been reported in many higher plants (Chitwood, 2002). *Tagetes erecta* cv. Crackerjack reduced the population of root-lesion nematodes, *P. penetrans* in carrots field (Kimpinski and Sanderson, 2004). The high yielding soybean cultivar "Padre" suppressed the population of reniform nematode, *R. reniformis* (Westphal and Scott, 2005). Also, *Arachis pintoi* reduced the galling of *Meloidogyne incognita* or tomato (Marban-Mendoza et al., 1992) and decreased *R. reniformis* numbers on coffee (Herrera and Marban-Mendoza, 1999). Castor bean, *Ricinus communis* has been successful in managing nematode populations (Hagan et al., 1998) and American joint-vetch,

Table 3. Nematode population density / g^{-1} root at harvest.

Treatments	Nematodes		
	Heli	*Meloi*^{z}	*Praty*
Weed-free *Mucuna*	24(4.6)^{y}a	39(5.8)a	35(5.5)a
Unweeded *Mucuna*	35(5.6)ab	68(6.1)b	56(7.4)b
Clean fallow	58(7.2)b	97(9.7)c	83(8.7)bc
Bush fallow	61(7.2)b	165(12.7)d	89(9.2)c
Grand mean	6.2	8.6	7.7
Lsd at P < 0.01	1.0	0.1	1.3

Values following different letters in a column are significant at P < 0.01.^{z}Juveniles; ^{y}ln (x + 1) transformed data used in ANOVA in parenthesis. *Heli: H. Multicintus; Meloi: Meloidogyne* spp.; *Praty: P. brachyurus.*

Table 4. Yield of pineapple.

Treatments	Yield (t/ha)
Weed free *Mucuna*	51a
Unweeded *Mucuna*	48a
Clean fallow	36b
Bush fallow	34b
Grand mean	42.1
Lsd at P <0.01	15

Values following different letters in a column are significant at P < 0.01.

Aeschynomene virginica and partridge pea; *Cassia fasciculata* significantly suppressed nematode population densities (Rodríguez-Kábana et al., 1991). Plants with the potential to suppress nematode populations by producing anti-helminthic compounds have been designated "antagonistic plants" (Pandey et al., 2003).

In the current study, *M. pruriens* established itself as an effective candidate in controlling nematode population build up and improving the yield of pineapple. At harvest, the weed free-*Mucuna* treatment reduced the populations of *H. multicintus, Meloidogyne* spp., *P. brachyurus* and *R. reniformis* by 95, 322, 56 and 119% over the bush fallow treatment, respectively. The effectiveness of *M. pruriens* was greatest at six months after the application of treatments as weed free-*Mucuna* treatment reduced the populations of the four nematode pests by 327, 732, 341 and 616% over the bush fallow treatment, respectively. Nematode population extracted from pineapple roots at harvest further demonstrated the nematicidal potential of *M. pruriens*. Populations of *Meloidogyne* spp., *P. brachyurus* and *H. multicintus* recovered from roots from weed free-*Mucuna* plots represented significant reduction of 323, 154 and 154% over the bush fallow treatment, respectively. Results were consistent with the findings of Quénéhervé et al. (1998) who used *M. pruriens* to effectively control *M. incognita* and *R. reniformis* in vegetables in France. It was observed from the results that, the effectiveness of *M. pruriens* on soil samples populations reduced with time. A probable explanation might be that, the active ingredients in *M. pruriens* diluted with time. It has also been reported that, the antagonistic activity of *Mucuna* might be due to the production of phytoalexins by the roots (Vargas et al., 1996). Some tropical legumes have demonstrated their ability to increase crop yields (Kumbhar et al., 2007) perhaps due to their potential to positively enhance the nitrogen economy of soils (Giller, 2001). Becker and Johnson (1998) reported that, *Mucuna* spp. *Canavalia ensiformis* and *Crotalaria anagyroides* significantly increased upland rice yield in the Ivory Coast. Also, the growth and yield of okra following *M. pruriens* resulted in a significant 48%

yield increase over the control and *Phaseolus vulgaris* treatments in Ghana (Osei et al., 2010). In addition to nematode suppression and yield improvement, some legumes have the potential to control insect pests. Rotenone extracted from the roots of tropical legumes, *Lonchocarpus* and *Derris* control insect pests such as aphids, beetles, fleas and lice by inhibiting cellular respiration (Buss and Park-Brown, 2009). In the current study, *M. pruriens* not only suppressed populations of plant parasitic nematodes but also increased the yield of pineapple significantly. Weed free-*Mucuna* treatment out yielded the bush fallow treatment by 50%.

Conclusions

Antagonistic plants have a significant role in integrated pest management. Where the antagonistic plant is a leguminous crop, capable of replenishing the fertility status of the soil, then the results of the biotrophic system (host plant–nematode interaction) becomes more rewarding. The fast growing *M. pruriens* is capable of controlling soil erosion, suppressing weed growth and providing hay for livestock.

More importantly, *Mucuna* is used for food in some African countries (Rachie and Roberts, 1974). The Ghanaian farmer should consider the incorporation of the versatile *M. pruriens* in his farming system. The cost effectiveness, environmental friendliness and product acceptability especially in the international trade render the system sustainable.

ACKNOWLEDGEMENTS

The authors acknowledge (GTZ) Gmbh, for co-funding this study with the West Africa Fair Fruit. Special thanks also go to the staff of the Ministry of Food and Agriculture for their collaboration in the study.

REFERENCES

Becker M, Johnson DE (1998). Legumes as dry season fallow in upland rice-based systems of West Africa. Biol. Fertil. Soils, 27:

358-367.

Buss EA, Park-Brown SG (2009). Natural products for insect pest management. Department of Entomology and Nematology, UF/IFAS, Florida Extension Service, ENY-350 (IN197). Available @http://edis.ifas.ufl.edu/in197.

Chitwood DJ (2002). Phytochemical based strategies for nematode control. Ann. Rev. Phytopathol., 40: 221–249.

Giller KE (2001). Nitrogen fixation in tropical cropping systems. CAB International, Wallingford, UK, p. 423.

Gogoe S, Dekpor A, Williamson S (2001). Prickly issues for pineapple pesticides. Pestic. News, 54: 4-5

Hagan A, Gazaway W, Sikora E (1998). Nematode suppression crops. Alabama Coop. Ext. Syst., ANR 856: 4.

HEII (2006). Implementation Guide MD2 pineapple sucker multiplication programme. AgSSIP/MoFA, Accra.

Kimpinski J, Sanderson K (2004). Effects of crop rotations on carrot yield and on the nematodes Pratylenchus penetrans and Meloidogyne hapla. Phytoprotect., 85: 13-17.

Kumbhar AM, Buriro UA, Oad FC, Chachar QI (2007).Yield parameters and N-uptake of wheat under different fertility levels in legume rotation. J. Agric. Tech., 3: 323- 333.

Marban-Mendoza N, Dicklow MB, Zuckerman BM (1992). Control of Meloidogyne incognita on tomato by two leguminous plants. Fundam. Appl. Nematol., 15: 97-100.

Osei K, Fening JO, Gowen SR, Jama A (2010). The potential of four non-traditional legumes in suppressing the population of nematodes in two Ghanaian soils. J. Soil .Sci .Environ. Manage., 1: 63-68.

Pandey R, Sikora RA, Kalra A, Singh HB, Pandey S (2003). Plants and their products act as major inhibitory agents. In: Trivedi, PC. (ed.) Nematode Manage., Plants, Scientific Publishers, Jodhpur, India. pp. 103–131.

Quénéhervé P, Topart P, Martiny B (1998). Mucuna pruriens and other rotational crops for control of Meloidogyne incognita and Rotylenchulus reniformis in vegetables in polytunnels in Martinique. Nematropica, 28: 19-30.

Rachie KO, Roberts LM (1974). Grain legumes of the lowland tropics. Adv. Agron., 26 : 1-132.

Rodríguez-Kábana R, Robertson DG, King PS, Wells L (1991). American jointvetch and partridge pea for the management of Meloidogyne arenaria in peanut. Nematropica, 21: 97-103.

Sarah JL (1986). Influence of Pratylenchus brachyurus on the growth and development of pineapple in the Ivory Coast. Rev. Nematol., 3: 308-309.

Spies BS, Caswell-Chen EP, Sarah JL, Apt WJ (2005). Nematode parasites of pineapple. In: Luc M, Sikora RA, Bridge J (eds.) Plant parasitic nematodes in subtropical and tropical agriculture (Second edition) CABI Wallingford, UK, pp. 709-731.

Trienekens JH (2003). Market induced innovations through international supply chain development. KLCT TR-207. A Working Paper.

Vargas R, Rodriquez A, Acosta N (1996). Components of nematode suppressive activity of velvet bean, Mucuna deeringiana. Nematropica, 26: 323.

Wardy W, Saalia FK, Asiedu MS, Budu AS, Sefa-Dede S (2009). A comparison of some physical, chemical and sensory attributes of three pineapples (Ananas comosus) varieties grown in Ghana. Afr. J. Food Sci., 3: 022-025.

Westphal A, Scott Jr AW (2005). Implementation of soybean in cotton cropping sequences for management of reniform nematode in South Texas. Crop Sci., 45: 233-239.

16

A review of on *Aleurodicus dispersus* Russel.
(spiralling whitefly) [Hemiptera: Aleyrodidae] in Nigeria

A. D. Banjo

Department of Plant science and applied zoology, Olabisi Onabanjo University, P. M. B. 2002, Ago-Iwoye, Ogun State, Nigeria. E-mail: adaba55@yahoo.co.uk.

The developmental biology of *Aleurodicus dispersus* Russel., have been investigated. It was found to have a cumulative developmental period of (23 - 41) days. The mean numbers of egg developing to adult have been found to be 138.1 per thousand eggs. The spread of the insect have been found to be connected to human traffics. The oviposition and feeding occurs simultaneously and occur more on their abaxial surface of host leaves. Rainfall and temperature play a prominent role on the abundance and seasonal fluctuation of the insect and infact, regulating their population. Presently, *A. dispersus* is found on arable as well as ornamental plants but rarely on gramminae. At present, *A. dispersus* is a minor pest with the potential of becoming a serious pest with the increasing global warming.

Key words: *Aleurodicus dispersus*, oviposition, abaxial surface, abundance, gramminae.

INTRODUCTION

Aleurodicus dispersus (Russel, 1965) otherwise known as spiralling whitefly (swf) is a small (1 - 2 mm long) insect as other whiteflies (Avidov and Harpaz, 1969) with a characteristic spiralling pattern of oviposition on the underside of leaves (Russel,1965). Spiralling whitefly is a polyphagous whitefly species of tropical or neotropical origin (Russell, 1965; Martin, 1987). It is a vector of plant pathogens as reported by Martin in 1987 when it was first observed on *Terminalia catappa* in Hawaii which later spread to infest other crops.

The Spiralling whiteflies had spread westward across the pacific and southeast Asia (Waterhouse and Norris 1989). Waterhouse and Norris (1989) further mentioned that the pest has been reported in Brazil, Ecuador, Peru, Philippines, Fiji, Maldives, Mariana Islands and Canary Islands etc. However, the first report of *A. dispersus* in Nigeria was in 1993 by Akinlosotu et al and other West African part (M'boob and Van oers, 1994). Although, it is possible that this insect occurred in West Africa much earlier than 1992 as was reported, but was confused with the cassava mealybug (Asiwe et al., 2002). The host range of this polyphagous insect increases day by day as it spreads to other part of the world (Banjo et al., 2003), some of which are *Hura crepitans, Psidium guajava* (Banjo et al., 2003) and the most important root crop in tropical African: cassava (*Manihot esculantum*) (FAO, 1996). The pest also attacks many trees, arable crops and ornamental plants though rarely on gramminae and

cowpea (Waterhouse and Norris, 1989). Altogether, *A. dispersus* has been reported on more than 27 plant families, 38 genera with over 100 species including citrus and ornamental plants (Russell, 1965; Cherry, 1980).

Rainfall and temperature are the major climatic factors which affect the population of *A. dispersus* as they affect the development of each of its six life stages (Banjo and Banjo, 2003) irrespective of the host type associated with the insect. Control of the Spiralling whiteflies include eradication of the growth of a low lying weed *Sida acuta* which serves as refuge for the population of this insect in the wetter season as too much rainfall affects adversely their population during such season (Banjo and Latunde-Dada, 1999). *Stenthonus* sp. which is a small dark beetle was found to prey on the pest serving more or less as a biological control for this plant vector (Banjo, 1998).

TAXONOMY

A. dispersus Russel belongs to the subfamily aleyrodicinae and family aleyrodidae (Mound and Halsey, 1978) of the order Hemiptera. The basis for the present generic classification was laid by Quaintance and Baker (1913-14) who divided the group into three subfamilies containing one, four and eighteen genera respectively, "since then large number of species and genera have been described in the oriental, Neotropical and Ethiopian

Table 1. Mean developmental (days) period of Swf in three clones of cassava.

	Egg	No.	First nymph	No.	Second nymph	No.	Third nymph	No.	Fourth nymph	No.
	6	57	3	125	4	97	4	73	6	39
	7	192	4	199	5	162	5	126	7	54
	8	311	5	78	6	56	6	47	8	67
	9	120	6	68	7	23	7	19	9	38
	10	35	7	12					10	10
ADP	6 - 10		9 - 17		13 - 24		17 - 31		23 - 41	
TNSWF	715		482		338		565		218	
X±SE days	8.10 ± 3.11		4.26 ± 0.48		5.01 ± 0.43		5.04 ± 0.43		7.75 ± 0.54	

ADP means Accumulated Developmental period. Source: Banjo et al. (2003).

regions. As a result, the present catalogue listed 1156 species in 126 genera (Mound and Halsey, 1978) which included *A. dispersus* named by Russel in 1965.

BIOLOGY

Spiralling whiteflies are small insect which feed on plants by sucking plant juices from the phloem through a slender stylet as other whiteflies do (Muniyappa, 1980). The pest has six life stages on all host plants (Banjo and Banjo, 2003) which are the egg, first, second and third nymphal stages, the fourth nymphal stage also known as the pupa and the adult.

The newly hatched larva has functional legs and moves about before settling to feed, the second instar larva flattens out on the leaf whose legs and antennae become vestigial (that is, it is sessile) and cottony secretion is sparse, the third instar larva resembles the previous instar in shape but slightly larger as found in other whiteflies (Bryne et al., 1990). The fourth instar (pupa) in its first stage feeds and exudes honey dew after which it becomes dormant and can be knocked off the leaf (Bryne et al., 1990). Young pupae are ventrally flattened but mature ones with ventral surface swollen and surrounded by a band of wax with atrophied legs and antennae (Avidov and Harpaz, 1969; Banjo and Banjo, 2003).

The developmental biology of *A. dispersus* was studied in the fields of International Institute of Tropical Agriculture (IITA) Research Farm in Ibadan, Nigeria on three cassava genotypes (TMS 30512,TMS 91934,TME 1) following the report of Banjo et al. (2003), the cumulative developmental period in all genotypes was twenty three to forty one (23 - 41) days. The above fact followed the summation of the incubation period (6-10 days), the nymph (crawler stage) which was 3 - 7 days, the second nymph stage 4 - 7 days and the third nymph which was 6 - 10 days. The Table 1 summarises better the result (Banjo et al., 2003). This result however differs in some respect from that of Palaniswami et al. (1995)

who studied the developmental biology of the SWF in India. The incubation period was longer in any of the three genotypes than that reported by Palaniswami et al. (1995) and the cumulative period lasted for longer days than twelve to fourteen days observed in India by Palzaniswami et al. These differences may be due to environmental or inherent factors resulting in ecotypes of SWF.

DEMOGRAPHY

Part of studies already done and reported in Nigeria on the spiralling whiteflies includes its population dynamics (demography), oviposition and feeding site preference. However, the general population of *A. dispersus* is set to rapid declination during rainy season as eggs are being washed away by rain (Banjo et al., 2003) as in other whiteflies (Van Lenterens et al., 1990). Following the report of Banjo and Banjo, (2003) on the life history of *A. dispersus* on some host plants of economic importance in South Western Nigeria, eight different plant species which had significant *A. dispersus* infestation were selected for the study and the number of SWF in each that developed to later stages were noted . The mean number of *A. dispersus* that developed to adult from every one thousand eggs was 138.1.

According to the result of this report, mortality of *A. dispersus* from egg to the first nymphal stage was high with average number of 660 per thousand of eggs laid and this correlated with report from other parts of the world by Van Lenteren (1978) as for other whiteflies.

Banjo and Adenuga, (2001) also reported that mortality rate is high from egg to first instar nymphal phase on all host plants selected for their investigation and both immature and adult whiteflies occur in specific distribution patterns, which appear to be at least partly due to reactions to differences in host plant and characteristics This report by Banjo and Adenuga (2001) correlated with that of Ekbom (1980) as found in spiralling whiteflies' counterpart.

THE SPREAD OF SPIRALLING WHITEFLIES

The spread of *A. dispersus* in Nigeria has been hypothesized by Asiwe et al. (2002) to be connected with human, in that the risk of spread increases with frequency of movement. When *A. dispersus* is introduced via human activity or other yet-unreported means, they are clumped where their most preferred host plant in that area are located like cassava (Asiwe et al., 2002) and ornamental plants like *Acalypha* sp. etc (Banjo and Latunde Dada, 1999) before they are spread to infest other plant host (Asiwe et al., 2002).

The aforementioned statements could be inferred from the investigation of Asiwe et al. (2002) on the spread of the spiralling whiteflies by monitoring the intensity and infestation of spiralling whiteflies in all research plots, the residential and administrative areas of the International Institute of Tropical Agriculture (IITA) Ibadan, Nigeria overtime. The result showed that the incidence of *A. dispersus* within IITA was localized and quite severe in a few blocks. The average monthly infestation was highest where improved cassava and perennial tree cassava were grown and around the residential and administrative blocks where preferred ornamental plants (including *Acalypha* sp., *Hibiscus* sp. etc) were abundant. According to Asiwe et al. (2002), the highly susceptible tree cassava plants in the area may have been the source of infestation and there was usually continuous movement of people and materials in and around all areas of high infestation. This could have increased the likelihood of cross infestation from foci of high infestation to those with lower pest densities.

SELECTION OF FEEDING AND OVIPOSITION SITES

The relationship between selection of oviposition sites and growth, survival and reproduction of offspring are central element in the evolution of host association between herbivorous insects and plant (Singer, 1986; Thompson, 1988). Generally for whiteflies, selection of suitable site for ovipositon is very important since the larval stages are completely sessile except for the early first crawler stage which can move to cover a very short distance under the leaf where the eggs was deposited (Hassell and Southwood, 1978).

According to Banjo et al. (2001) oviposition and feeding occur simultaneously on the same leaves. They further reported that the occurrence of nymphs and adult of *A. dispersus* may be due to a number of factors as mentioned by Van lenterens and Noldus (1990) which includes thinness of lower cuticle, proximity of phloem to lower surface, presence of stomata and protection from rain. Others include negative phototaxis, positive geotaxis and dorsal position of anus and methods of ejecting excreta. This selection phase is mediated in all whiteflies

by visual olfactory or gustatory stimuli (Van lenterens and Noldus (1990).

A study specifically on the oviposition site preference of the spiralling whiteflies in Nigeria was carried out by Banjo et al. (2001) using cassava leaf surface and strata within canopy. The following conclusions were made via the result got;

1) Spiralling whiteflies oviposition was an insect biological response which could not be modified by changing position of leaf surface as oviposition was confined to the lower (abaxial) leaf surface of the inverted plant irrespective of the position of the plant either normal standing position or inverted position.
2) Leaf canopy level has little effect on oviposition or feeding preference at the lower or upper strata of the plant canopy among the three genotypes tested and
3) No genotype preference for oviposition among the three cassava genotypes used (Banjo et al., 2001).

Also in the behavioural studies of feeding and oviposition pattern of *A. dispersus* by Banjo et al. (2003), using three cassava genotypes, it was reported that at high infestation zone in the wild *Manihot* field, egg spirals started to be laid on 59 of the leaves within 24 h and within two days (48 h), all the matured leaves had egg spirals. Hence, oviposition commences within 48 h of adults emergence and usually on the lower surface of the older leaves (Banjo et al., 2003).

EFFECT OF SEASONAL CHANGES

Rainfall, temperature (and other weather factors) cause seasonal fluctuation and are important regulating factors of many tropical insects (Delinger, 1986) but it is mostly their combined effect evapo-transpiration that is more important (Asiwe et al., 2002). When it rains heavily, many small insects get dislodged from plant surfaces by the combined effect of wetness and the kinetic energy of the rain drops as well as strong winds. The orientation of the leaves on the plant and consequently the position of the insect on the plant would be critical (Asiwe et al., 2002).

A. dispersus population as most tropical insects is affected by the climatic conditions which dictate the season (Banjo et al., 2003, Banjo and Latunde Dada, 2001; Asiwe et al., 2002). A period of moderate rainfall combined with high day temperature which usually occur between April and May, following the onset of rain after the very dry months (December and January) in Nigeria and other tropical regions favours high population of the spiraling whiteflies (Banjo and Banjo, 2003). However the population is at optimum at the drier months of November, December and January (Banjo et al., 2003). During wetter season (June and July) conversely, the population of spiraling whiteflies declines gradually as all the stages of life especially eggs of the insect are washed away by heavy rain couple with wind that is normally

Table 2. Canopy levels showing percentage proportion of leaves infested in different host plant.

Host plants	Canopy level 1	Canopy level 2	Canopy level 3
Terminalia catappa	9.91 ± 2.4	3.6±1.9	0
Ipomea carnea	22.4 ± 23.5	86. ±59.1	35.6±8.7
Acalypha hispida	24.7 ± 0.4	73.3±7.9	50.65±.9
Acalypha wilkesiana	7.5 ± 2.5	22.43±.7	16.24±.6
Sida acuta	8.41±11.8	80.11±0.5	58.02±0.1
Psidium guajava	33.1±6.3	63.01±3.0	92.1±11.2
Ficus exasperata	8.4 ± 2.3	55.22±.7	39.9±1.4
Manihot sp.	0	60.82±5.2	31.9±21.5

Source: Banjo and Adenuga (2001).

associated with such rain (Banjo et al., 2003; Asiwe et al., 2002; Banjo and Latunde Dada, 1999). During this wetter month, low lying wild weed, *Sida acuta* serves as refuge for the spiralling whiteflies to form a relic population after which it is expected to reinfest the taller and cultivated plants when favourable conditions return in the drier months of the year (Banjo and Latunde Dada, 1999).

A. *dispersus* as plant pest

Whiteflies generally are important as vector of plant pathogens primarily but not exclusively, in the tropics and sub-tropics (Muniyappa, 1980) and feed by sucking plant juices from the phloem through a slender stylet. The disease transmitted by white flies includes cotton leaf curl, tobacco leaf curl and cassava mosaic (Costa, 1976). Viruses transmitted by whiteflies are heterogeneous group inducing diverse diseases. Symptoms of infection are mosaic, leaf and vein discolouration and tissue distortion such as curling and crinkling (wrinkling) (Costa, 1969). However, *A. dispersus* is mostly commonly implicated as a vector and has been associated with more than 25 different diseases and feeds on a larger number of plant species (Russel, 1965; Costa, 1969; Muniyappa, 1980). The transmission characteristics are similar in gross details to persistent transmission of viruses by aphids, in that acquisition time can be as long as several hours (Costa, 1969).

Specific host preference

A. *dispersus* has been recorded on more than 27 plants families, 38 genera with over 100 species including citrus and ornamentals (Russell, 1965; Cherry, 1980). In Nigeria, however, Akinlosotu et al. (1993) had reported that *Anacardium occidentale*, *Annona sp.*, *Cocos nucifera* and *Psidiuim guajava* were among the host range of the pest.

Asiwe et al. (2002) also reported that investigation done on the spread and host range of A. *dispersus* in selected states of South-western Nigeria where the outbreak was severe in 1993 revealed that spiralling whiteflies attack many trees and arable crops, ornamental plants but rarely on Gramminae and cowpea, *Vigna unguiculata*.

A further investigation reported by Banjo and Adenuga (2001) on infestation rate as index of plant preference on eight plant species of societal and economic values was carried out and the canopy level preference was determined by zoning the various hosts into three equal zones depending on the height including the ground level to the first one third of the total height (lower canopy), the middle canopy and then the top canopy level which is the terminal to the one third of the total height of the plant. The result of this investigation showed that spiralling whiteflies were not evenly distributed on all host plants and the older leaves of the first canopy level were more preferred (especially for *Psidium guajava*) and by active flight, newly emerged adults leaves the older lower leaves to the next upper younger leaves.

It is noteworthy here to mention that in all the host plants used for this investigation, almost all the spiralling whiteflies were found on the underside (abaxial surface) of the leaves which presumably protect the spiralling whitefly from rain, wind and direct effect of sunlight (Banjo and Adenuga, 2001). The analysis of the result was summarized (Table 2).

When eight different plant species were used to asses the host plant preference of the spiralling whitefly using egg count fortnight method, *Psiduim guajava* has again the highest egg count followed by *Terminalia catappa*, *Acalypha sp* and *Ficus exasperata*. Others with relatively less egg count include *Manihot esculenta*, *Musa sp*, *Bauhinia monandra* and *Sida acuta* (Banjo and Latunde-Dada, 1999).

However, it is noteworthy that attention of the spiralling whiteflies shifted to low lying weed like *S. acuta* during wetter season from where they reinfest the taller and cultivated plants when favourable conditions return (Banjo and Latunde-Dada, 1999).

Ecological and economic effect

A. dispersus is a pest of crops and ornamental plants in two ways, through their debilitating effect in sucking plant sap and through the introduction of various diseases (Nakahara, 1978; Mound, 1973). The spiralling whitefly is a phloem sap feeder and its direct consumption of transportable carbohydrate and other nutrients carried in phloem reduces productivity of host plants by competing for available nutrients and causing premature leaf shedding (Bryne et al., 1990) in its extensive host range of over 100 species of which guava, banana, mango, grape and tropical almond are part of the most preferred hosts (Nakahara, 1978).

A. dispersus also excrete honey dew as other whiteflies do which can cover the surface of leaves and serves as a medium for the growth of sooty mold (Bryne et al., 1990). These interfere with photosynthetic process by not allowing enough light to reach the cytochrome tissues of the leaves. The sooty mold may also increase thermal absorption and raise leaf temperature, thus in turn reduces leaf efficiency and may even cause premature death of tissue (Bryne et al., 1990).

However, cassava which is the most concerned among the host of the pest because of its status as the most important root crops in tropical Africa (Otoo, 1988) is not significantly affected by the spiralling whiteflies even at high infestation probably due to perennial nature of the crop which the pest can not withstand (Banjo et al., 2003).

MANAGEMENT AND CONTROL

The present status of A. dispersus as a plant pest in Nigeria has not called essentially for extensive management as it was regarded as a minor pest for the most concerned host crop in Nigeria i.e. cassava; Manihot esculanta (Banjo et al., 2004, , Banjo et al., 2003, Asiwe et al., 2002).

In the investigation on the growth indices and yield of three genotypes of cassava according to Banjo et al. (2004), yield loss was not significantly different between infested cassava genotypes and those not infested. Also, there were no significant differences between the growth indices (leaf area index, crop growth rate, harvest index, net assimilation rate and total biomass used in assessing yield). Banjo et al. (2004) further reported that the cassava genotype used seemed to undergo compensatory growth, even at high infestation level. The compensatory growth type occur due to pruning effect caused by the feeding of the pest which causes suppression of growth in one organ and increases the size or weight of others (Banjo et al., 2004). Although it may build up and achieve a major pest status in some localities (Banjo, 1998).

However, Stenthonus species (a small dark beetle) was found to prey on spiralling whiteflies, nymphs and pupae (Banjo et al. 2004) as predator which is more or less a natural biological control of A. dispersus. The efficacies of some biological control agents (such as Encasia haitensis and Encasia guadloupea vigigani) was reported to have being under investigation in Nigeria and Ghana (Neuenschwender, 1994) but has not been reported fully. Physical control of A. dispersus following the report by Banjo and Latunde-Dada (1999), Banjo et al. (2003) and Banjo et al. (2004) that population of spiralling whiteflies during wetter season, declines drastically as sporadic rainfall washed off the eggs and nymphs from the host leaves, Banjo et al. (2004) therefore concluded that intensive spraying of the underside of leaves with water will only reduce the population of this pest.

Nevertheless, removal of low lying weed Sida acuta on uncultivated lands as crop protection measure was recommended by Banjo and Latunde-Dada (1999) following the observation that population of A. dispersus form a relic under the low lying weed during wetter season from which they are expected to reinfest the taller and cultivated plants when favourable condition returns in the drier months (Banjo and Latunde-Dada, 1999).

Asiwe et al. (2002) however recommended the use of insecticide when the infestation becomes severe and causes economic damage to crops, if the source of infestation can be identified since the effect of distance on the spread of the insect is unambiguous

Conclusion

A. dispersus has not really reached a pest status in Nigeria (Banjo et al., 2004; Banjo, 1998) but regarded as only a minor pest of cassava (Banjo, 1998). However as Bardner and Fletcher (1974) put it, pest assessment studies frequently show that crops vary greatly between sites and between years on their response to natural infestation by similar reaction of individual plants of the same crops or genotype. Also Le Clerg (1970) stated that neither pest population nor crop losses were static and these change from year to year in a given location. Therefore the status and the effect of spiralling whiteflies on plant may change from year to year and form a serious pest of plants especially cassava (Banjo and Latunde-Dada, 1999, Banjo et al., 2004).

Moreover, as suggested by Le Clerg (1970), professsionals (in Nigeria) are urged to reinvestigate overtime and space, the status of A. dispersus as a plant host (Banjo et al., 2001; Banjo et al., 2004) for at least three years at a number of locations (Banjo et al., 2004). Also, because of rapidly changing cultural practices, information on this pest should be updated perhaps every five years (Banjo et al., 2004).

REFERENCES

Akinlosotu TA, Jackal LEN, Ntonifor NN, Hassan AT, Agyakwa CW,

Odebiyi JA, Akingbohungbe AE, Russel HW (1993). Spiralling whitefly, *Aleurodicus dispersus* in Nigeria. FAO plant protection bulletin 41(2): 127-129.

Asiwe JAN, Dixon AGO, Jackal LEN, Nukenine EN (2002). Investigation on the the spread of the spiralling whitefly (*A. dispersus*, Russell) and field evaluation of elite cassava population for genetic resistance. A research article in AJRTC 5 (1): 12-17

Avidov Z, Harpaz I (1969). Plant pests of Israel. Israel University Press, Jerusalem. p. 549.

Banjo AD (1998). Population changes, yield loss assessment and physiological consequences of the infestation of spiralling whitefly (*A. dispersus russel*) on cassava (*Manihot esculanta* Crantz). Ph.D. thesis, university of Ibadan, Ibadan Nigeria.

Banjo AD, Latunde Dada IL (1999). An assessment of host plant preference of the spiralling whitefly (*A. dispersus*) in Ago-Iwoye, Nigeria. J. Crop Res. 17(3): 390-394

Banjo AD, Adenuga FM (2001). Infestation rate, Vertical distribution pattern and population dynamics of *A. dispersus* (the spiralling whitefly) on selected host plants in Southwest Nigeria. J. Sci. Eng. Technol. 8(4): 3594-3603

Banjo AD, Hassan AT, Dixon AGO, Ekanayake IJ, Jackal LEN (2001). Preliminary evaluation of resistance in cassava genotypes to spiralling whitefly. Afr. J. Sci. Technol. 1(1- 2): 194-196.

Banjo AD, Hasssan AT, Jackal LEN, Ekanayake IJ, Dixon AGO (2001). Oviposition preference of the spiralling whitefly (*A. dispersus* Russel) on cassava leaf surface and strata within canopy. Afr. J. Sci Technol 2(1-2): 190-193.

Banjo AD Banjo FM (2003). Life history and the influences of agroclimatological factors on the spiralling whitefly (*A. dispersus* Russel) (homoptera: aleyrodidae) on some host plants of economic importance in south-western Nigeria. J. Crop Res. 26(1):140-144.

Banjo AD, Hassan AT, Jackal LEN, Dixon AGO, Ekanayake IJ (2003). Developmental and Behavioural study of spiralling whitefly (*A. dispersus*) on three cassava (*Manihot esulanta* crantz) genotypes. J. Crop Res . 26(1): 145-149.

Banjo AD, Hassan AT, Ekanayake IJ, Dixon AGO, Jackal LEN (2004). Effect of *Aleurodicus dispersus* Russel (Spiralling whitefly) on growth indices and yield of three genotypes of cassava (*Manihot esculanta* crantz). J. Res. Crops. 5(2-3): 252-260

Bardner R, Fletcher KF (1974). Insect infestations and their effects in growth and yield of field crops: a review. Bull. Entomol. Resour. 64: 141-60

Bryne DN, Bellows TS, Parella MP (1990). Whiteflies in agricultural system In: Whiteflies- their bionomics, pest status and management. Gerling, D. (ed). Wimborne U.K. Intercept. pp. 227-61

Cherry RH (1980). Host plant preference of the whitefly *A.dispersus* Russell. Florida entomologist 63: 222-225

Costa AS (1969). Whiteflies as vectors. In viruses vectors and vegetation. k. Maramoruskh. (ed.) John Wiley and sons, New York.111pp.

Costa AS (1976). Whitefly transmitted plant disease. Ann Rev Phytopathol 14: 429-449.

Delinger DL (1986). Dormancy in Tropical Insects, Ann Rev Entomol, 31: 239-264.

Ekbom BS (1980). Some aspects of the population dynamics of *Trialeuriodes vaporariorum* and *Encarsia formosa* and their impotance for biological control 10BC/UPRS Bull. 3 (3): 25-34

Evans LT (1993). Crop evaluation, adaptation and yield. Cambridge University Press. Cambridge. 500pp.

FAO (1996). Food Outlook March/April. p. 19.

Gates DM (1993). Climatic change and its biological consequences. Sinauer, Massachussets. 280p.

Le clerg EL (1970). Field experiments for assessment of crop losses. In:FAO manual on evaluation and prevention of losses by pest and disease. pp. 2(1): 1-6

M'Boob SS, Van Oers CCCM (1994). Spiralling whitefly (Aleurodicus dispersus), a new problem in Africa. FAO plant protection Bulletin 42(1-2): 59-62.

Martin JH (1987). An identification guide to common whitefly species of the world. Trop. Pest Manage. 33(4): 298-322.

Mound LA, Halsey SH (1978). Whitefly of the world: A systematic catalogue of the aleyrodidae (Hemiptera) with host plant and natural enemy data. British museum (Natural history), Chicheser. 321p.

Muniyappa V (1980). Whiteflies. In vectors of plant pathogens (K. F. Harris and K. Maramorosch, (ed.) Academic Press, New York. pp 39-85.

Nakahara L (1978). Hawaii Cooperative Economist Pest Report. State of Hawaii October 20, National Oceanic and Atmospheric Administration (1980-1991). Climatological Data – Hawaii and Pacipic 1: 26-77.

Neuenschwender P (1994). Spiralling whitefly, Aleurodicus dispersus Russel, a recent invader and new cassava pest. Afr. J. Crop Sci. 2(40): 419-421.

Otoo JA (1988). IITA Afro wide cassava improvement program In : In praise of cassava, Hahn, N.D. (ed.) UNICEF/IITA Pub. pp. 67-75

Palaniswami MS, Pillar KS, Nair RR, Mohandas C (1995). A new cassava pest in India, Cassava Newsl. 19: 6-7.

Quaintance AL, Baker AC (1913) Classification of the Aliyrodidae part I Tech. Ser. Bur. Entomol. U.S. 27: 93.

Russel LM (1965). A new species of Aleurodicus Douglas and two close relatives (Homoptera: Aleyrodidae). Florida Entomol 48(1): 47-55.

Singer MC (1986). The Definition and Measurement of Oviposition Preference in plant feeding insect. In insect-plant relation. Miller J. Miller TA (eds) Springer, New York. pp. 65-94.

Thompson JN (1988). Evolutionary ecology of the relationship between oviposition preference and performance of offspring in phytophagous insects. Entomol. Exp. Appl. 4: 13-14

Van Lenterens JC, Noldus LPJJ (1990). Whitefly plant relationship, behavioural and ecological aspects. In: whiteflies: their Binomics, pest status and management, Gerling D (1978) (ed). Intercept Pub. Ltd.UK. pp. 47-89.

Vetten HJ, Allen DJ (1983). Effect of environment and host on vector biology and incidence of two whiteflies in the spread of legume ir Nigeria. Ann. Appl. Biol. 102: 219-27.

Waterhouse DF, Norris KR (1984). *Aleurodicus dispersus* Russel Hemiptera: Aleyrodidae. Spiralling whitefly. Pages 13-22. in: Biological control, pacific prospects – suppl. 1. ACIAR, Canberra.

Weems HV Jr. (1971). *Aleurodicus dispersus* Russel. Homoptera Aleyrodidae, a possible vector of the lethal yellowing disease of coconut palms. Florida division of plant industry. Enthomol Circular N0. 111. 2.

17

Studies on some economic traits and biological characters of regular and reciprocal cross between a multivoltine and bivoltine race of the silkworm *Bombyx mori*

M. S. Doddaswamy[1], G. Subramanya[1]* and E. Talebi [1, 2]

[1]Department of Sericulture Science, University of Mysore, Manasagangotri, Mysore, Karnataka, India.
[2]Department of Animal Science, Islamic Azad University, Darab, Fars, Iran.

In order to understand the genetics of cross breeding system between multivoltine and bivoltine race, a hybridization experiment was conducted by crossing females of multivoltine Pure Mysore race (PM) with males of a bivoltine race C_{108} and its reciprocals. The F_1 and F_2 progenies of the above crosses were derived in order to study the economic traits and biological characters. The results of the analysis of the thirteen economic traits in the F_1 progeny between the regular and reciprocal crosses clearly indicated that wherever PM is participated as female partner the values of the economic traits are on the higher side than the reciprocals. In regard to the studies on the biological characters, the F_1 progeny of a cross of females of PM with males of C_{108} produced non-diapause eggs exhibiting 23 - 24 days of larval duration and spinning light green cocoons. On the other hand a cross between females of C_{108} with males of PM produced only diapause type of eggs having larval duration of 22 days which is shorter by 1 - 1.5 days compared to regular cross. The F_2 progenies involving both the crosses produced 3:1 ratio diapause: non-diapause eggs and green: white cocoons indicating Mendelian pattern of inheritance for voltinism and cocoon colour. The authors in the present investigation discussed the importance of hemizygous and heterozygous individuals of F_1 progeny and its utility in genetics and breeding of silkworm *Bombyx mori*.

Key words: Regular and reciprocal crosses, *Bombyx mori,* Economic traits, biological characters.

INTRODUCTION

The sericultural industry involved in the production of silk occupies an important place in the Indian economy and provides gainful occupation to lakhs of people. Even though India is one of the oldest silk producing countries, majority of the silk produced is by multivoltine × bivoltine hybrids. It is estimated that nearly 80% of the silk in India is produced by multivoltine × bivoltine hybrids where multivoltine races are used as female parent. The above combinations of hybrids have performed uniformly in the rearer's house exhibiting moderate productivity. The average yield is now estimated as 50 Kg/100 disease

free layings from multi- bi hybrids. Contrary to this, in the sericultural states of India the hybrids of bivoltine females with males of multivoltines are not commercially exploitted. The main reason attributed is that the contributions of bivoltines by virtue of its maternal inheritance may result in regular crop losses. As a result, fifty percent of male population of multivoltines and other fifty percent of female population of bivoltines are not properly utilized resulting in the revenue loss running to several corers of rupees. Thus, the genetic potentialities of the silkworm breeds are not fully exploited for silk production in our country. Understanding the genetic capabilities of multivoltine × bivoltine hybrids as well bivoltine × multivoltine hybrids is of utmost practical importance and it is worth while to examine the prospects of developing a thorough knowledge of the above crossing system in the

*Corresponding author. E-mail: subramanyag2000@yahoo.

light of the basic principles of genetics and understanding of basic biological characters. In the present paper the results of such differences for some economic traits and biological characters are discussed.

MATERIALS AND METHODS

The pure races of Mulberry Silkworm *Bombyx mori Viz.*, multivoltine Pure Mysore (PM), bivoltine race C_{108} formed the materials for the present investigations. These races were drawn from the germplasm bank maintained in the Department of Studies in Sericultural Science, University of Mysore, Mysore, India. Parental seed cocoons of the above said two races were collected and layings of the pure races were prepared adopting the method described by Tazima (1962). After incubation of eggs at $25 \pm 1\,^{\circ}C$ and relative humidity of $80 \pm 5\%$, two layings of each of the two pure races were selected. The larvae hatched from each layings were reared separately under uniform laboratory conditions as described by Narasimhanna and Krishnaswamy (1972). The larvae were fed with M_5 variety of mulberry (*Morus indica*) leaves.

For assessing the comparative performance of the pure races, thirteen economic characters namely Fecundity, Hatching percentage, Weight of fifth instar larvae, Larval duration, Yield/10000 larvae brushed by number and weight, Cocoon weight, Shell weight, Shell percentage, Filament length, Denier, Renditta and Pupation rate were analysed in three different seasons of the year. The parental seed cocoons of the above races, after the analysis were utilized in the preparation of the hybrid layings by crossing females of Pure Mysore with males of bivoltines and their reciprocals. The Duncan system of statistical model for one-way classification was employed following the method of Snedecor and Cochran (1967) for the data obtained during of course hybridization programme.

RESULTS AND DISCUSSION

The mean values of the economic traits of two parents viz; Pure Mysore and C_{108} and their F_1 hybrids along with statistical data for three seasons of the year are presented in Tables 1 - 3. Based on the results it is clear that the two races revealed differential performances in three seasons. The multivoltine Pure Mysore exhibited longest larval duration of 661 h in Post-monsoon season compared to 570 h in C_{108}. In addition quantitative traits like cocoon weight, shell weight and shell ratio are significantly higher ($P < 0.05$) in the bivoltine C_{108} compared to multivoltine race. This could be ascribed to the differences in the voltinism and racial specificity as pointed out by Murakami (1989a and b). Based on the findings on the performance of the F_1 hybrids in three different seasons of the year, it is obvious that F_1 hybrids of reciprocal crosses also spun cocoons of tightly formed shells, uniform growth and uniform crop performance compared to popular crosses where multivoltine Pure Mysore is used as female parent. In all the three seasons (Tables 1 - 3) analysis of the various economic traits in the regular and reciprocal crosses of F_1 hybrids revealed significant differences ($P < 0.05$) for eleven out of thirteen economic traits analysed. Based on the results it is obvious that Pure Mysore wherever participated as

female partner with bivoltines, the value of economic traits are on the higher side than the reciprocals. The above findings tally with the findings of Benchamin et al. (1983) and Tazima (1988) who have observed similar results in their studies on multivoltine × bivoltine and bivoltine × multivoltine hybrids. It is also important that several multivoltine x bivoltine hybrids are commercially exploited in addition to the F_1 hybrids derived from a cross between Pure Mysore females with bivoltine males (Radhakrishna et al., 2001; Ravindra Singh et al., 2005; Umadevi et al., 2005). In the present study the F_1 hybrids of multivoltine × bivoltine analysed in three seasons clearly indicated uniform larval growth period from I-V instars while, in the reciprocal crosses the larval growth period is normal from I-IV instars and V instar larval period is reduced by 25 - 28 h. Similar results were recorded by Tazima (1988) utilizing multivoltine × bivoltine hybrid and he opined that such differences in larval duration are due to Lm^e genes. On the other hand Murakami (1994) opined that a reduction in the larval duration in reciprocal crosses in a multivoltine × bivoltine hybrid is due to the role played by X chromosome. Analysis of the genetic characteristics for voltinism and cocoon colour during F_1 and F_2 generation in both the crosses revealed interesting results (Table 4, Figures 1 and 2). In the crosses wherever Pure Mysore female is utilized with bivoltine race the F_1 eggs are non-diapause and F_2 eggs are diapause type, but the eggs laid by the F_2 moths revealed a typical Mendelian pattern of inheritance exhibiting 3:1 diapause to non-diapause type pattern. In the reciprocal crosses the F_1 progenies are diapause type and F_2 progenies are non-diapause type and F_2 moths laid 3:1 diapause to non-diapause eggs (Figure 1). Tazima (1988) while reporting on the improvement of multivoltine Pure Mysore race revealed that dormancy is always dominant over non-dormancy. Similar result was observed by Murakami (1989 a) utilizing Cambodge as a female parent. The results of the authors in the present investigation supports the findings of Tazima (1964) and Murakami (1989 a). Based on the data on cocoon colours (Figure 2) it is clear that the F_2 progenies also revealed 3:1 segregation in a typical Mendelian pattern (three green and one white) and is in conformity with the findings of Murakami (1988 and 1994). Schematic representations of the two different crosses are shown in Figure 3. Based on the segregation of the characters it is important that breeders should take cognescence of hemizygous and heterozygous populations to understand the viability features of F_1 hybrids. The above results are in conformity with findings of Subramanya and Murakami (1994) who have proposed that hemizygous and heterozygous population play an important role in the expression of quantitative and qualitative traits and also basic biological characters such as voltinism and moultinism, larval growth, cocoon colour and shape of the cocoons.

Thus, based on the overall performance of economically important parameters multivoltine female × bivoltine male

Table 1. Mean values of the thirteen economic traits of two parental races and their F$_1$ hybrids in Pre-monsoon season

Traits / Races	Fecundity	Hatching %	Larval duration (h)	Larval Wt (gm)	Yield / 10000 No	Wt in Kg	Cocoon Wt (gm)	Shell Wt (gm)	Shell %	Filament length (m)	Denier	Renditta	Pupation Rate
PM	444.33 ± 20.21[c]	91.42± 2.51[b]	648 ± 13.85[a]	19.78 ± 0.319[d]	8947 ± 61.61[a]	9.730 ± 0.026[d]	1.115 ± 0.020[d]	0.160 ± 0.018[d]	14.35 ± 1.34[b]	382 ±1069[d]	1.91 ± 0.040[b]	11.20 ± 0.134[a]	90.80 ± 0.333[a]
C108	545.66 ± 3.48[a]	95.41 ± 0. 344[ab]	535 ±2.51[b]	42.16 ± 0.272[a]	8052.66 ± 34.70[c]	16.430 ± 0.017[a]	2.076 ± 0.031[a]	0.383 ± 0.002[a]	18.48 ± 0.140[a]	1032 ±2.64[a]	2.256 ± 0.032[a]	7.95 ± 0.017[c]	92.10 ± 0.387[a]
PM × C108	530.33 ± 1.45[b]	96.47 ± 0.315[a]	525 ± 13.98[b]	34.03 ± 0.442[b]	8066 ± 33.32[c]	13.920 ± 0.057[b]	1.739 ± 0.033[b]	0.315 ± 0.004[b]	18.11 ± 0.292[a]	728.33 ± 9.70[b]	2.06 ± 0.017[b]	10.25 ± 0.053[b]	92.15 ± 0.713[a]
C108 × PM	592 ±5.19[a]	93.190 ± 0.213[ab]	513 ± 13.74[b]	30.60 ± 0.303[c]	8254 ± 28.82[b]	11.236 ± 0.014[c]	1.405 ± 0.026[c]	0.241 ± 0.008[c]	17.14 ± 0.504[a]	590.66 ± 7.68[c]	2.16 ± 0.061[a]	10.44 ± 0.057[b]	90.97 ± 0.260[a]
F-value	33.818	3.098 NS	26.752	741.153	102.385	7715.44	215.07	84.79	6.45	1082.87	14.120	321.40	2.47NS

Note: The values are derived from three replicates ± SE. The values with the same letters are not statistical significant (P > 0.05) when subjected to DMRT (Duncan's Multiple Range Test).

Table 2. Mean values of the thirteen economic traits of two parental races and their F$_1$ hybrids in Monsoon season.

Traits / Races	Fecundity	Hatching %	Larval Duration (h)	Larval Wt (gm)	Yield / 10000 No	Wt in Kg	Cocoon Wt (gm)	Shell Wt (gm)	Shell %	Filament Length (m)	Denier	Renditta	Pupation Rate
PM	452 ±7.02[c]	96.51 ± 0.545[a]	657 ±16.29[a]	28.27 ± 0.433[d]	9182.66 ± 50.42[a]	9.756 ± 0.040[d]	1.088 ± 0.008[d]	0.153 ± 0.008[c]	14.01 ± 0.592[b]	351 ±19.35[c]	2.08 ± 0.084[b]	11.68 ± 0.268[a]	93.14 ± 0.531[a]
C108	538.33 ± 2.33[b]	96.52 ± 0.439[a]	577.33 ± 2.03[b]	46.29 ± 0.260[a]	8118.33 ± 57.85[b]	15.92 ± 0.072[a]	2.008 ± 0.038[a]	0.368 ± 0.014[a]	18.29 ± 0.365[a]	1025.33 ± 8.37[a]	2.35 ± 0.044[a]	8.20 ± 0.052[c]	92.25 ± 0.465[ab]
PM × C108	533.33 ± 5.78[b]	96.81 ± 0.833[a]	555.66 ± 7.51[bc]	34.91 ± 0.167[b]	8432.66 ± 216.20[b]	13.59 ± 0.551[b]	1.640 ± 0.056[b]	0.285 ± 0.012[b]	17.36 ± 0.176[a]	684 ± 13.08[b]	2.11 ± 0.064[b]	10.34 ± 0.104[b]	92.19 ± 0.366[a]
C108 × PM	586.33 ± 8.51[a]	94.75 ± 0.847[a]	534.66 ± 2.73[c]	31.10 ± 0.261[c]	8351 ±56.65[b]	12.35 ± 0.068[c]	1.505 ± 0.009[c]	0.271 ± 0.007[b]	18 ±0.463[a]	671.33 ± 7.31[b]	2.17 ± 0.060[ab]	10.45 ± 0.038[b]	90.92 ± 0.336[b]
F-value	77.29	1.86NS	34.31	711.13	15.20	83.44	122.60	67.52	21.31	453.59	3.307NS	96.29	4.45

Note: The values are derived from three replicates ± SE. The values with the same letters are not statistical significant (P > 0.05) when subjected to DMRT.

male represent good hybrid combinations, but bivoltine female × multivoltine male also could be exploited with more scientific care.

ACKNOWLEDGEMENTS

Two of the authors wish to express sincere thanks to University Grants Commission for providing the necessary funds. We wish to thank Chairman, Department of Studies in Sericultural

Table 3. Mean values of the thirteen economic traits of two parental races and their F_1 hybrids in Post-monsoon season.

Traits / Races	Fecundity	Hatching %	Larval Duration (hrs)	Larval Wt (in gms)	Yield / 10000 No	Yield / 10000 Wt in Kg	Cocoon Wt (in gms)	Shell Wt (in gms)	Shell %	Filament Length (meters)	Denier	Renditta	Pupation Rate
PM	498.66 ± 14.11[b]	97.77 ± 0.181[a]	661 ±12.70[a]	20.35 ± 0.115[c]	9424.66 ± 46.08[a]	11.13 ± 0.031[c]	1.195 ± 0.005[c]	0.165 ± 0.004[c]	13.803 ± 0.329[c]	394.33 ± 5.61[c]	1.86 ± 0.037[c]	10.99 ± 0.080[a]	93.39 ± 0.394[ab]
C108	567.66 ± 2.03[a]	95.85 ± 0.228[b]	570.33 ± 1.20[b]	43.25 ± 0.431[a]	9527.33 ± 32.57[a]	18.45 ± 0.934[a]	2.06 ± 0.025[a]	0.388 ± 0.004[a]	18.84 ± 0.232[a]	1018.66 ± 6.49[a]	2.5 ± 0.046[a]	8.20 ± 0.028[c]	93.32 ± 0.497[ab]
PM × C108	562 ±7[a]	97.75 ± 0.383[a]	555.66 ± 6.12[bc]	32.21 ± 0.535[b]	9295 ±68.55[a]	16.91 ± 0.229[ab]	1.84 ± 0.032[b]	0.323 ± 0.010[b]	17.56 ± 0.284[b]	693.33 ± 12.99[b]	2.123 ± 0.048[b]	10.28 ± 0.129[b]	93.63 ± 0.268[b]
C108 × PM	576 ±3.21[a]	96.68 ± 0.263[b]	528.66 ± 10.52[c]	31.48 ± .583[b]	9314.33 ± 146.99[a]	16.30 ± 0.318[b]	1.77 ± 0.020[b]	0.304 ± 0.004[b]	17.16 ± 0.297[b]	688.66 ± 8.21[b]	2.153 ± 0.032[b]	10.33 ± 0.075[b]	92.08 ± 0.509[b]
F-value	19.107	11.46	42.34	424.26	1.57NS	39.26	247.17	198.45	55.67	839.94	39.67	197.79	2.64NS

Note: The values are derived from three replicates ± SE. The values with the same letters are not statistical significant ($P > 0.05$) when subjected to DMRT.

Table 4. Segregation of egg colour, cocoon colour in the F_2 progenies of regular hybrids and its reciprocals.

Hybrids →	Types of eggs (in numbers)			Cocoon colours (in numbers)		
	Diapause	Non-diapause	Total	Green	White	Total
PM♀♀ × C108♂♂	68370 74.70%	23151 25.30%	91521	3332 73.02%	1231 26.98%	4563
F2moths	3	1		3	1	
C108♀♀ × PM♂♂	69819 73.39%	25317 26.61%	95136	3543 73.35%	1287 26.65%	4830
F2moths	3	1		3	1	

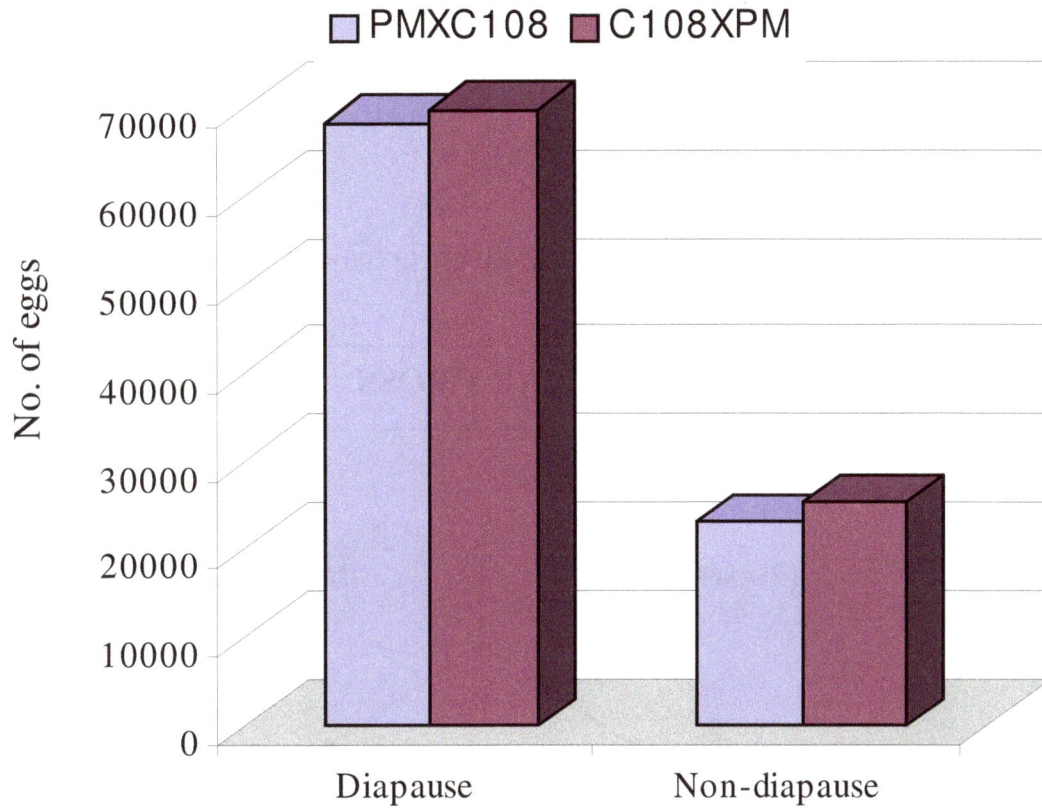

Figure 1. Diapausing and Non-diapausing egg segregation in F_2 progenies of PM × C_{108} and C_{108} × PM hybrids.

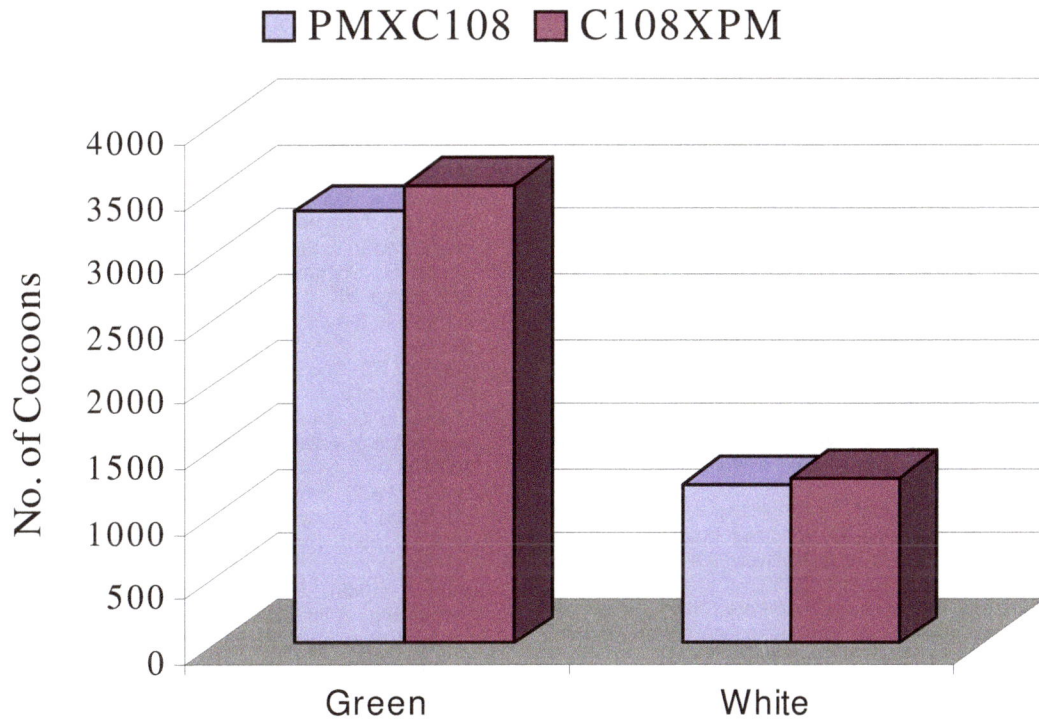

Figure 2. Green and White cocoon segregation in F_2 progenies of PM × C_{108} and C_{108} × PM hybrids.

$$x^m y \;_{♀♀} \quad \times \quad x^2 x^2 \;_{♂♂}$$

(Multivoltine)   (Bivoltine)

$$x^2 y \qquad\qquad\qquad x^m x^2$$

(hemizygous)   (heterozygous)

A. REGULAR CROSS

$$x^2 y \;_{♀♀} \quad \times \quad x^m x^m \;_{♂♂}$$

(Bivoltine)   (Multivoltine)

$$x^m y \qquad\qquad\qquad x^m x^2$$

(hemizygous)   (heterozygous)

B.RECIPROCAL CROSS

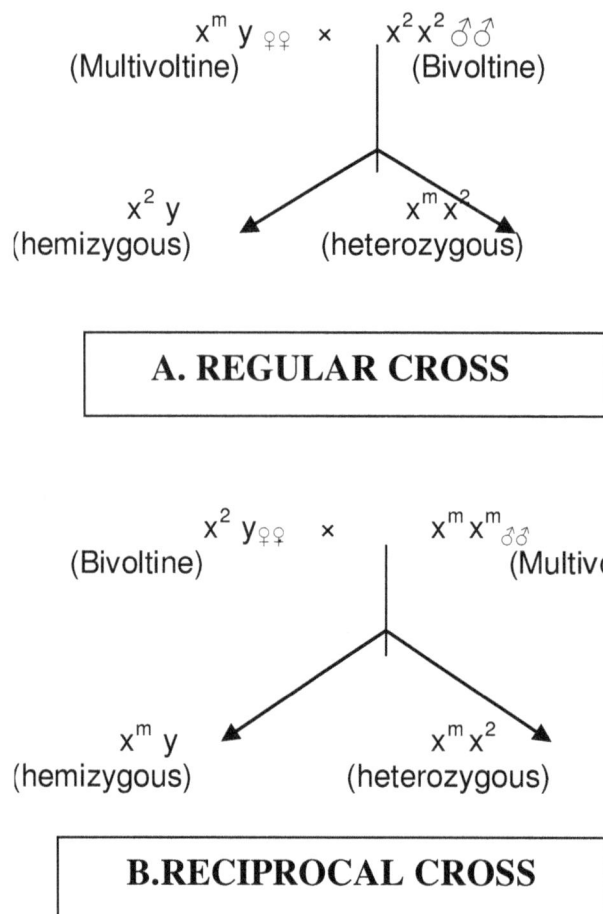

Figure 3. Schematic representation of the parental and F_1 genotypes in a cross between multivoltine females × bivoltine males and its reciprocal cross.

Science, University of Mysore, Manasagangotri, Mysore for extending the facilities.

REFERENCES

Benchamin KV, Jolly MS, Benjamin DAI (1983). Study on the reciprocal crosses of multivoltine × bivoltine with special reference to the use of bivoltine hybrid as a parent. National Seminar on Silk Research and Development, Bangalore March 10-13.

Murakami A (1988). Ecogenetic studies on voltinism in the tropical mulberry silkworm, *Bombyx mori*. Proceedings of the International Congress, on Tropical Sericulture Practices, Central Silk Board. pp. 11-23.

Murakami A (1989b). Genetic studies on tropical races of silkworm (*Bombyx mori*) with special reference to cross breeding strategy between tropical and temperate races 2. Multivoltine silkworm strains in Japan and their origin. JARQ 23(2): 127-133.

Murakami A (1994). Growth phenomena in *Bombyx mori* (L.) with a special reference to genetic factors responsible for growth acceleration and moultinism. Indian J. Seric. 33(1): 12-14.

Murakami A, Ohtsuki Y (1989 a). Genetic studies on tropical races of silkworm (*Bombyx mori*) with special reference to cross breeding strategy between tropical and temperate races 1. Genetic nature of the tropical multivoltine strain Cambodge. JARQ 23(1): 37-45.

Narasimhanna MN, Krishnaswamy S (1972). Improved techniques of silkworm rearing. Paper presented at seminar on sericulture, 1-11,CSRTI, Mysore.

Radhakrishna PG, Sekarappa BM, Gururaj CS (2001). Seasonal response of the new multi-bivoltine hybrids of the silkworm, *Bombyx mori* L. Indian J. Seric. 40(2): 174-176.

Ravindra S, Sharma SD, Raghavendra RD, Chandrashekaran K, Basavaraja HK, Kariappa BK (2005). Line x Tester and heterosis analysis in the silkworm, *Bombyx mori* L. Indian J. Seric. 44(1): 92-99.

Snedecor WG, Cochran CW (1967). Statistical methods. Oxford and IBH Pub, India.

Subramanya G, Murakami A (1994). Climatic differential phenotypic expression of voltine genes in *Bombyx mori* L. Indian J. Seric. 33(2): 103-109.

Tazima Y (1962). Silkworm egg: *scientific brochure-1*, pp. 1-49, Central Silk Board Bombay, India.

Tazima Y (1964). The Genetics of the Silkworm: *Logos Press*, London pp. 103-117

Tazima Y (1988). A view on the improvement of Mysore breeds. Proceedings of Int. Conress, on tropical sericulture 1-5, Feb 18-23.

Umadevi K, Sudha VN, Prakashmurthy DP, Raghavendra RD, Ravindra S, Premalatha V, Kariappa BK (2005). Evaluation of new multivoltine x bivoltine hybrid "Cauvery" (BL67 x CSR 101) of the silkworm, *Bombyx mori* L. Indian J. Seric. 44(1): 131-133.

18

Estimation of virulence type and level of soybean cyst nematode field populations in response to resistant cultivars

J. W. Zheng[1,2] and S. Y. Chen[1]*

[1]Southern Research and Outreach Center, University of Minnesota, Waseca, MN 56093; USA.
[2]Institute of Biotechnology, College of Agriculture and Biotechnology, Zhejiang University, Hangzhou 310029, P. R. China.

The use of resistant cultivars is the most common practice in managing the soybean cyst nematode (SCN), *Heterodera glycines.* Currently, most commercial SCN-resistant soybean cultivars were developed from a single source of resistance, plant introduction (PI) 88788. The effect of crop sequences including rotations of SCN-susceptible soybean 'sturdy' with SCN-resistant soybean 'freeborn' carrying resistance derived from PI 88788, soybean 'Pioneer 9234' carrying resistance derived from PI 548402 (peking), and nonhost corn was studied at two field sites in southern Minnesota, the United States. Parasitic ability of SCN measured as a female index (FI) on PI 88788 and freeborn increased with the number of years freeborn was planted. After more than 5 years of freeborn, either in monoculture or rotation with other soybean cultivars and corn, the SCN population changed from the original race 3 (HG type 0 or 7) to race 1 (HG type 2.7 or 2.5.7). After 10 years of freeborn, the changed nematode population reproduced freely on the once resistant cultivar (FI > 60). There was no selection pressure from the use of PI 88788-resistance on SCN populations that can overcome peking resistance. Planting 3 or fewer years of Pioneer 9234 had no noticeable effect on the virulence phenotype of the SCN population. This study suggests that more cultivars from resistance sources other than PI 88788 are urgently needed for effective management of the nematode in Minnesota and other regions in the world.

Key words: *Heterodera glycines, HG* type, host-parasite relationship, race, soybean cyst nematode, virulence phenotype.

INTRODUCTION

The soybean cyst nematode (SCN), *Heterodera glycines* Ichinohe, is a major soybean yield-limiting factor (Wrather and Koenning, 2006). Use of resistant cultivars and crop rotation are the most effective means to manage the nematode (Niblack, 2005; Niblack and Chen, 2004). However, the effectiveness of using resistant cultivars may depend on the interaction between the cultivars and

nematode populations. Breeding and deployment of resistant cultivars is challenging due to the genetic variability of SCN (Niblack and Riggs, 2004), the selection pressure on SCN when resistance is used (Anand and Shumway, 1985; Luedders and Dropkin, 1983; McCann et al., 1982; Riggs et al., 1977; Triantaphyllou, 1975; Young, 1994; Young and Hartwig, 1992; Young et al., 1986) and the linkage of yield-suppressive factors with SCN-resistance genes (Kopisch-Obuch et al., 2005; Mudge et al., 1996). A number of soybean lines have resistance to SCN, but only a few of them have been used in breeding commercial soybean

*Corresponding author. E-mail: chenx099@umn.edu.

cultivars (Shannon et al., 2004). Most (> 90%) commercial SCN-resistant cultivars in the North Central United States were derived from crosses with PI 88788 and only a few cultivars were from PI 548402 (peking) and PI 437654 ('cystx' or 'hartwig' resistance) (Shier, 2011). Changes in virulence phenotypes, measured as races (Riggs and Schmitt, 1988) or HG types (Niblack et al., 2002) of SCN populations following the use of resistant soybean cultivars have been reported from a number of surveys.

Nationwide surveys in the late 1980s and early 1990s suggested that the frequency of SCN populations that could develop (FI ≥ 10) on resistant cultivars was higher in the southern than in the northern USA, probably due to longer use of resistant soybean cultivars in the southern states (Anand et al., 1994; Kim et al., 1997). In Tennessee, most SCN populations could not develop well on PI 88788 in the 1970s, but after cultivars derived from PI 88788 for two decades, the prevalent virulence phenotypes in the state became HG type 1.2 and 2, which can develop on the resistance source PI 88788 (Young, 1990, 1998b). In North Carolina, the frequency of races 2 and 4, which can develop on both peking and PI 88788, increased from 28 to 62% over a span of approximately 10 years from the 1980s to 1990s (Koenning and Barker, 1998). The increase of the frequency of virulent populations has also been reported recently from central states in the USA (Hershman et al., 2008; Mitchum et al., 2007; Niblack et al., 2006). In Illinois, 70% of populations were virulent on PI 88788 (HG type 2) in 2004 to 2005, whereas 64% of populations collected in 1989 to 1990 were HG type 0, which does not develop on PI 88788 (Niblack et al., 2006; Sikora and Noel, 1991).

In Missouri, the percentage of populations that develop well on PI 88788 increased from 58% in 1998 (Niblack et al., 2003) to 78% in 2005 (Mitchum et al., 2007). In Kentucky, 60% of the 20 populations collected in 2006 to 2007 were able to develop on PI 88788, contrasting to that most common HG type in late 1980s was HG type 0 (race 3) (Hershman et al., 2008). The increase of virulent populations of SCN in Illinois, Missouri and Kentucky presumably resulted from the extensive use of soybean cultivars derived from PI 88788 in the past two decades. Use of SCN-resistant cultivars in managing the nematode has a relatively short history in Minnesota compared with the southern states. The frequency of virulence phenotypes in Minnesota did not change significantly from 1997 to 2002 and most populations (more than 80%) did not develop well on PI 88788 and peking (Zheng et al., 2006).

A recent survey conducted in 2007 to 2008 showed that virulence level of the SCN populations has increased over years in Minnesota, with 72.6% of the SCN populations in the state were virulent to PI 88788 soybean, and 12.1% virulent to Peking (Chen et al., 2010). It is expected that the frequency of populations that can develop on PI 88788 and Peking will further increase in the future with longer and more extensive use of SCN-resistant cultivars in Minnesota. Knowing the virulence phenotypes of SCN in fields and speed of change following the use of resistant cultivars is important for strategically deploying and breeding resistant cultivars and making management recommendations to farmers. The objective of this study was to estimate virulence types and levels of SCN field populations in response to crop sequences including non-host crop and SCN-resistant soybean cultivars in the northern climate.

MATERIALS AND METHODS

Field sites and soil sample collection

Soil samples were collected from the plots at two field sites in Waseca County, Minnesota. Site A was initiated in 1993 in a field in New Richland for study of the effect of tillage and row spacing on SCN populations (Chen, 2007; Chen et al., 2001b). The soil at Site A was a Webster clay loam (fine-loamy, mixed, mesic, Typic Endoaquoll) with 37.4% sand, 32.4% silt, 30.2% clay, 7.3% organic matter and pH 7.8 measured in 1998. The initial SCN population was HG type 0 (race 3). SCN-resistant soybean cultivar had not been used in the field before the experiment. At this site, nine combinations of the SCN-susceptible 'sturdy', resistant 'freeborn' carrying resistance from PI 88788 (R1) and 'Pioneer 9234' carrying resistance from peking (R2) soybean were rotated annually with the nonhost corn (Table 1). In the fall of 2002 when the tillage study was terminated, soil samples were collected from the no-till plots of each crop sequence and composited. The nematode population density was increased on the susceptible soybean sturdy and maintained in the greenhouse until they were used for HG type analysis. Site B was established in 1996 in a field near Waseca to study the effect of crop sequences including corn and soybean rotations on the SCN population density and soybean yield (Chen et al., 2001a).

The soil at Site B was a Webster clay loam (fine-loamy, mixed mesic Typic Endoaquoll) with 22% sand, 46% silt, 32% clay, 9.9% organic matter and pH 7.8. The initial SCN population was HG type 0 or 7 (Race 3). SCN-resistant soybean cultivar had not been used in the field before the experiment. At this site, the SCN-resistant soybean freeborn was included in rotations. In spring 2004, soil samples were collected from the four replicated plots each of the five selected crop sequence treatments (Table 1) and composited by treatment. The nematode population density was increased on sturdy in the greenhouse for analysis of HG types. In addition, soil samples were collected in the spring of 2007 from five crop sequences containing various numbers of years of freeborn (Table 2). The nematode population density was increased on sturdy and parasitic ability of the populations was analyzed on peking, PI 88788 and freeborn with Lee 74 as the control.

Bioassay

Newly formed females on the roots were collected and eggs were released from the cysts. The eggs were used as inoculums for tests of HG types and races of SCN following the procedures used in previous studies (Niblack et al., 2002; Riggs and Schmitt, 1988) with some modification (Zheng et al., 2006). The HG type soybean

Table 1. HG types and races of *Heterodera glycines* from different crop sequences in Minnesota fields.

Field site	Population	Crop sequence[b]	Female index on HG type indicators[a]							FI on picket	Females/plant on Lee 74	HG type	Race
			1 (peking)	2 (PI 88788)	3 (PI 90763)	4 (PI 437654)	5 (PI 209332)	6 (PI 89722)	7 (PI 548316)				
A	N1	C-R$_1$-C-R$_1$-C-R$_1$-C-R$_1$	0	12.1	0	0	6.2	0	13.4	0	114	2.7	1
	N2	C-R$_1$-C-R$_1$-C-R$_2$-C-R$_2$-C-R$_1$	0	2.1	0	0	4.9	0	1.5	0	98	0	3
	N3	C-S-C-S-C-S-C-S-C-S	0	3	0	0	0.4	0	3.4	0	65	0	3
	N4	C-S-C-S-C-R$_2$-C-R$_2$-C-R$_2$	0	0.4	0	0	1.7	0	7.5	0	60	0	3
	N5	C-R$_1$-C-S-C-R$_2$-C-R$_1$-C-R$_2$	0	2.2	0	0	0.9	0	3.8	3.9	77	0	3
	N6	C-R$_1$-C-S-C-R$_2$-C-S-C-R$_1$	0	2.5	0	0	1.8	0	7.1	2.4	81	0	3
	N7	C-R$_1$-C-S-C-R$_1$-C-S-C-R$_2$	0	0.7	0	0	3.4	0	18.6	0	109	7	3
	N8	C-R$_1$-C-S-C-R$_1$-C- R$_1$-C-R$_2$	0	5.3	0	0	10.1	0	17.2	4.9	103	5.7	3
	N9	C-R$_1$-C-S-C-R$_1$-C- R$_2$-C-R$_1$	0	2.8	0	0	3.7	0	14.7	1.4	71	7	3
B	W1	S-S-S-S-S-S-S-S	0	1	0	0	1.9	0	5.2	0	102	0	3
	W2	S-C-S-C-S-C-S-C	0	4.1	0	0	3.4	0	14.5	0	127	7	3
	W3	R$_1$-C-S-C-R$_1$-C-S-C	0	1.2	0	0	3.1	0	3.5	0	122	0	3
	W4	R$_1$-C- R$_1$-C-R$_1$-C-R$_1$-C	0	4.2	0	0	5.3	0	16.1	0	137	7	3
	W5	R$_1$-R$_1$-R$_1$-R$_1$-R$_1$-R$_1$-R$_1$-R$_1$	0	22.3	0	0	27.8	0	19.9	1.7	116	2.5.7	1

[a] Female index (FI) = 100 × number of females on indicator/number of females on Lee 74. [b] The letters represent the sequence of crops from 1993 to 2002 at Site A and 1996 to 2003 at Site B: C = corn; S = susceptible soybean 'sturdy'; R$_1$ = SCN-resistant soybean 'freeborn' (PI 88788 source of resistance – SR); and R$_2$ = SCN-resistant soybean 'Pioneer brand 9234' (peking SR).

Table 2. Parasitic ability of *Heterodera glycines* populations after planting the SCN-resistant soybean 'freeborn' for various years during 1996 to 2006 in a Minnesota field (Site B) infested with an original population of HG type 0 (Race 3).

Population	Crop sequence[b]	Female index[a]			Females/plant on Lee 74
		Peking	PI 88788	Freeborn	
W1-07	SSSSSSSSSCS	0.3	2.5	15.9	174
W3-07	RCSCRCSCRCS	0.2	8.2	15.6	239
W4-07	RCRCRCRCRCR	1.2	13.5	34.3	145
W5-07B	RRRRRRRRSCS	0.7	25.5	39.7	207
W5-07A	RRRRRRRRRCR	1.6	48.7	70.9	158

[a] Female index (FI) = 100 × number of females on indicator/number of females on Lee 74. [b] The letters represent the sequence of crops from 1996 to 2006: C = corn; S = susceptible soybean 'sturdy'; R = SCN-resistant soybean 'freeborn'.

indicator lines PI 548402 (peking), PI 88788, PI 90763, PI 437654, PI 209332, PI 89772, and PI 548316 (cloud) and the susceptible cultivar Lee 74 were obtained from USDA soybean Germplasm Collection (Urbana, Illinois, USA) for the study. In addition, "Pickett" was included for SCN race determination. The soybean seeds were germinated in moist, sterilized germination paper in Petri dishes at 29 ℃ in an incubator. After 48 h, the seedlings with root radicles 2 to 3 cm long were selected for the tests. A cone-tainer (4 cm diameter and 13.5 cm high) was filled with autoclaved sandy loam soil to half and 2,000 eggs in 2.5 ml water were added. Additional soil was placed in the cone-tainer to approximately 1 cm from the top. A hole was made at the center to a depth of 3 cm with a glass stick (0.5 cm diameter). A soybean seedling was placed in the hole. Another suspension of 2,000 eggs in 2.5 ml water was added near the seedling and the seedling was covered with additional soil to about 1 cm depth. Each soybean line was replicated five times (five cone-tainers). All of the cone-tainers for each SCN population were inserted into autoclaved sand in a container (35 × 31 × 15 cm). The cone-tainers were maintained in the greenhouse at 25 to 28°C and watered daily.

After 30 days, the seedlings with soil were separately removed from the cone-tainers and soaked in water in 1 L beakers for at least 30 min. The soybean plants were gently removed from the beakers and the soil on the roots was gently washed off. The females on the roots were counted directly with the aid of a magnifier in the 2005 assay. In the 2007 assay, the females were extracted from the roots and then counted. The Female index (FI) was determined for each soybean line by dividing the mean female count on the indicator line (or other assay soybean) by the mean female count on the susceptible check 'Lee 74', and expressed as percentage. The populations were classified to HG type and races (Niblack et al., 2002; Riggs and Schmitt, 1988).

RESULTS

At Site A, two resistance sources (PI 88788 and peking) were used during 1993 to 2002. The ten years of corn/freeborn annual rotation (5 years of freeborn; SCN population N1) resulted in higher FI (12.1) on PI 88788, the source of resistance for freeborn, compared with the susceptible soybean annually rotated with corn (N3) and all other sequences involving two resistant sources (all other SCN populations) (Table 1). The virulence phenotype of N1 was changed from HG type 0 (race 3) to HG type 2.7 (Race 1) with the ten years of corn/freeborn annual rotation. In the rest of the crop sequences, whether corn was rotated with the susceptible cultivar or with two different sources of resistance, there was no significant change of virulence phenotypes except that the FI was more than 10 on PI 209332 for N8, and PI 548316 for N7, N8, and N9 SCN populations (Table 1). At Site B, 8 years of SCN-resistant freeborn monoculture from 1996 to 2003 increased parasitic ability of the SCN population (W5) on PI 88788; the FI on PI 88788 following the freeborn monoculture was 22.3, contrasting to 1 of the population from plots following susceptible soybean monoculture (W1) (Table 1). The virulence phenotype of the W5 population was changed from HG type 0 (race 3) to HG type 2.5.7 (Race 1), which is virulent to freeborn and probably other commercial

soybean cultivars with the PI 88788 source of resistance. The parasitic ability of the W5 population also increased on the soybean lines PI 209332 and PI 548316 compared with the population from the monoculture of susceptible soybean (W1) (Table 1). However, there was no noticeable change of parasitic ability in the plots of corn/resistant soybean annual rotation (W4) or resistant soybean/corn/susceptible soybean rotation (W3) within the 8 years (Table 1).

In 2007, soil samples were taken from plots with different number of years of the resistant cultivar freeborn at Site B, and the FI was analyzed on freeborn, peking and PI88788. The FI of the SCN population from the plots with monoculture of susceptible soybean (W1-07) was only 0.3 and 2.5 on peking and PI 88788, respectively, confirming that the initial population was HG type 0 and the FI of this population on freeborn was 15.9, indicating freeborn was moderately resistant to HG type 0 (Race 3) (Table 2). In the sequences with freeborn more than 5 years, the FI was more than 10 on PI 88788 and it increased with increasing number of years of freeborn (Figure 1). After 10 years of freeborn, the FI increased to 48.7 on PI 88788 and 70.9 on freeborn, respectively and the resistant cultivar freeborn became susceptible to the changed nematode population. There was no noticeable change of FI on peking after the ten years of use of PI 88788 source of resistance (Table 2).

DISCUSSION

A number of previous studies with field plot experiments in southern and central USA have demonstrated that planting an SCN-resistant cultivar selected SCN populations with increased parasitic ability of the nematode population on the selecting cultivar and other cultivars with similar resistance. In Tennessee, Young and coworkers conduced extensive long-term experiments in two fields initially infested with populations virulent to cultivars carrying peking resistance (Race 14 or HG type 1) and found that planting soybean cultivars with PI 88788 source of resistance increased parasitic ability of the SCN populations on the selecting soybean cultivars and PI 88788, while planting cultivars with peking source of resistance did not significantly change the level of virulence (Young, 1994, 1998a; Young and Hartwig, 1992; Young et al., 1986). Similar results were obtained in Missouri and Arkansas in fields infested with HG type 1 (Race 14) (Anand et al., 1995; Young et al., 1986). In another study in a Mississippi field initially infested with HG type 0 (race 3), which is avirulent to the current common sources of resistance, after 10 years of various crop sequences, parasitic ability of the nematode populations on peking increased in the sequences with 6 or more years of planting a cultivar with peking source of resistance (Young and Hartwig, 1988). The present study

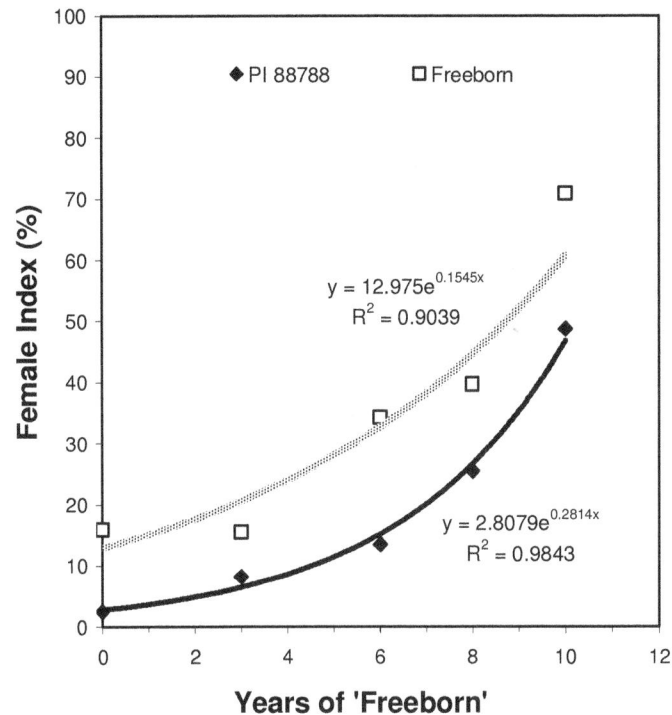

Figure 1. Relationship between parasitic ability (female index) of *Heterodera glycines* on PI 88788, and 'freeborn' after the use of the SCN-resistant soybean freeborn for various years during 1996 to 2007 in a Minnesota field (Site B) infested with an original population of race 3 (HG type 0 or 7).

demonstrated that use of freeborn carrying PI 88788 resistance had a significant effect on virulence phenotype of SCN populations in the northern climate in Minnesota fields initially infested with HG type 0 (race 3) populations.

The parasitic ability of SCN on PI 88788 increased with increasing number of years of planting the resistant cultivar. After 5 years, the original population of HG type 0 (Race 3) became a population that was able to overcome the resistance of PI 88788 (FI > 10). After 10 years, freeborn that was moderately resistant (FI ≈ 15) to the original population became susceptible (FI > 60) to the resulting population. Selection pressure may depend on a number of factors including resistance level of hosts, virulence types and level of initial nematode population, and climate and soil conditions that may affect nematode parasitism and survival. For developing a comprehensive model to predict the virulence types and level of SCN populations in response to the use of resistant cultivars, additional data from long-term field experiments are needed. Nevertheless, the present study provided approximate estimation of virulence level of SCN populations following the use of the resistant cultivars. Our results may be useful for estimating the change of virulence of SCN populations following the use of a

cultivar with similar resistance in fields infested with HG type 0 (Race 3) in the similar climatic conditions such as in the Northern USA and Northeastern China.

In some fields in Minnesota, SCN populations are still HG types with low parasitic ability on PI 88788. Similarly, most populations in Northeastern China are probably still HG type 0 (Race 3) (Liu et al., 1997) because there has been limited use of SCN-resistant cultivars. The results may also be useful for estimating change of virulence of SCN populations in other recently infested regions such as Iran where most of populations are HG type 0 or 7 (Race 3) (Maafi et al., 2008). Based on our study, if a cultivar carrying PI 88788 resistance and having similar level of resistance as in freeborn is used in the fields infested with HG type 0 (Race 3) populations in the northern climates, it may take approximately 20 years for the cultivar to become susceptible to the resulting populations if the cultivar is annually rotated with a nonhost, a common practice in the Northern USA and Northeastern China, and there is no other source of resistance in the rotation. The results of this field plot study agree with the statewide surveys (Chen et al., 2010; Zheng et al., 2006). The frequency of virulence phenotypes on PI 88788 in Minnesota did not differ significantly from 1997 to 2002 (Zheng et al., 2006), but significant increase occurred after 2007 (Chen et al., 2010) because use of SCN-resistant cultivars in many fields by 2007 in the state was approximately for 5 years of soybean or 10 years of soybean-corn annual rotation, and most of the cultivars carried PI 88788 resistance. We could not detect any significant change of virulence phenotype of SCN following the use of a cultivar (Pioneer 9234) carrying resistance derived from peking for 3 or fewer years. However, change of virulence phenotypes on peking cultivars is also expected because it will pose a selection pressure on the nematode populations (Riggs et al., 1977; Triantaphyllou, 1975; Young, 1994; Young and Hartwig, 1988, 1992; Young et al., 1986).

Soybean is a major crop and in most cases it is annually rotated with corn in Minnesota. One year of non-host corn or other crops is insufficient in lowering SCN population density to a level where there is limited damage to soybean (Chen et al., 2001a). Other practices have yet limited success. Consequently, the SCN management largely relies on the use of resistant cultivars. However, with the current resistant cultivars, most from the single source PI 88788, the use of resistant cultivars will become ineffective in the foreseeable future. Efforts are needed to develop commercial cultivars with other sources of resistance for long-term effective management of the nematode (Young, 1998b). Our results indicate that planting PI 88788-resistance freeborn had little or no selection pressure towards SCN populations that can overcome peking resistance because the resistances of the two sources did not correlate (Colgrove and Niblack, 2008;

Niblack et al., 1993; Young, 1994; Zheng et al., 2006). Consequently, rotation of cultivars from PI 88788 with the cultivars from peking is a good practice to slow down the change of virulence. However, additional sources of resistance may be needed for long-term effective management with the use of resistant cultivars.

Conclusion

In this study, field experiments were conducted in Minnesota to determine the virulence types and levels of SCN in response to the use of SCN-resistant cultivars. Planting 3 or fewer years of Pioneer 9234 carrying peking resistance had no noticeable effect on the virulence phenotype of the SCN population. There was no selection pressure from the use of PI 88788-resistance on SCN populations that can overcome peking resistance. Parasitic ability of SCN on PI 88788 and freeborn increased with the number of years freeborn was planted. After more than 5 years of freeborn, either in monoculture or rotation with other soybean cultivars and corn, the SCN population changed from the original HG type 0 or 7 (Race 3) to HG type 2.7 or 2.5.7 (Race 1). After 10 years of freeborn, the resistant cultivar was susceptible (FI > 60) to the changed nematode population. It may take approximately 20 years of corn-soybean annual rotation for a cultivar carrying similar resistance as in freeborn to become susceptible to the resulting SCN populations if nonhost corn does not affect SCN virulence, and there is no other source of resistance in the rotation.

This study suggests that more cultivars from resistance sources other than PI 88788 are urgently needed for effective management of the nematode in Minnesota and other regions in the world.

ACKNOWLEDGEMENTS

This research was supported by Minnesota Soybean Producers check-off funding through the Minnesota Soybean Research and Promotion Council and the Minnesota Agricultural Experiment Station. The authors thank D. Miller for reviewing the manuscript prior to the submission, and C. Johnson, W. Gottschalk and J. Ballman for technical assistance.

REFERENCES

Anand SC, Koenning SR, Sharma SB (1995). Effect of temporal deployment of different sources of resistance to soybean cyst nematode. J. Product. Agri., 8: 119-123.

Anand SC, Sharma SB, Rao Arelli AP, Wrather JA (1994). Variation in parasitic potential of Heterodera glycines populations. Crop Sci., 34: 1452-1454.

Anand SC, Shumway CR (1985). Response of secondary selection on soybean cyst nematode reproduction on resistant soybean lines Crop Protection, 4: 231-234.

Chen S, Potter B, Orf J (2010). Virulence of the soybean cyst nematode has increased over years in Minnesota. J. Nematol. 42: in press (Abstract).

Chen SY (2007). Tillage and crop sequence effects on Heterodera glycines and soybean yields. Agron. J., 98: 797-907.

Chen SY, Porter PM, Reese CD, Stienstra WC (2001a). Crop sequence effects on soybean cyst nematode and soybean and corn yields Crop Sci., 41: 1843-1849.

Chen SY, Stienstra WC, Lueschen WE, Hoverstad TR (2001b). Response of Heterodera glycines and soybean cultivar to tillage and row spacing. Plant Dis., 85: 311-316.

Colgrove AL, Niblack TL (2008). Correlation of female indices from virulence assays on inbred lines and field populations of Heterodera glycines. J. Nematol., 40: 39-45.

Hershman DE, Heinz RD, Kennedy BS (2008). Soybean cyst nematode, Heterodera glycines, populations adapting to resistant soybean cultivars in Kentucky. Plant Dis., 92: 1475.

Kim DG, Riggs RD, Robbins RT, Rakes L (1997). Distribution of races of Heterodera glycines in the Central United States. J. Nematol., 29 173-179.

Koenning SR, Barker KR (1998). Survey of Heterodera glycines races and other plant-parasitic nematodes on soybean in North Carolina. J Nematol., 30: 569-576.

Kopisch-Obuch FJ, McBroom RL, Diers BW (2005). Association between soybean cyst nematode resistance loci and yield in soybean. Crop Sci., 45: 956-965.

Liu XZ, Li JQ, Zhang DS (1997). History and status of soybean cyst nematode in China. Int. J. Nematol., 7: 18-25.

Luedders VD, Dropkin VH (1983). Effect of secondary selection on cyst nematode reproduction on soybeans. Crop Sci., 23: 263-264.

Maafi ZT, Salati M, Riggs RD (2008). Distribution, population density race and type determination of soybean cyst nematode, Heterodera glycines, in Iran. Nematol., 10: 919-924.

McCann J, Luedders VD, Dropkin VH (1982). Selection and reproduction of soybean cyst nematodes on resistant soybeans. Crop Sci., 22: 78-80.

Mitchum MG, Wrather JA, Heinz RD, Shannon JG, Danekas G (2007). Variability in distribution and virulence phenotypes of Heterodera glycines in Missouri during 2005. Plant Dis., 91: 1473-1476.

Mudge J, Concibido V, Denny R, Young N, Orf J (1996). Genetic mapping of a yield depression locus near a major gene for soybean cyst nematode resistance. Soybean Genetics Newslett., 23: 175-178.

Niblack TL (2005). Soybean cyst nematode management reconsidered Plant Dis., 89: 1020-1026.

Niblack TL, Arelli PR, Noel GR, Opperman CH, Orf JH, Schmitt DP Shannon JG, Tylka GL (2002). A revised classification scheme for genetically diverse populations of Heterodera glycines. J. Nematol. 34: 279-288.

Niblack TL Chen SY (2004). Cropping systems. In Biology and management of the soybean cyst nematode. Eds. Schmitt DP Wrather JA, Riggs RD, Schmitt & Associates of Marceline, Marceline MO. pp. 181-206.

Niblack TL, Colgrove KB, Colgrove AC (2006). Soybean cyst nematode in Illinois from 1990: Shift in virulence phenotype of field populations J. Nematol., 38: 285.

Niblack TL, Heinz RD, Smith GS, Donald PA (1993). Distribution density, and diversity of Heterodera glycines in Missouri. J. Nematol. 25: 880-886.

Niblack T4 Riggs RD (2004). Variation in virulence phenotypes. In Biology and management of the soybean cyst nematode. Eds Schmitt DP, Wrather JA, Riggs RD. Schmitt & Associates of Marceline, Marceline, MO. pp. 57-72.

Niblack TL, Wrather JA, Heinz RD, Donald PA (2003). Distribution and virulence phenotypes of Heterodera glycines in Missouri. Plant Dis. 87: 929-932.

Riggs RD, Hamblen ML, Rakes L (1977). Development of Heterodera glycines pathotypes as affected by soybean cultivars. J. Nematol., 9 312-318.

Riggs RD, Schmitt DP (1988). Complete characterization of the race scheme for *Heterodera glycines*. J. Nematol., 20: 392-395.

Shannon JG, Arelli PR, Young LD (2004). Breeding for resistance and tolerance. In Biology and management of the soybean cyst nematode. Eds. Schmitt DP, Wrather JA, Riggs RD. Schmitt & Associates of Marceline, Marceline, MO., pp. 155-180.

Shier M (2011). Soybean sarieties with soybean cyst nematode resistance. University of Illinois Extension. Web/URL: https://netfiles.uiuc.edu/tjw/www/cover.htm.

Sikora EJ, Noel GR (1991). Distribution of *Heterodera glycines* races in Illinois. J. Nematol., 23: 624-628.

Triantaphyllou AC (1975). Genetic structure of races of *Heterodera glycines* and inheritance of ability to reproduce on resistant soybeans. J. Nematol., 7: 356-364.

Wrather JA, Koenning SR (2006). Estimates of disease effects on soybean yields in the United States 2003 to 2005. J. Nematol., 38: 173-180.

Young LD (1990). Survey of soybean cyst nematode races in Tennessee. J. Nematol., 22: 672-675.

Young LD (1994). Changes in the *Heterodera glycines* female index as affected by ten-year cropping sequences. J. Nematol., 26: 505-510.

Young LD (1998a). Influence of soybean cropping sequences on seed yield and female index of the soybean cyst nematode. Plant Dis., 82: 615-619.

Young LD (1998b). Managing soybean resistance to *Heterodera glycines*. J. Nematol., 30: 525-529.

Young LD, Hartwig EE (1988). Selection pressure on soybean cyst nematode from soybean cropping sequences. Crop Sci., 28: 845-847.

Young LD, Hartwig EE (1992). Cropping sequence effects on soybean and *Heterodera glycines*. Plant Dis., 76: 78-81.

Young LD, Hartwig EE, Anand SC, Widick D (1986). Responses of soybeans and soybean cyst nematodes to cropping sequences. Plant Dis., 70: 787-791.

Zheng JW, Li YH, Chen SY (2006). Characterization of the virulence phenotypes of *Heterodera glycines* in Minnesota. J. Nematol., 38: 383-390.

19

Efficacy of different benzimidazole derivatives on microsporidiosis of lamerin breed of the silkworm, *Bombyx mori* L.

Shabir Ahmad Bhat[1]*, B. Nataraju[2] and and Ifat Bashir[3]

[1]Temperate Sericulture Research Institute Mirgund, Sher-e-Kashmir University of Agricultural Sciences and Technology, Kashmir, Srinagar-190001, India.
[2]Central Sericultural Research and Training Institute, Mysore-570 008, India.
[3]Sericulture Development Department Jammu and Kashmir, Srinagar, India.

Six benzimidazole derivatives *viz.,* Metronidazole, Albendazole, Tinidazole, Ornidazole, Mebendazole and Satinidazole at three concentrations (0.25, 0.50 and 1.00%) were screened against microsporidiosis of Lamerin breed of the silkworm, *Bombyx mori in vitro* and *in vivo*. All drugs were found significantly effective in minimizing the microsporidiosis. Further more, mebendazole were found more effective at all concentrations and emerged as the most efficient and promising drug for minimizing the microsporidiosis to the significant extent in Lamerin breed of the silkworm, *B mori.*

Key words: Microsporidiosis, benzimidazole, *Bombyx mori*, lamerin breed.

INTRODUCTION

Chemotherapy was attempted by many researchers to control pebrine in insects in general and in silkworms in particular. Sahay et al. (2005), Hayasaka (1991), and Baig (1994), attempted to asses the effect of chemicals or drugs on microsporidians. Griyaghey (1976) studied the effect of chemotherapic agents on Pebrine of Tassar silkwom, *Antheraea mylitta* and found an effective way to control the microsporidiosis. Schmah and Benini (1998) reported the effectiveness of benzimidazole derivatives in controlling the microsporidiosis of fishes. The scientific literature shows there are no report as such on the effects of benzimidazole derivatives on the microsporidiosis of Lamerin breed of the silkworm, *B. mori*. So in the present study the effect of benzimidazole derivatives on the microsporidiosis of Lamerin breed is investigated.

MATERIALS AND METHODS

Six benzimidazole derivatives *viz.,* Metronidazole (M/s R.P.L.

Pharmaceuticals Pvt. Ltd., New Delhi, India), Albendazole, (M/s Cipla Meditab Specialties Pvt. Ltd., Goa, India), Tinidazole, (M/s Kopran Pvt. Ltd. Khalapur), Ornidazole, (M/s Vapicare Pharma Pvt. Ltd. India), Mebendazole, (M/s Cipla Ltd. by Meditab Specialties Pvt. Ltd., Goa, India) and Satinidazole, (M/s Nicholas Piramal India Pvt. Ltd. Auragabad, India) were screened for antimicrosporidian activity against Lamerin microsporidian spores.

In vitro efficacy of benzimidazole derivative against microsporidia

These drugs were screened *in vitro* for anti microsporidian activity. The microsporidian spores (1×10^7 spores/ml) were suspended in one ml of drugs of concentrations *viz.,* 0.25, 0.50 and 1.00% for 30 min respectively. The incubated spores were centrifuged at 300 rpm for 5 min and the sediment was collected washed properly in distilled water by repeated centrifugation and were subjected to viability test.

Toxicity test of drugs

The drugs were screened for their toxicity at 1% dosage by feeding to 2nd instar silkworm larvae daily once for three days along with the mulberry leaves and the toxicity symptoms were observed up to 10 days.

In another set of experiment, one day old 3rd instar Lamerin

*Corresponding author. E-mail: bhat_shabir2003@yahoo.com.

Table 1. *In vitro* screening of different of benzimidazole derivatives against microsporidia.

S/No.	Name of the drug	Chemical formula	Concentrations (%)	Effective
1	Metronidazole	$C_6H_9N_3O_3$	0.25	-
			0.50	-
			1.00	+
2	Albendazole	$C_{12}H_{15}N_3O_2S$	0.25	-
			0.50	-
			1.00	+
3	Tinidazole	$C_8H_{13}N_3O_4S$	0.25	-
			0.50	-
			1.00	-
4	Ornidazole	$C_7H_{10}C_1N_3O_3$	0.25	-
			0.50	-
			1.00	+
5	Mebendazole	$C_{16}H_{13}N_3O_3$	0.25	+
			0.50	+
			1.00	+
6	Satranidazole	$C_{18}H_{29}C_1N_3\text{-}H_3PO_4$	0.25	-
			0.50	-
			1.00	-
			1.00	-
	Control		-	-

- Ineffective; + Effective.

silkworm larvae were classified into four batches. Each batch which consisted of three replications of 100 larvae was per orally inoculated with Lb_{ms} spores (1×10^7 spores/ml). The provisionally infected larvae were treated with these drugs of 0.25, 0.50 and 1.00% concentration along with mulberry leaves once on alternate days till the onset of spinning. A control batch (without any treatment) was also kept for comparison.

RESULTS AND DISCUSSION

Observations on the *in vitro* studies of microsporidia spores after incubating in different concentrations of benzimidazole derivative were recorded and presented in Table 1. Among six drugs tested, mebendazole were found effective against microsporidian spores at all the concentrations. Rest of the drugs viz., metronidazole, albendazole, ornidazole and satinidizole were found effective only at 1%. However, tinidizole was not found effective against the microsporidian spores at any of the concentrations tested. The results of the *in vivo* observations showed that among tested benzimidazole derivatives, mebendazole (0.25 to 1.00%), albendazole (1.00%) were found effective in reduction of larval mortality to an extent of 100% (Table 2). However, the

other drugs at concentrations ranging 0.25 to 1.00% reduced the mortality from 78.56 to 96.46%. At 1% concentration orinidazole reduced the larval mortality by 96.46%, followed by metronidazole 92.92% and satranidazole 92.92%. The percent of infected moths was significantly low in all treatments as compared to the inoculated control. It is recorded 59.38% in inoculated control was reduced to 22.44 to 30.04% in treated batches (Table 3).

Several therapeutic drugs have been identified as antimicrosporidia agents to control *Nosema. bombycis* infection in silkworm (Baig, 1994). In the present study benzimadizole derivatives were found effective in preventing larval mortality and also suppressing infection at moth stage as compared to control. A similar observation was made by Joythi et al. (2005) stated that when carbendazimin given 48 h post inoculation of *N. bombycis* reduced the infection but could not eliminated the infection at moth stage completely. Chandra et al. (1995) also reported 3 to 4% bavistin treatment twice a day reduced *N. bombycis* infection in silkworms. Analogues of benzimidazole and benlate, bavistin, derosal, Fumidil - B or fumagillin, methylthiophanate, ethyl thiophanate and anisomycin have been reported to

Table 2: Efficacy of benzimidazole derivatives in larval mortality.

Treatment	% mortality and disease reduction due to treatment			
	0.25%	0.50%	1.00%	Control
Metronidazole	1.33 ± 0.57 (85.74)	1.00 ± 0.00 (89.28)	0.66 ± 0.57 (92.92)	
Albendazole	1.33 ± 0.57 (85.74)	0.33 ± 0.57 (96.46)	0.00 ± 0.57 (100.0)	
Tinidizdazole	1.66 ± 0.57 (82.20)	1.66 ± 1.15 (82.20)	1.00 ± 1.00 (89.28)	
Ornidizole	2.00 ± 1.00 (78.56)	1.33 ± 0.57 (85.74)	0.33 ± 0.57 (96.46)	9.33 ± 0.57
Mebendazole	0.00 ± 1.00 (100.00)	0.00 ± 0.57 (100.0)	0.00 ± 0.57 (100.0)	
Satranidizole	1.66 ± 0.57 (82.20)	2.00 ± 1.00 (78.56)	0.66 ± 0.57 (92.92)	

Each value is mean±SD of three replications; values within parenthesis indicate % reduction in mortality.

Table 3. Efficacy of benzimidazole derivatives on suppression of infection at moth stage.

Treatment	% disease suppression at moth stage due to treatment			
	0.25%	0.50%	1.00%	Control
Metronidazole	25.18 ± 2.88 (59.59)	24.56 ± 0.66 (58.63)	23.40 ±1.54 (60.59)	
Albendazole	27.36 ± 1.06 (53.92)	22.76 ± 1.19 (61.67)	22.94 ± 0.51 (61.36)	
Tinidizdazole	25.56 ± 5.24 (56.95)	28.41 ± 1.98 (52.15)	27.44 ± 1.45 (53.78)	59.38 ± 4.78
Ornidizole	23.60 ± 3.55 (60.25)	30.04 ± 1.44 (49.41)	24.14 ± 1.81 (59.34)	
Mebendazole	22.44 ± 1.02 (62.20)	22.51 ± 2.63 (62.09)	23.79 ± 0.92 (59.93)	
Satranidizole	24.58 ± 2.52 (58.60)	24.05 ± 2.86 (59.49)	27.05 ± 1.61 (54.44)	

Each value is mean±SD of three replications; values within parenthesis indicate % reduction in infection.

be effective against different microsporidians. With observations of earlier studies, it is confirmed that benzimidazole derivatives are effective way to control microsporidiosis. Though the treatment controlled larval / moth mortality but the infection was again observed at moth stage. The present study is with agreement (Joythi et al., 2005; Brook et al., 1978) where they stated that chemical treatment arrested the further multiplication of microsporidians during feeding stage although they could not eliminate the disease completely and the development of some spore stages again started during non- feeding stage where feeding of chemicals/ drugs is impossible. Hence, for the complete elimination of microsporidiosis is only possible by feeding mulberry drug fortified leaves to silkworms continuously rather alternatively.

REFERENCES

Baig M (1994). Studies on *Nosema bombycis* N - A pathogen of silkworm *Bombyx mori* L. Ph.D Thesis, University of Mysore, Mysore.

Chandra AK, Sahai RK, Bhattacharya J, Krishna N, Sen SK, Sarachandra B (1995). Efficacy of carbendazimin as an antimicrosporidial agents and its influence on the growth and cocoon characters of silkworm, *Bombyx mori* L.. Insect Sci. Appl., 16: 233-235.

Griyaghey UP, Jolly MS, Kumar P (1976). Studies on the thermic control of microsporodiosis of the tropical Tasar silkworm, *Antherea mylitta* D. Ind. J. Seric., 14: 27-30.

Hayasaka S (1991). Inhibitory effects of antimicrobial chemicals on the sporogenesis of *Nosema bombycis* infecting larvae of silkworm, *Bombyx mori* L. Acta Series Entomol., 4: 53-54.

Jyothi NB, Patil CS, Dass CMS (2005). Action of C arbedazmin on the development of *Nosema bombycis* Naegeli in silkworm *Bombyx mori* L, J. Appl. Entmol., 129: 205-210.

Sahay A, Singh GP, Roy DK, Sahay DN, Suryanarayana N (2005). Effect of chemotherapy on Pebrine in Tasar silkworm, *Antherea mylitta* D. The 20th congress of the International Sericulture Commission 15th-18th December, 2005 Banagalore, India. Non-Mulberry silkworm, 2(3): 126-130.

20

Nematicidal effect of *Acacia nilotica* and *Gymnema sylvestris* against second stage juveniles of *Meloidogyne incognita*

Nighat Sultana[1]*, Musarrat Akhter[2], Muhammad Saleem[3] and Yousaf Ali[1]

[1]Pharmaceutical Research Center, Pakistan Council of Scientific and Industrial Research (PCSIR) Laboratories Complex, Karachi-75280, Pakistan.
[2]Food and Marine Resources Research Centre, Pakistan Council of Scientific and Industrial Research (PCSIR) Laboratories Complex, Karachi-75280, Pakistan.
[3]Department of Chemistry, the Islamia University of Bahawalpur, Pakistan.

The nematicidal effect of different extracts of *Acacia nilotica* leaves/seeds and *Gymnema sylvestre* leaves were tested against *Meloidogyne incognita* larvae at the concentrations of 1, 0.5, 0.25 and 0.125%, up to three days. All extracts showed nematicidal mortality. Nematicidal mortality was 100% with the use of 1% concentration of leaves ethyl alcohol extract of *G. sylvestre* and ethyl acetate leaves extracts of *A. nilotica* after 2 days. Qualitative analysis of the phytochemicals of alcohol extracts revealed the presence of carbohydrates, saponins, triterpene saponins belonging to oleanane and dammarene, phytosterols, phenols, flavonoids and tannins in all the plants. Quantitative analysis showed that, the crude saponin was the major phytochemical constituent present in highest percentage followed by crude oleanane and dammarene triterpene acids in both of two plants. It is suggested that both of these two plants possess nematicidal properties that could be developed and used as natural nematicides for nematode control.

Key words: *Acacia nilotica*, nematicidal activity, ellagic acid, oleanolic acid 3-glycosides, quercetin 3-glycosides, 3,19-dihydroxy-12-ursen-28-oic acid.

INTRODUCTION

Plant parasitic nematodes constitute one of the most important pest groups of the economic crops, especially in developed and developing countries of the world. The use of the plants and plant products is one of the promising methods for nematode controls. They are cheap, easy to apply, produce no pollution hazards and have the capacity to structurally and nutritionally improve the soil health. In view of these facts, investigations have been undertaken by various groups of scientists (Decker et al., 1981; Gommers et al., 1981; Qamar et al., 1995; Nogueira et al., 1996) which shows an effective control of root-knot nematodes. In the present article, studies on the nematicidal activity of different extracts, fractions isolated from the air dried aerial parts (leaves and seeds)

of *Acacia nilotica* and *Gymnema sylvestre* (Imoto et.al., 1991) are described. *A. nilotica* belongs to Leguminosae and sub-family, Mimosoideae (Brenan JPM et.al., 1983). It is an economically and medicinally important plant. Its different parts are used for different purposes. Inner bark contains (18 to 23%) tannin, used for tanning and dyeing leather black. Young parts produce a very pale tint in leather, notable goat hides. Pods were used by the ancient Egyptians. Young bark used as fiber, twigs esteemed for tooth brushes. Trees tapped for gum arabic. Because of its resins, it resist insects and water and trees are harvested for timber for boat making, posts, buildings, water pipes, well planking, plows, cabinet work, wheels, mallets and other implements wood yield excellent firewood and charcoal (Duck, 1981).

Medicinally, *A. nilotica* was used in large number of diseases in different parts of world. For example, Zulu take bark for cough, chipi use root for tuberculosis, Masai are intoxicated by the bark and root decoction, said to

*Corresponding author. E-mail: nighat2001us@hotmail.com.

impart coverage even ephrodisia and root is said to cure impotence. Astringent bark used for diarrhea, dysentery and leprosy. According to hear well the gum or bark is used for cancers and tumors of ear, eye and testicles and in duration liver and spleen. It is also said to be used for cancer, colds, congestion cough, diarrhea, dysentery, fever, gallbladder, hemorrhage, leucorrhoea, ophthalmia, sclerosis, small pox and tuberculosis. Bark, gum leaves and pods are used medicinally in West Africa. Different parts are strongly astringent due to diarrhea. Other preparations are used for cough, gargle, toothache, ophthalmia and syphilitic ulcers. In Tonga, root is used to treat tuberculosis. In Lebanon, the resin is mixed with orange flower infusion for typhoid convalescence. Masai use the bark decoction as a nerve stimulant. In Italian Africa, the wood is used to treat small pox. Egyptian Nubians believe that diabetics may eat unlimited as long as they also consume powdered pods (Duke, 1983). Extracts are inhibitory to at least four species of pathogenic fungi (Umodkar et al., 1977). The aqueous extract of fruit rich in tannin (18 to 23%) has shown algicidal activity against chroccoccus, closteruim, coelastrum, cosmarium, cyclotella, euglena, microcystis, Pediastrum, rivularia, spirogyra and spirolina (Ayoub, 1983).

No phytochemical investigation has so far been carried out on this specie. Keeping in view the pharmacological significance of the plant, phytochemical studies were undertaken on the constituents of the aerial parts of the plant in this laboratory two years earlier, which resulted in the isolation and characterization of various sugars, include Androst-5-ene-3,17-diol (Chaubal et al., 2003), 2-O-L-Arabinofuranosyl-L-arabinose (Chalk, 1968), 3-O-L-Arabinopyranosyl-L-arabinose (Verkerk et al., 1998), dehydrodigallic acid (Ishimatsu et al., 1989), 3,5-Dihydroxy-4',7-dimethoxyflavone (da Silva et al., 2000), 3-(3,4-dihydroxyphenyl)-2-propenal (Demin et al., 2004), 3,19-dihydroxy-12-ursen-28-oic acid (Zheng et al., 2004), ellagic acid (Press et al., 1969), 24,25-epoxytirucall-7-ene-3,23-diol, (Liu et al., 2001) 3,3',4',5,5',7-hexahydroxyflavan (Miketova, 1998), N-(4-(4-Hydroxybenzoylamino) butyl)-1,3-dimethyllumazine-6-carboxamide (Voerman et al., 2005), 2-hydroxy-9,12,15-octadecatrienoic acid (Bohannon et al., 1975), nilocitin (Lee et al., 1989), 2',3,3',4',5,6,7,8-Octahydroxyflavone (Chauhan et al., 2000), oleanolic acid bisdesmosides (Domon, 1984), oleanolic acid 3-glycosides (Nihei et al., 2005), 3,4,8,9,10-Pentahydroxy-6H-dibenzo(b,d)pyran-6-one (Nawwar et al., 1984), 3,3',4,4',9-Pentahydroxy-7,9'-epoxylignan (Dobner et al., 2003), 3,3',4',5,7-Pentahydroxyflavan (Gao et al., 2004), 1,3,7,11,12-Pentahydroxy-14-meliacen-28-oic acid (Torto et al., 1995), Quercetin 3-glycosides (Aqil et al., 1999), 3,4',5,7-Tetrahydroxy-8-prenylflavone (Shin et al., 2002) and Tirucall-7-ene-3,23,24,25-tetrol (Vieira et al., 1998).

Acacia species is a rich source of gallic and ellagic acid (Nighat et al., 2010). *G. sylvestre* leaves contain triterpene saponins belonging to oleanane and dammarene classes. Oleanane saponins are gymnemic acids (Kennady et.al., 1989) and gymnemasaponins while dammarene saponins are gymnemasides. Besides this, other plant constituents are flavones, anthraquinones, hentri-acontane, pentatriacontane, α and β- chlorophylls, phytin, resins, d-quercitol, tartaric acid, formic acid, butyric acid, lupeol, β-amyrin related glycosides and stigmasterol. Leaves of this species yield acidic glycosides, anthro-quinones and their derivatives (Mukherjee et al., 1995; Chakravarthi et al., 1981; Glase et al., 1984; Gupta, 1961; Imoto et al., 1991; Kennedy 1989; Shanmugasundaram et al., 1983; Stocklin et al. 1969b; Yoshikawa et al., 1989a; Yoshikawa et al., 1989b; Yoshikawa et al., 1992a; Mukherjee et al., 1995; Anil e al., 1994).

MATERIALS AND METHODS

General experimental procedures

Visual examination is not completely reliable for determining mortality; therefore, final viability determinations on tomato plants were carried out according to Feldmesser et al. (1983). The inoculated tomato seedlings were examined after three weeks to determine the viability of the nematode inocula expressed as roo infection. Five replicates of each treatment were carried out and the results are reported as an average of the five replicates. Infections were evaluated on an arbitrary basis (the root-knot index) by assigning values of 0 = no infection, 1 = 1 to 25% of the roots galled, 2.0 = 26 to 50% galled, 3.0 = 51 to 75% galled and 4.0 = 76 to 100% root infection.

Plant material

The aerial parts of *A. nilotica* and *G. sylvestris*, (15 kg) were collected from Karachi, in June 2010.

Extraction and isolation

Air-dried aerial parts of *A. nilotica* (15 kg dry weight) and *G sylvestris* were dried and extracted with EtOH (100 L). The EtOH extract was concentrated to a gum (813 g), dissolved in distilled water and extracted thoroughly with petroleum ether (40 L). The hexane soluble portion was dried (74.96 g). The remaining aqueous layer was acidified with acetic acid to pH 3 and then, extracted with CHCl₃. The CHCl₃ soluble portion was dried (83.84 g). The remaining aqueous layer was basified with NH₄OH to pH 12 and extracted with CHCl₃ (35 L). The CHCl₃ soluble portion was dried (79.84 g). The acidic chloroform soluble portion was dried as a crude mixture. Extraction of aqueous extract with ethyl acetate (5 L) yielded an impure mixture to afford E.AcS extract.

Nematicidal activity

Experiments were performed under laboratory conditions at 28 + 2°C. Fresh egg masses collected from stock culture maintained on tomato root tissue were kept in water for egg hatching. The larvae emerged after 48 h from egg masses incubated at 30°C and were used as test species for larval mortality studies. The movements of nematodes were checked by touching them with needle.

For the nematicidal activity of leave extract, stock solutions (30 mg/ml) of different fractions of *A. nilotica* in ethyl alcohol (AL-AS.), ethyl acetate (AL-E.AcS.), methanol (AL-MS.), chloroform (AL-CS.) and n-haxane (AL-HS) were prepared. To determine nematicidal effect of various fractions, 100 freshly hatched second stage juveniles were taken in 5 ml tap water.

Measured amounts of stock solution were added to make dilution of 1, 0.5, 0.25 and 0.125%. Standard nematicide *Azadirachta indica* (0.05%) was taken for comparison and tap water taken as control. After 24 and 48 h exposure with various *A. nilotica* fractions, the larvae were counted for mortality and non-mortality under stereoscopic microscope. The deaths of nematodes were confirmed by keeping them in tap water for 24 h. The percent mortality was worked out from an average of three replicate. The result of percent mortality in different fraction of *A. nilotica* and *G. sylvestris* after 24 and 48 h of leaf extract were given in Tables 1 to 4.

RESULTS AND DISCUSSION

The nematicidal activity of different fractions of leaves extract of *A. nilotica* and *G. sylvestre* (AL-E.Al., AL-E.Ac., AL-Meth., AL-Chl., AL-Hex.) and different fractions of seeds extract (AS-AS., AS-E.Ac., AS-MS., AS-CS., AS-HS) of *A. nilotica* after 24 and 48 hours were tested against a root-knot nematode (*Meloidogyne incognita*).

The total alcohol soluble extract, MeOH soluble, Et-acetate soluble, chloroform soluble and pet. ether soluble extracts of the aerial parts of *A. nilotica* (AL-AS, AL.CS, AL.E-AcS, and AL-PS) (AL-AS, AL.CS, AL.E-AcS and AL-PS) and *G. sylvestre* (GL-AS, GL.CS, GL.E-AcS and GL-PS) were screened for its nematicidal activity against nematodes freshly hatched second stage juveniles of *M. incognita* (root-knot nematode). Negative results were obtained for the nematicidal activity of pet. ether extract of *A. nilotica* leaves. This is the first reported on the nematicidal activity of these compounds and any part of *A. nilotica* and *G. sylvestre.* Thus, both of these plants extract might be beneficial as a potent nematode inhibitor under specified conditions.

The crude ethyl acetate leaves extract of *A. nilotica* (AL- E.AcS) showed 70% mortality at 1.0% concentration after 24 h, while 90% mortality at 1.0% concentration after 48 h, whereas, the pet. ether soluble fraction (AL-PS) showed 65% mortality and the chloroform soluble fraction AL-CS showed 45% mortality at the same concentration after 48 h. Conventional nematicide *Azadirachta indica* showed 88% mortality after 48 h. The alcoholic soluble (AL-AS), pet.ether soluble (AL-PS) and chloroform soluble (AL-CS) fractions showed 50, 40 and 38% mortality, respectively after 24 h of *M. incognita* larvae.

From the crude leaf extract, fraction AL-AS shows maximum mortality (70%), while fraction AL-CS, shows minimum mortality (38%) at 1% concentration after 24 h. Similarly, fraction AL-E.AcS shows maximum mortality (90%), while fraction (AL-CS.) shows minimum mortality (45%) at 1% concentration after 48 h. Other fractions that is, AL-.AS, AL-MS, AL-HS, have percent mortality in between them at the same concentration. The percent mortality of all fractions gradually decreases with decreasing concentration of fraction from 1 to 0.0125%

after 24 and 48 h. Similarly, the nematicidal activity of the crude alcoholic leaves extract, as well as crude alcoholic leaves extract fractions (GL -AS, GL.CS, GL.E-AcS and GL-PS) of *G. sylvestre* were tested against a root-knot nematode (*M. incognita*).

The crude ethyl acetate extract of *G. sylvestre* (GL-AS) showed 90% mortality at 1.0% concentration after 24 h, while 95% mortality at 1.0% concentration after 48 h, whereas, the pet. ether soluble fraction (GL-PS) showed 10% mortality and the chloroform soluble fraction (GL-CS) showed 18% mortality at the same concentration after 48 h. Conventional nematicide *A. Indica* showed 88% mortality after 48 h. The alcoholic soluble (GL-AS), pet.ether soluble (GL-PS) and chloroform soluble (GL-CS) fractions showed 90, 2.0 and 15% mortality, respectively after 24 h of *M. incognita* larvae. Negative results were obtained for the nematicidal activity of pet. ether extract of *G. sylvestre* leaves. The direct antinemic action shown by GL –AS and its fractions GL -CS, GL -PS in the *in vitro* investigation against second-stage juveniles of *M. incognita* is presented in Tables 1 to 2.

The nematicidal activity of alcohol extract of *G. sylvestris leaves* showed 99, 91, 75 and 56% of death with the use of 1, 0.5, 0.25 and, 0.125% concentrations, respectively, after 3 days. The third day, 0.5% concentration killed more than 91% of the larvae. However, 100% mortality was observed only in 1% concentration alone (Table 1). Only the highest concentration (1%) of all two plants extract showed 95% mortality (Table 3). Among the two plants, the leaves alcohol extract of *G. sylvestris* was found more lethal than other extracts.

The crude methanol extract of *A. nilotica* seeds (AS-MS) showed 85% mortality at 1.0% concentration after 24 h, while 92% mortality at 1.0% concentration after 48 h, whereas, the pet. ether soluble fraction (AS-PS) showed 50% mortality and the chloroform soluble fraction (AS-CS) showed 70% mortality at the same concentration after 48 h. Conventional nematicide *A. Indica* showed 88% mortality after 48 h. The alcoholic soluble (AS-AS), pet.ether soluble (AS-PS) and chloroform soluble (AS-CS) fractions showed 80, 30 and 65% mortality, respectively, after 24 h of *M. incognita* larvae. The percent mortality was worked out from an average of three replicate. The result of percent mortality in different fraction of *A. nilotica* after 24 and 48 h of leaf extract were given in Tables 3 and 4.

The direct antinemic action shown by AS-AS and its fractions AS-CS, AS-PS in the *in vitro* investigation against second-stage juveniles of *M. incognita* is presented in Tables 1 and 2. It has also been observed that, different fractions of leaf extract at different fraction concentration show same percent mortality after 24 and 48 h. AL-AS at 0.5% concentration and AL-E.AcS at 0.125% concentration, AL-CS and AL-HS at 0.25 and 0.125% concentration, respectively, AL-HS and AL-MS at 1 and 0.125% concentration, respectively. AL-AS at 1% concentration and AL-MS at 0.5% concentration, have

Table 1. Nematicidal activity of different fractions of leaves extract isolated from *A. nilotica* and *G. sylvestris* on the larval mortality of *M. incognita* (Root knot nematode). Percent mortality/concentration after 24 h.

Concentration↓ (%)	Fractions	Ethanol		E. Acetate		Methanol		Chloroform		Hexane	
		AL (%)	GL	AL (%)	GL	AL (%)	GL	AL (%)	GL	AL (%)	GL
1		50	90	70	80	68	12	38	15	40	2
0.5		42	88	66	60	50	5	28	5	28	-
0.25		22	70	58	50	46	2	10	4	17	-
0.125		16	50	42	44	40	-	05	-	10	-

Control = 2%. AL-AS= *A. nilotica* leaves alcohol soluble; AL.CS= *A. nilotica* leaves chloroform soluble; AL.E-AcS=*A. nilotica* leaves ethyl acetate soluble; AL-PS= Acacia leaves pet.ether soluble; GL-AS= *G. sylvestris* leaves alcohol soluble; GL.CS= *G. sylvestris* leaves chloroform soluble; GL.E-AcS= *G. sylvestris* leaves ethyl acetate soluble; GL-PS= *G. sylvestris* leaves pet.ether soluble; GL= *G. sylvestris* leaves; AL = *A. nilotica* leaves.

Table 2. Nematicidal activity of different fractions of leaves extract isolated from *A. nilotica* and *G. sylvestris* on the larval mortality of *M. incognita* (Root knot nematode). Percent mortality/concentration after 48 h.

Concentration ↓ (%)	Fractions	Ethanol		E. ac.		Methanol		Chloroform		Hexane	
		AL (%)	GL (%)	AL (%)	GL (%)	AL (%)	GL	AL (%)	GL	AL (%)	GL
1		62	99	90	95	85	15	45	18	65	10
0.5		46	91	78	70	62	7	30	7	30	8
0.25		25	75	70	62	50	2	15	5	20	4
0.125		18	56	50	53	42	2	12	5	15	-

Control = 3%. AL-AS = *A. nilotica* leaves alcohol soluble; AL.CS = *A. nilotica* leaves chloroform soluble; AL.E-AcS = *A. nilotica* leaves ethyl acetate soluble; AL-PS = *A. nilotica* leaves pet.ether soluble; GL-AS= *G. sylvestris* leaves alcohol soluble; GL.CS= *G. sylvestris* leaves chloroform soluble; GL.E-AcS= *G. sylvestris* leaves ethyl acetate soluble; GL-PS= *G. sylvestris* leaves pet.ether soluble; GL= *G. sylvestris* leaves; AL = *A. nilotica* leaves.

Table 3. Nematicidal activity of different fractions of seed extract isolated from *A. nilotica* on the larval mortality of *M. incognita* (Root knot nematode). Percent mortality/concentration after 24 h.

Concentration ↓(%)	Fractions →	AS-AS. (%)	AS-E.AcS. (%)	AS-MS. (%)	AS-CS. (%)	AS-HS. (%)
1		80	60	85	65	30
0.5		60	50	70	60	26
0.25		58	48	58	56	20
0.125		38	40	38	42	12

Control = 2%. AS-AS = *A. nilotica* seeds alcohol soluble; AS.CS = *A. nilotica* seeds chloroform soluble; AS.E-AcS =*A. nilotica* seeds ethyl acetate soluble; AS-PS = *A. nilotica* seeds pet.ether soluble.

Table 4. Nematicidal activity of different fractions of seed extract isolated from *A. nilotica* on the larval mortality of *M. incognita* (Root knot nematode). Percent mortality/concentration after 48 h.

Concentration ↓ (%)	Fractions →	AS-AS. (%)	AS-E.Acs. (%)	AS-MS. (%)	AS-CS (%)	AS-PS. (%)
1		90	70	92	70	50
0.5		75	57	80	67	-
0.25		48	50	63	60	25
0.125		40	40.2	40	45	15

Control = 5%. AS-AS = *A. nilotica* seeds alcohol soluble; AS.CS = *A. nilotica* seeds chloroform soluble; AS.E-AcS = *A. nilotica* seeds ethyl acetate soluble; AS-PS = *A. nilotica* seeds pet.ether soluble.

same percent mortality after 24 h. Similarly, AL-CS at 0.25% concentration and AL-HS at 0.125% concentration, AL-AS at 1% concentration and AL-MS at 0.5% concentration, have same percent mortality after 48 h.

Different fractions at same concentration also showed same percent mortality. AL-CS and AL-HS at 0.5% concentration showed same percent mortality after 24 and 48 h. Conventional nematicide *A. indica* showed 88%

mortality at the 0.5% concentration used in the present studies. It was noted that, at all the concentrations, all the tested fractions exhibited significant larval mortality are against the test nematode but the activity decreases with a decrease in concentration in all the cases (Tables 1 to 4).

REFERENCES

Anil K I, Nazaam PA, Joseph L, Vijay Kumar NK (1994). Response of "Gurmar" for in vitro propagation (Eng.Recd 1996,6 ref). 42(6): 365-368.

Antunes AS, Da Silva BP, Parente JP (2000). Flavonol glycosides from leaves of Costus spiralis. Fitoterapia, 71: 507-510.

Aqil M (1999). Euphorbianin, a new glycoside from Euphorbia hirta Linn. Global. J. Pure. Appl. Sci., 5: 371-373.

Ayoub SMH (1983). Algicidal properties of A. nilotica, Fitoterapia, 53(5-6), 175-8.

Bohannon MB, Kleiman R (1975). Lipids, Unsaturated C18 a-Hydroxy Acids in Salvia nilotica. pp. 10-703.

Brenan JPM (1983). Manual on the taxonomy of Acacia species, present taxonomy of four species of Acacia (A. albida, A. senegal, A. tortilis) FAO, Rome, Italy, 42P.

Chakravarthi D, Debnath NB (1981). Isolation of Gymnemagenin, the Sapogenin from Gymnema Sylvestre R.Br. (Asclepiadaceae). J. Institution. Chem. India., 53: 155-158.

Chalk RCJF, Stoddart WA, Szarek JK, Jones N (1968). Isolation of two arabinobioses from Acacia nilotica gum. Can. J. Chem., pp. 46-2311.

Chaubal R, Mujumdar AM, Puranik VG, Deshpande VH, Deshpande NR (2003). Isolation and X-ray study of an anti-inflammatory active androstene steroid from Acacia nilotica. Planta. Med., 69: 287-288.

Chauhan D, Singh J, Siddiqui IR (2000). Isolation of two flavonol glycosides from the seeds of Acacia nilotica.. Indian. J. Chem., 39: 719-722.

Da Silva Antunes A (2000). Fitoterapia, 71: 507-510.

Decker H (1981). Phytonematology, Amerind Publishing Company Pvt. Ltd., New Delhi.pp, 19 – 540.

Demin P, Rounova O, Grunberger T, Cimpean L, Sharfe N, Roifman CM (2004). Tyrenes: synthesis of new antiproliferative compounds with an extended conjugation. Bioorg. Med. Chem., 12: 3019-3026.

Dobner MJ, Ellmerer-Müller EP, Schwaiger S, Batsugh O, Narantuya S, Stütz M (2003). New lignan, benzofuran and sesquiterpene derivatives. Helv. Chim. Acta., 86: 733-738.

Domon B , Hostettmann K (1984). New Saponins from Phytolacca dodecandra I' Herit.. Helv. Chim. Acta, 67: 1310-1315.

Duke JA (1983a). Medicinal plants of the Bible, trado-Medical books, Owerri, NY. Gao, S, 2004. Planta. Med., 70: 1128-1134

Glaser D, Hellekant, Gwalior, Brouwer JN (1984). Van der wel. Happy, Effects of Gymnemic Acid and on sweet taste perception in primates, Chem. Sci., 8: 367- 374.

Gommers FJ, Nematol J (1981). Biochemical interactions between nematodes and plants and their relevanc,e to control. Helminth. Plant. Nematol. pp. 50: 9.

Gupta SS (1961). Inhibitory effect of Gymnema sylvestre (Gurmar) on adrenaline induced hyperglycemia in rats, Indian. J. Med. Sci., 15: 883-887.

Imoto T, Miyasaka A, Ishima R, Akasaka K (1991). A novel peptide isolated from the leaves of Gymnema Sylvestre I. Charactorization and its supressive effect on the neural responses to sweet taste stimuli in the rat. Comparative. Biochem. Physiol., 100: 309-314.

Ishimatsu M , Tanaka T, Nonaka G, Nishioka I, Nishizawa M, Yamagishi T (1989). Isolation and characterization of novel diastomeric ellagitannins. Chem. Pharm. Bull., pp. 37-1735.

Kennady LM (1989). Gymnemic Acids, specificity and comperitive inhibitation. Chem. Senses, 14: 853-858.

Le Houerou HN (1980). Chemical composition and nutritional value of browse in tropical West Africa. In H.N. Lettouerou (ed), Browse in Africa, the current state of knowledge, ILCA, Ethiopia, pp. 26-289.

Lee SHT , Tanaka G, Nonaka, Nishioka I (1989). Sediheptulose digallate from Cornus officinallis,. Phytochemistry, pp. 28- 3469.

Liu H, Heilmann J, Rali T, Sticher O (2001). New tirucallane-type triterpenes from Dysoxylum variabile.. J. Nat. Prod., 64: 159-163.

Miketova P, Schram KH, Whitney JL, Kerns EH, Valcic S, Timmermann BN, Volk KJ (1998). Mass spectrometry of selected components of biological interest in green tea extracts. J. Nat. Prod., 61: 461-467.

Mukherjee PK, Rajesh Kumar M, Saha K, Giri SN, Pal M, Saha BP (1995). Preparation and evaluation of Tincture of Gymnema Sylvestre (Family- Asclepiadaceae) by Physico- Chemical, TLC and Spectroscopic characteristics. J. Scientific. Industrial. Res., 55(3): 178-181.

Nawwar MAM, Hussein SAM, Merfort I (1984). NMR Spectral Analysis of Polyphenolics from Punica granatum,. Phytochem., pp. 23-2966, 36-793.

Nighat Sultana, Musarrat Akhter, Zakia Khatoon (2010). Nat. Prod. Res., 24(5): 407-415.

Nihei KI (2005). J. Agric. Food. Chem., 53: 608-613.

Nogueira MA, De Oliveira JS, Ferraz S (1996). Phytochemistry, pp. 42-997.

Press RE , Hardcastle D (1969). J. Appl. Chem., pp. 19-247.

Qamar F, Kapadia Z, Khan SA, Badar Y (1995). Nematicidal natural products from the aerial parts of Lantana camara. Pak. J. Sci. Ind. Res., pp. 38-319.

Shanmugasundaram KR, Panneerselvam C, Samudram P, Shanmugasundaram ERB (1983). Enzyme changes and glucose utilisation in diabestic rabbits; the effect of Gymnema Sylvestre, J. Agric. Food. Chem. 17: 704-708.

Shin HJ, Kim HJ, Kwak JH, Chun HO, Kim JH, Park H, Kim DH, Lee YS (2002). A prenylated flavonol, sophoflavescenol: a potent and selective inhibitor of cGMP phosphodiesterase,. Bioorg. Med. Chem. Lett., 12: 2313-231.

Stocklin W (1969b). Chemistry and Physiological properties of Gymnemic acid, the anti-saccharine principle of the leaves of Gymnema Sylvestre, J. Ethnopharmacol. 7: 205-234.

Torto B (1995). A limonoid from Turraea floribunda. Phytochemistry., pp. 40-239, 42-1235.

Umodkar CV, Begum S, Nehemiah KMA (1977). Inhibitory effect of Acacia nilotica extract in pectolytic enzyme production by some pathogenic fungi, Indian. Phytopath. Pull., 29(4): 469-70.

Verkerk R, Dekker M, Jongen WMF (1998). Natural Toxicants in Food,. Natural Toxicants in Food, Watson R, Sheffield Academic Press, pp. 29-53.

Vieira IJC, Rodrigues-Filho E, Vieira PC, Silva MFGF, Fernandes JP (1998). Quassinoids and protolimonoids from Simaba cedron. Fitoterapia, 69: 88-90.

Voerman G, Cavalli S, van der Marel GA, Pfleiderer W, van Boom JH, Filippov DV (2005). 1,3-Dimethyllumazine derivatives from Limnatis nilotica. J. Nat. Prod., 68: 938-941.

Yoshikawa K, Amimoto K, Arihara School, Matsuura K (1989a). Structure Studies of new anti-sweet constituents from Gymnema Sylvestre. Trtrahedron Lett., 30: 1103- 1106.

Yoshikawa K, Amimoto K, Arihara S, Matsuura K (1989b). Gymnemic Acid V, 6 and 7 from Gurmar, the leaves of Gymnema Sylvestre R.Br. Chem. Pharmaceutical Bullitin, 37: 852-854.

Yoshikawa K, Arihara S, Matsuura K, Miyase,T (1992a). Demmarane Saponins from Gymnema Sylvestre, Phytochem.,, 31: 237-241.

Zheng Q, Koike K, Han LK, Okuda H, Nikaido T (2004). New biologically active triterpenoid saponins from Scabiosa tschiliensis.. J. Nat. Prod., 67: 604-613.

21

Botanicals for the management of insect pests in organic vegetable production

Mochiah M. B.[1]*, Banful B.[2], Fening K. N.[1], Amoabeng B. W.[1], Offei Bonsu K.[1], Ekyem S.[1], Braimah H.[1] and Owusu-Akyaw M.[1]

[1]Entomology Section, CSIR-Crops Research Institute, P. O. Box 3785, Kumasi, Ghana.
[2]Department of Horticulture, College of Agriculture and Natural Resources, Kwame Nkrumah University of Science and Technology, Kumasi, Ghana.

This study was conducted to evaluate the efficacies of some botanical products on insect populations associated with two vegetables; Eggplant and Okra. Two field experiments were conducted on-station at the CSIR-Crops Research Institute, Kwadaso, Kumasi, and on-farm at Eatwell farm at Agona-Mampong all in the Ashanti region of Ghana from December 2009 to March 2010 and September to December 2010, respectively. Seven botanical treatments were applied viz; Ecogold (10 ml/l of water); Alata soap (5 g/l of water); Garlic (30 g/litre of water); Neem oil (3 ml/l of water); Papaya leaves (92 g/l of water); Wood ash (10 g/plant stand) and control (no botanical). The experimental set up was a Randomized Complete Block design (RCBD) with three replications. Parameters studied included insect pest numbers and their natural enemies, number of days to 50% flowering, plants height at flowering (cm), number of fruits per plant, fruit damage, and mean weight of fruits (g). Major insect pest recorded on the two vegetables included aphids (*Aphis gossypii*), flea beetles (*Podagrica* spp), white flies (*Bemisia tabaci*), fruit borers (*Earias* sp), cotton strainers (*Dysdercus superstitiosus* (F.)), variegated grasshoppers (*Zonocerus variegatus* L.), *Urentius hysterricellus* (Richter) and shoot and fruit borers (*Leucinodes orbonalis* Gn). The natural enemies of pests of the two vegetables identified were the ladybird beetles, (*Cheilomenes* sp) and predatory spiders (*Araneae*). There were significant percentage reductions in pests for all the botanicals applied (P< 0.05) on both the eggplant and okra plants compared to the control. Generally, plants to which the botanicals were applied produced the highest mean weight of fruits, translating into mean percentage increases in fruit weight ranging between 21 and 59% on both the eggplant and okra plants compared to the control in both growing periods. It is concluded that botanicals such as Ecogold, Alata soap, exotic garlic, neem oil, papaya leaves and wood ash could be effectively considered as pest management options to reduce insect pest populations and increase eggplant and okra productivity.

Key words: Organic vegetables, botanicals, insect pests, beneficial insects.

INTRODUCTION

Okra (*Abelmoschus esculentus* (L.) Moench) and eggplant (*Solanum intergrifolium* L.) are vegetables, which are consumed nearly on a daily basis in many households in Sub-Saharan Africa and therefore are considered important components of the diet (El-Shafie, 2001; Timbilla and Nyarko, 2004; Bennette-Lartey and Oteng-Yeboah, 2008; FAO, 2008; Obopile et al., 2008) even though they are exotic to Africa (FAO, 2000; 2003).

The fruit of *Solanum intergrifolium* L. (Solanaceae) is a good source of vitamins A and C, potassium, phosphorus, calcium and, dietary fibre (USDA Nutrient database, 2008) and is known to possess medicinal properties as well. The fruit of the eggplant can be eaten raw or served as a baked, grilled, fried or boiled vegetable and can be used in stews or as a garnish (FAO, 2003).

In Ghana, the fruits of eggplant and okra are produced and marketed primarily by small-scale farmers who are distributed throughout the country. A broad range of market participants are involved in trading these vegetables.

*Corresponding author. E-mail: mochiah63@yahoo.com.

They are particularly a source of cash for rural households in the southern and central parts of Ghana (Obeng-Ofori et al., 2002). In spite of its usage throughout the year, the country can meet the domestic demand only during the rainy season. In the dry season, the lack of irrigation facilities together with the higher incidence of pests relative to the rainy season drastically reduces total production of these vegetables. These vegetables are usually attacked by insect pests, mostly caterpillars that cause extensive damage to parts of the plants affecting their yield and marketability (Zehnder et al., 1997; Sibanda et al., 2000; FAO, 2000, 2003). For okra, the most significant pest damage is caused by the flea beetle (*Podagrica* spp) while for eggplant white flies (*Bemisia tabaci* [Genn.]), transmitter of viral diseases, and thrips (*Thrips tabaci* Lind.) represent the most economic important pests (Owusu-Ansah et al., 2001). Over time, chemical control has been practiced by farmers for higher gains (Gerken et al., 2001), but these pests can become resistant to chemical insecticides very quickly. Moreover, the misuse of chemical insecticides in terms of quantity applied or in dangerous combinations (Obeng-Ofori et al., 2002) have created a myriad of problems which include pest resistance, resurgence of pests, pesticide residues, destruction of beneficial fauna and environmental pollution (Sibanda et al., 2000; AVRDC, 2003b). Under such debilitating circumstances, interest in organic farming has been growing and therefore exploring alternative options to control pests of okra and eggplant is a fundamental means of supporting the smallholder farmer to diversify into organic production and be able to tap into the high profits associated with organic products. Furthermore, the export market continues to impose tight restrictions (including zero tolerance) on many widely used pesticides. Large-scale growers, who produce for the export market, are looking for alternatives to synthetic pesticides. One viable alternative is the use of botanical insecticides which is also considered eco-friendly. Hence this study aims at evaluating different botanicals and their efficacies in the management of insect pests in organic vegetable production.

MATERIALS AND METHODS

Study sites

Studies were conducted in the dry season and minor rainy season. In the dry season the study was conducted on-station at the Crops Research Institute, Kwadaso, Kumasi from December 2009 to March 2010 while in the minor rainy season, the study was conducted on-farm at Eatwell Farms located at Agona-Mampong from September to December 2010. Both sites have ferric acrisols as the dominant soil type.

Land preparation and transplanting

The land was cleared and root stumps removed after weeding prior to sowing of seeds. Ploughing and harrowing were performed on the land before beds were made. The vegetables sown were Okra (*A. esculentus* (L.) Moench) and egg plant (*S. intergrifolium* L.).

Eggplant seedlings from the nursery were transplanted to the main experimental plot on 11th January, 2010 and 2nd September, 2010 in the dry and minor rainy seasons, respectively, while the okra seed were directly sown on 12th January, 2010 and 2nd September 2010 in the dry and minor rainy seasons, respectively. Uniform seedlings of height 15 cm with 3 to 5 leaves were transplanted. The planting distances were 70 cm × 30 cm for okra on plots that measured 4.2 m × 4.0 m whereas for eggplant, the plant spacing was 80 cm × 80 cm on plots that measured 4.8 m × 4.0 m. Watering was done twice a day (morning and evening) especially in the dry season.

Experiment design

There were two experiments, one each for okra and eggplant. In each experiment, the design was a Randomized Complete Block (RCBD) consisting of seven treatments replicated three times. The treatments were Ecogold (10 ml/l of water); Alata soap (5 g/l of water (Alata soap at 0.5%W/V); exotic garlic (30 g/l of water); neem oil (3 ml/l of water); papaya leaves (92 g/l of water); wood ash (10 g/plant stand) and control (no botanical but sprayed with only water). Both Eco-gold and neem oil are commercial products manufactured by PAKS Agro Division.

Preparation and application of garlic bulb extract, papaya leaves extract and wood ash

Garlic bulb extract

The outer layers of the matured garlic were peeled off. 200 g of garlic were mixed with 1 L of water and ground with a blender to obtain garlic juice. This juice was thoroughly mixed with additional 1 L of water. The mixture was then sieved to obtain a uniform extract.

Papaya leaves extract

92 g of papaya leaves were collected and ground using local mortar and pestle. 1 L of water was then added and left to stay for 20 to 24 h. 1 L of water was later added to the mixture/extract which was sieved to obtain a uniform extract. 10 ml of fish oil and liquid soap were added to the garlic and papaya leaves extracts to improve their delivery and to allow them to stick unto the surface of the leaves of the plants.

Wood ash

Wood ashes were collected and sieved using a fine wire mesh such that sieved materials were just like powder which was sprinkled uniformly and moderately on both the lower and upper surfaces of the vegetable leaves.

Spraying of botanical extracts as well as the wood ash commenced 14 days after transplanting and 21 days after planting of the eggplants and okra, respectively, using a knapsack sprayer. 2 L of the extracts to 3 L of water was put in the knapsack to form the spraying mixture before application. Both garlic and papaya extracts were applied bi-weekly but application was repeated within the week whenever there were heavy rains a day or two after application in the minor growing season. The bi-weekly spraying schedule continued until about 14 days to harvesting. Since some of these botanicals do have some systemic effect in plants, spraying was done as other contact insecticides, ensuring thorough spray coverage and targeting the undersides of the leaves where pests tend to cluster and hide.

Application of nutrients

In each season (dry and minor rainy season), poultry manure was applied two times each at a rate of 50 g per plant in a ring. The first application was done ten days after transplanting and the second application was 25 days later.

Data collected

Data collection started 21 days after transplanting. Data on numbers of insects pests, natural enemies, days to 50% flowering, plant height at flowering (cm), number of fruits per plant, fruit damage and yield (mean fruit weight, g) were recorded from the two central rows of each plot which had 24 and 14 plants for okra and eggplant, respectively. The assessment of the numbers of various insect pest species and natural enemies were done by carefully examining each vegetable plant; leaf by leaf to collect any insects from the under-surface of the leaves. The insect pests collected from each plot were identified and counted. The insects' population data were collected every 7 days until harvest, between 06:30 and 09:00 h. Mean % increase in fruit weight was calculated using the formula thus:

$$\text{Mean \% increase in fruit weight} = \frac{(X_2 - X_1)}{(X_2)} \times 100$$

Where, X_2 = protected yield and X_1 = unprotected yield

Statistical analysis

Data were analysed by analysis of variance (ANOVA), using the general linear model (GLM) procedure of SAS Version 9 (SAS, 2005). Number of insects were log (x+1) transformed. Treatment means separation was carried out with the Student Newman Keul's (SNK) test and the probability of treatment means being significantly different was set at $P < 0.05$.

RESULTS

Insect pests of okra and eggplant

Insects from five major orders (Homoptera, Lepidoptera, Heteroptera, Orthoptera and Coleoptera) were found associated with the okra and eggplant.

Okra (*A. esculentus* (L.) Moench): Insect pests recorded on the okra included aphids (*A. gossypii*), Flea beetles (*P. uniformis*) (Figure 1), White flies (*B. tabaci*), cotton strainers (*D. superstitiosus* (F.) (Figure 2), variegated grasshoppers (*Zonocerus variegatus* L.) (Figure 3) and fruit borer (*Earias* spp) (Figure 4).

Eggplant (*S. intergrifolium* L.): Insect pests recorded on the eggplant included aphids (*A. gossypii* Glover) (Figure 5), whiteflies (*B. tabaci*), *Urentius hysterricellus* (Richter) (Figure 6), shoot and fruit borers (*Leucinodes*

orbonalis Guenée) (Figure 7), cotton strainers (*D. superstitiosus*) and variegated grasshoppers (*Z. variegatus* L.).

Effect of botanicals on insect pest populations

Generally, insect pest populations were lower and ranged between 12 and 16% in the minor rainy growing season of 2010 between September and December compared to the dry season of (December 2009 to March 2010 for the two crops (Tables 1, 2, 3 and 4). Botanical treated plots were observed to have reduced insect pest populations compared to the control plots. For Okra, the percentage reduction in pests as a result of botanical applications ranged between 49.2 and 62.3%, 68.1 and 76.8%, 52.9 and 66.8%, 54.9 and 76.9%, 42.8 and 63.3% as well as 52.9 and 63.3% for *A. gossypii, P. uniformis, D. superstitiosus B. tabaci, Z. variegatus* and *Erias* sp respectively. Similarly for eggplant, the percentage reduction in pests as a result of the same botanical applications ranged between 52.4 and 68.9%, 33.1 and 66.9%, 48.3 and 73.5%, 49.2 and 67.4%, 46.1 and 68.4% as well as 54.9 and 66.0 for *A. gossypii, U. hysterricellus, D. superstitiosus B. tabaci, Z. variegatus* and *L. orbonalis* sp respectively. Among the botanicals, there were significant differences in the number of *P. uniformis, D. superstitiosus* and *Z. variegatus*. Plots sprayed with botanicals such as EcoGold 999 Plus, neem oil and exotic garlic significantly reduced the populations of *P uniformis* (by 76.8, 75.8 and 72.7%), *D. superstitiosus* (66.8, 66.1 and 59.5%) and *Z. variegatus* (63.3, 61.3 and 55.9%) than plots that were sprayed with Alata soap (68.1, 57.5 and 48.6%), papaya leaf extracts (68.1, 56.8 and 50.3%) and wood ash (68.1, 52.8 and 42.8%) respectively, for the same pest populations (Tables 1 and 2).

Natural enemies of okra and eggplant insect pests

The natural enemies of pests of both okra and eggplant are the ladybird beetles, (*Cheilomenes* sp) (Coleoptera Coccinellidae) (Figure 8) and predatory spiders (Araneae) (Figure 9). The mean numbers of the various natural enemies are presented in Tables 3 and 4. Generally for the two seasons and the two crops, *Cheilomenes* sp and predatory spiders' populations were highest on the control plots, even though the differences compared to the botanical treatments were not significant (Tables 3 and 4).

Crop growth, fruit yield and crop damage assessment

In both dry and minor rainy seasons for the two vegetable crops, there were significant differences in plant height at flowering (cm), number of fruits per plant, mean fruit weight and fruit damage, among the botanical treatments

Figure 1. *Podagrica* sp on okra.

Figure 2. *Dysdercus* sp. mating on Okra.

(P<0.05). On Okra, Ecogold, Alata soap, neem oil and garlic were observed to have performed better in terms of plant height compared with the wood ash and the Control. In terms of number of fruits, all the botanicals performed better than the control. Ecogold performed best in terms of fruit weight compared to the Control which gave the poorest in both the dry and rainy seasons.

Similarly for the eggplant, all the botanical treated plants performed better in terms of plant height compared with the Control in both the dry and rainy seasons. However, Ecogold-treated plants gave the highest number

Figure 3. *Z. variegatus* mating on okra.

Figure 4. *Earias* sp (larva) on okra.

nof fruits while the Control recorded the least in the minor rainy season of 2010. Ecogold-treated plants produced the heaviest okra fruits for both dry (336 g) and minor rainy (341 g) seasons. On the other hand, garlic extract-treated plants produced the heaviest fruits of eggplant for both dry (267 g) and minor rainy (287 g) seasons. For both okra and eggplant, the control plants yielded the least fruits for both dry and minor rainy sea-sons, (Tables 5 and 6). Ecogold, garlic extract and neem extract - treated plots recorded the least mean number of fruit damage, whilst the control plots recorded the largest number of fruit damage (Tables 5 and 6). The highest

Figure 5. Aphids attack on egg plant.

Figure 6. *U. hysterricellus* attack on eggplant.

(a)   **(b)**

Figure 7. Stem and fruit borer (*L. orbonalis*) (a) inside stem (b) borer hole on fruit of eggplant.

Table 1. Effect of botanical treatments on numbers of major insect pests of okra: (a) minor season, 2009 and (b) major season, 2010 at Kwadaso.

Botanical	Mean ± (SE) number of insects/plant					
	A. gossypii	*P. uniformis*	*D. superstitiosus*	*B. tabaci*	*Z. variegatus*	*Earias* spp
Minor season, 2009						
EcoGold 999 Plus	7.01 ± 0.23[a]	6.05 ± 0.11[a]	5.02 ± 0.23[a]	2.50 ± 0.38[a]	5.14 ± 0.33[a]	4.12 ± 0.38[a]
Alata soap	8.23 ± 0.15[a]	8.32 ± 0.12[b]	6.42 ± 0.10[b]	2.62 ± 0.21[a]	7.21 ± 0.21[b]	4.82 ± 0.21[a]
Papaya leaves	9.06 ± 0.23[a]	8.32 ± 0.21[b]	6.52 ± 0.12[b]	5.01 ± 0.04[a]	7.15 ± 0.13[b]	5.32 ± 0.04[a]
Neem oil	7.68 ± 0.59[a]	6.32 ± 0.55[a]	5.12 ± 0.42[a]	4.61 ± 0.23[a]	5.42 ± 0.02[a]	4.18 ± 0.23[a]
Exotic garlic	8.14 ± 0.01[a]	7.12 ± 0.23[a]	6.12 ± 0.26[a]	4.62 ± 0.18[a]	6.17 ± 0.21[a]	4.32 ± 0.08[a]
Wood ash	8.19 ± 0.75[a]	8.32 ± 0.16[b]	7.12 ± 0.15[b]	4.72 ± 0.03[a]	8.02 ± 0.23[b]	5.12 ± 0.07[a]
Control	18.56 ± 0.41[b]	26.12 ± 0.12[c]	15.10 ± 0.03[c]	11.1 ± 0.34[b]	14.0 ± 0.58[c]	11.2 ± 0.31[b]
P	0.0012	0.0013	0.0001	0.0001	0.0001	0.0001
Major season, 2010						
EcoGold 999 Plus	5.42 ± 0.28[a]	5.02 ± 0.21[a]	3.02 ± 0.23[a]	2.50 ± 0.38[a]	4.12 ± 0.33[a]	3.12 ± 0.38[a]
Alata soap	6.22 ± 0.18[a]	7.42 ± 0.13[b]	5.42 ± 0.11[b]	2.62 ± 0.21[a]	6.11 ± 0.28[b]	3.62 ± 0.21[a]
Papaya leaves	7.16 ± 0.18[a]	7.52 ± 0.22[b]	5.52 ± 0.02[b]	3.52 ± 0.04[a]	6.15 ± 0.12[b]	4.52 ± 0.04[a]
Neem oil	5.68 ± 0.48[a]	5.12 ± 0.45[a]	3.12 ± 0.42[a]	2.51 ± 0.23[a]	4.42 ± 0.08[a]	3.12 ± 0.23[a]
Exotic garlic	7.12 ± 0.01[a]	6.12 ± 0.28[a]	4.12 ± 0.28[a]	2.52 ± 0.18[a]	5.17 ± 0.28[a]	3.32 ± 0.18[a]
Wood ash	7.18 ± 0.83[a]	8.12 ± 0.18[b]	6.12 ± 0.13[b]	3.12 ± 0.08[a]	7.02 ± 0.28[b]	4.12 ± 0.08[a]
Control	13.56 ± 0.81[b]	23.1 ± 0.05[c]	12.12 ± 0.07[c]	7.12 ± 0.36[b]	13.02 ± 0.58[c]	9.12 ± 0.36[b]
P	0.0014	0.0023	0.0001	0.0001	0.0001	0.0001

Means within a column followed by the same letter do not differ significantly from each other (P > 0.05; SAS, PROC GLM, SNK).

mean percentage increase in okra fruit weight was recorded from plants spayed with Ecogold (51%) while Alata soap gave the lowest (22%). Similarly, the highest mean percentage increase in eggplant fruit weight was recorded

Table 2. Effect of botanical treatments on numbers of major insect pests of egg plant (a) Minor season, 2009 at Kwadaso and (b) Major season, 2010 at Agona-Mampong.

	(a) Minor season, 2009					
	Mean ± (SE) number of insects/plant					
Botanical	A. gossypii	U. hysterricellus	D. superstitiosus	B. tabaci	Z. variegatus	L. orbonalis
EcoGold 999 Plus	3.40±0.23[a]	3.02 ± 0.21[a]	3.42 ± 0.53[a]	2.32±0.34[a]	4.12±0.33[a]	3.12± 0.38[a]
Alata soap	4.23±0.11[a]	5.32 ± 0.18[b]	5.52 ± 0.31[b]	2.72±0.21[a]	6.11± 0.28[b]	3.62± 0.21[a]
Papaya leaves	5.19±0.13[a]	5.52 ± 0.25[b]	5.78 ± 0.12[b]	3.62±0.14[a]	6.15± 0.12[b]	4.52± 0.04[a]
Neem oil	3.68±0.31[a]	3.42 ± 0.45[a]	3.89 ± 0.42[a]	2.71±0.21[a]	4.42± 0.08[a]	3.12± 0.23[a]
Exotic garlic	5.21±0.01[a]	4.32 ± 0.28[a]	4.45 ± 0.38[a]	2.82±0.12[a]	5.17± 0.28[a]	3.32± 0.18[a]
Wood ash	5.04±0.33[a]	6.10 ± 0.17[b]	6.67 ± 0.23[b]	2.92±0.01[a]	7.02± 0.28[b]	4.12± 0.08[a]
Control	10.94±.51[b]	9.12 ± 0.18[c]	12.89 ± 0.17[c]	7.12±0.36[b]	13.02±0.58[c]	9.12± 0.36[b]
P	0.0012	0.0011	0.0001	0.0001	0.0001	0.0001
	(b) Major season, 2010					
EcoGold 999 Plus	4.41 ± 0.26[a]	3.92 ± 0.21[a]	3.23 ± 0.41[a]	2.13 ± 0.01[a]	3.98 ± 0.33[a]	1.52 ± 0.38[a]
Alata soap	5.23 ± 0.18[a]	4.32 ± 0.16[b]	5.22 ± 0.21[b]	2.11 ± 0.02[a]	6.01 ± 0.28[b]	1.62 ± 0.21[a]
Papaya leaves	6.17 ± 0.19[a]	4.52 ± 0.15[b]	5.13 ± 0.12[b]	3.10 ± 0.11[a]	6.12 ± 0.10[b]	2.22 ± 0.04[a]
Neem oil	4.68 ± 0.51[a]	3.42 ± 0.35[a]	4.79 ± 0.42[a]	2.24 ± 0.01[a]	4.42 ± 0.08[a]	1.12 ± 0.23[a]
Exotic garlic	6.22 ± 0.11[a]	3.32 ± 0.18[a]	4.12 ± 0.38[a]	2.23 ± 0.10[a]	4.17 ± 0.28[a]	1.32 ± 0.18[a]
Wood ash	6.08 ± 0.13[a]	7.10 ± 0.19[b]	6.01 ± 0.23[b]	2.21 ± 0.11[a]	6.02 ± 0.28[b]	2.12 ± 0.08[a]
Control	10.44 ± 0.51[b]	8.12 ± 0.15[c]	9.89 ± 0.17[c]	5.12 ± 0.08[b]	10.02 ± 0.58[c]	6.12 ± 0.36[b]
P	0.0010	0.0013	0.0001	0.0001	0.0001	0.0001

Means within a column followed by the same letter do not differ significantly from each other (P > 0.05; SAS, PROC GLM, SNK).

from plants sprayed with exotic garlic (59%) while Alata soap produced the lowest (21%). However, in terms of the number of days to 50% flowering, the plants treated with botanicals were similar compared with the Control. .

DISCUSSION

Populations of flea beetles (*Podagrica* spp.) and whiteflies (*B. tabaci*) in this present study were suppressed with the use of garlic bulb extract, a finding supported by Purseglove (1969) and Prasterink (2000) that garlic and onion acts as a repellent for insects. Garlic produces a pungent alliaceous compound, allyl-epropyl-disulphide, which probably is responsible for its pest repellent attribute. Population of the shoot and fruit borer (*L. orbonalis*) in the present study was significantly reduced by the application of the botanicals and therefore very few fruits were observed to be hollow and filled with frass, the indicative damage by the fruit borer. Among the botanicals, fruit borer populations were consistently lower with the application of Ecogold, garlic extract and neem extract. Similar results from studies in Asia and parts of Africa lend support to these findings (Fiscian, 1999; AVRDC, 2003a; FAO 2003).

The use of botanicals such as neem and other bio-pesticides to control insect pests of vegetables is gaining

attention (Obeng-Ofori and Ankrah, 2002; Coulibaly et al., 2007). The botanicals provide alternative means of insect pest management which conserves the ecosystem (Zehnder et al., 1997; 2011; Obeng-Ofori and Ankrah, 2002; Dively et al., 2003). Worldwide, the non-pesticide management (NPM) of crops is becoming popular among vegetable growers since it endeavours to keep manage-ment of insect pests and crop cultivation costs to a minimum and avoid dependency on manufactured inputs by utilizing materials that are readily available to farmers, in this case, the adoption and use of botanicals. In the present study, the populations of natural enemies mainly ladybird and huntsman spider were similar on both the botanically treated plants and the control, suggesting that the botanicals did not have any adverse effect on the natural enemies. This finding is significant against the background that several studies on biopesticides (Sibanda et al., 2000; Navon, 2000; Owusu-Ansah et al., 2001; Obeng-Ofori and Ankrah, 2002), do not clearly indicate the effect biopesticides on natural enemies of insect pests. In light of this, the present study is the first report of the non-harmful attributes of EcoGold 999 Plus, neem oil and garlic bulb extract as botanicals on ladybird and huntsman spider as natural enemies of white flies and aphids. The presence of the ladybird beetle on both the Control and Botanicals-treated plants in the present study, was indicative that the white flies were still present

Table 3. Effect of botanical treatments on numbers of natural enemies on okra minor season (a) 2009 at Kwadaso and (b) 2010 at Agona-Mampong.

Botanical	Minor season, 2009 at Kwadaso		Minor season, 2010 at Agona-Mampong	
	Mean ± (SE) number of beneficial insects		Mean ± (SE) number of beneficial insects	
	Lady bird beetles (*Cheilomenes* sp)	Predatory spiders (*Araneae*)	Lady bird beetles (*Cheilomenes* sp)	Predatory spiders (*Araneae*)
EcoGold 999 Plus	9.42 ± 0.28	8.01 ± 0.21	8.42 ± 0.38	7.01 ± 0.11
Alata soap	8.22 ± 0.18	7.42 ± 0.13	8.22 ± 0.10	6.82 ± 0.10
Papaya leaves	8.16 ± 0.11	7.42 ± 0.22	8.16 ± 0.11	6.92 ± 0.12
Neem oil	7.68 ± 0.48	6.82 ± 0.45	7.68 ± 0.21	6.82 ± 0.35
Exotic garlic	7.82 ± 0.01	7.12 ± 0.28	8.82 ± 0.01	7.12 ± 0.17
Wood ash	6.88 ± 0.83	7.22 ± 0.18	6.88 ± 0.03	6.52 ± 0.10
Control	10.16 ± 0.61	9.12 ± 0.45	8.96 ± 0.31	8.12 ± 0.45
P	0.6061	0.7043	0.4521	0.5610

Means within a column followed by the same letter do not differ significantly from each other (P > 0.05; SAS, PROC GLM, SNK).

Table 4. Effect of botanical treatments on numbers of natural enemies on egg plant (a) Major season, 2010 at Kwadaso and (b) Minor season, 2009 at Agona-Mampong.

Botanical	Major season, 2010 at Kwadaso		Minor season, 2009 at Agona-Mampong	
	Mean ± (SE) number of beneficial insects		Mean ± (SE) number of beneficial insects	
	Lady bird beetles (*Cheilomenes* sp)	Predatory spiders (*Araneae*)	Lady bird beetles (*Cheilomenes* sp)	Predatory spiders (*Araneae*)
EcoGold 999 Plus	7.12 ± 0.28	6.71 ± 0.01	6.14 ± 0.18	6.01 ± 0.02
Alata soap	6.67 ± 0.18	7.11 ± 0.12	5.63 ± 0.16	5.49 ± 0.11
Papaya leaves	6.56 ± 0.11	7.22 ± 0.01	5.61 ± 0.11	5.64 ± 0.02
Neem oil	5.68 ± 0.48	6.82 ± 0.45	4.89 ± 0.21	5.82 ± 0.05
Exotic garlic	5.82 ± 0.01	7.12 ± 0.28	4.82 ± 0.01	5.62 ± 0.25
Wood ash	6.58 ± 0.03	7.22 ± 0.20	5.87 ± 0.03	5.52 ± 0.21
Control	7.16 ± 0.61	8.01 ± 0.65	7.16 ± 0.52	6.41 ± 0.45
P	0.6132	0.7147	0.7132	0.6578

Means within a column followed by the same letter do not differ significantly from each other (P > 0.05; SAS, PROC GLM, SNK).

on the treated plants albeit not at the economic threshold. Fruit yields from okra and eggplant were higher on Botanically-treated plants than the Control plants, probably due to the reduced fruit damage on the treated plants. Furthermore, for both okra

Figure 8. *Cheilomenes* sp on okra plant.

Figure 9. Predatory spider on okra plant.

Table 5. Effect of insecticide treatments on growth and yield per plant of Okra. (a) Minor season, 2009 at Kwadaso and (b) Major season, 2010 at Agona-Mampong.

Botanical	Mean no. of days to 50% flowering	Mean plant height at flowering (cm)	Mean no. of fruits	Mean no. of fruits damaged	Mean weight of fruits (g)	Mean % increases in fruit weight
(a) Minor season, 2009 at Kwadaso						
EcoGold 999 Plus	33.5 ± 0.1[a]	62.1 ± 2.6[b]	9.8 ± 2.8[b]	1.1 ± 0.3[a]	336.1±1.8[c]	50.5 ± 1.8[c]
Alata Soap	34.1 ± 0.2[a]	61.9 ± 3.1[b]	7.6 ± 0.2[b]	1.3 ± 0.2[a]	215.7±4.0[b]	22.9± 2.1[b]
Papaya leaves	33.5 ± 0.6[a]	54.5 ± 2.2[ab]	6.2 ± 1.4[b]	1.5 ± 0.1[a]	219.3±5.8[b]	24.2± 3.4[b]
Neem oil	33.7 ± 0.7[a]	60.1 ± 2.5[b]	8.3 ± 1.4[b]	1.1 ± 0.2[a]	318. ± 5.2[c]	47.8±2.2[c]
Exotic garlic	34.1 ± 0.6[a]	60.7 ± 3.1[b]	7.7 ± 1.9[b]	1.3 ± 0.1[a]	234. ± 7.1[b]	29.1±3.1[b]
Wood ash	33.6 ± 0.1[a]	49.2 ± 2.6[a]	7.4 ± 1.8[b]	1.1 ± 0.1[a]	224. ± 3.8[b]	25.8± 2.4[b]
Control	34.1 ± 0.2[a]	49.5 ± 3.3[a]	4.5 ± 0.6[a]	3.2 ± 0.2[b]	166.3±7.8[a]	0.0
P	0.9987	0.0010	0.0001	0.0001	0.0001	0.0001
(b) Major season, 2010 at Agona-Mampong						
EcoGold 999 Plus	33.6 ± 0.2[a]	63.1 ± 2.5[b]	10.2 ± 2.8[b]	1.3 ± 0.1[a]	341.3 ± 1.6[c]	49.9 ± 1.2[c]
Alata Soap	34.3 ± 0.3[a]	62.9 ± 3.0[b]	8.7 ± 0.2[b]	1.5 ± 0.2[a]	219.8 ± 3.9[b]	22.2 ± 2.5[b]
Papaya leaves	33.6 ± 0.6[a]	53.7 ± 2.3[ab]	7.5 ± 1.4[b]	1.6 ± 0.2[a]	224.7 ± 3.8[b]	23.9 ± 2.4[b]
Neem oil	33.8 ± 0.8[a]	57.8 ± 2.4[b]	9.1 ± 1.2[b]	1.4 ± 0.2[a]	324.3 ± 4.1[c]	47.3 ± 3.1[c]
Exotic garlic	34.3 ± 0.5[a]	58.8 ± 3.3[b]	8.7 ± 1.7[b]	1.3 ± 0.2[a]	238.2 ± 5.0[b]	28.2 ± 3.0[b]
Wood ash	33.5 ± 0.2[a]	49.1 ± 2.5[a]	8.6 ± 1.5[b]	1.3 ± 0.1[a]	229.2 ± 2.2[b]	25.4 ± 2.1[b]
Control	34.2 ± 0.1[a]	49.4 ± 3.2[a]	5.2 ± 0.4[a]	3.4 ± 0.3[b]	171.0 ± 5.3[a]	0.0[a]
P	0.9481	0.0031	0.0001	0.0001	0.0001	0.0001

Means within a column followed by the same letter do not differ significantly from each other (P > 0.05; SAS, PROC GLM, SNK).

and eggplant, the reduction in population of the foliage defoliators (*P. uniformis* and *Z. variegatus* L.) implied the availability of more undamaged foliage which through photosynthesis resulted in increased dry matter accumulation in the fruits. Zehnder et al. (1996; 1997) also reported increased marketable yields of vegetables treated with garlic and red pepper.

Conclusion

The current study presents an array of botanicals

that could significantly reduce pest populations and conveniently maintain ecological balance with their natural enemies on okra and eggplants. However, for expected result as observed in this study we agree with Zehnder et al. (1997) and Ahmed et al. (2009) and caution users that the application frequency of these botanical materials should be based on climatic variations in particular regions in order to achieve the desired results. These botanicals when applied in the right dosages and frequencies could offer large-scale growers, who produce for both the local and export markets and are looking for alternatives to

synthetic pesticides to manage insect pest of vegetables at relatively less cost and at no harm to the consumer and environment. This will eventually make vegetable cultivation a profitable business and improve the livelihood of growers.

ACKNOWLEDGEMENTS

We thank Messrs. Augustine Agyekum, Anthony Gyimah and Adama Amadu for their assistance with data collection and compilation; and Dr. K. Osei for his inspiration and technical support in

Table 6. Effect of insecticide treatments on growth and yield per plant of egg plant. (a) Minor season, 2009 at Kwadaso and (b) Major season, 2010 at Agona-Mampong.

(a) Minor season, 2009 at Kwadaso

Botanical	Mean no. of days to 50% flowering	Mean plant height at flowering (cm)	Mean no. of fruits	Mean no. of fruits damaged	Mean weight of fruits (g)	Mean % increases in fruit weight
EcoGold 999 Plus	64.8 ± 0.1[a]	64.1 ± 2.2[b]	10.4 ± 2.3[a]	1.2 ± 0.3[a]	215.7 ± 4.0[bc]	49.4 ± 2.1[c]
Alata soap	64.8 ± 0.2[a]	64.2 ± 2.1[b]	8.3 ± 0.2[a]	1.5 ± 0.2[a]	166. ± 2.9[b]	34.6 ± 2.3[b]
Papaya leaves	64.5 ± 0.1[a]	64.1 ± 2.2[b]	7.4 ± 1.4[a]	1.3 ± 0.1[a]	143. ± 2.8[b]	24.1 ± 2.1[b]
Neem oil	65.0 ± 0.4[a]	64.5 ± 2.0[b]	7.2 ± 1.5[a]	1.3 ± 0.2[a]	216. ± 3.0[b]	49.5 ± 2.3[c]
Exotic garlic	65.1 ± 0.3[a]	64.7 ± 3.1[b]	6.7 ± 1.9[a]	1.4 ± 0.1[a]	266. ± 4.0[d]	59.0 ± 2.4[d]
Wood ash	64.9 ± 0.1[a]	62.7 ± 2.6[a]	6.2 ± 0.6[a]	1.5 ± 0.1[a]	138. ± 2.7[b]	21.4 ± 2.2[b]
Control	65.4 ± 0.2[a]	62.5 ± 3.3[a]	4.7 ± 0.3[b]	2.7 ± 0.4[b]	109.2 ± 1.6[a]	0.0a
P	0.7869	0.0019	0.0001	0.0001	0.0001	0.0001

(b) Major season, 2010 at Agona-Mampong

Botanical	Mean no. of days to 50% flowering	Mean plant height at flowering (cm)	Mean no. of fruits	Mean no. of fruits damaged	Mean weight of fruits (g)	Mean % increases in fruit weight
EcoGold 999 Plus	64.6 ± 0.2[a]	64.2 ± 2.0[a]	11.4 ± 2.7[c]	0.7 ± 0.1[a]	236.5 ± 4.0[c]	49.8 ± 2.6[c]
Alata soap	64.6 ± 0.1[a]	64.3 ± 1.9[a]	8.7 ± 0.2[ab]	0.8 ± 0.2[a]	178.2 ± 2.9[b]	33.4 ± 2.2[b]
Papaya leaves	64.3 ± 0.1[a]	64.3 ± 2.1[b]	9.1 ± 1.4[b]	0.6 ± 0.1[a]	166.9 ± 2.8[b]	28.9 ± 2.3[b]
Neem oil	65.1 ± 0.1[a]	64.5 ± 1.8[b]	9.3 ± 1.4[b]	0.7 ± 0.2[a]	228.4 ± 3.0[b]	48.0 ± 2.7[c]
Exotic garlic	65.2 ± 0.2[a]	64.7 ± 2.1[b]	9.8 ± 1.9[c]	0.7 ± 0.1[a]	287.3 ± 3.9[c]	58.7 ± 3.1[d]
Wood ash	64.7 ± 0.1[a]	62.4 ± 2.2[a]	8.6 ± 1.8[ab]	0.8 ± 0.1[a]	159.8 ± 2.6[b]	25.7 ± 2.1[b]
Control	65.2 ± 0.1[a]	62.5 ± 2.1[a]	5.8 ± 0.6[a]	2.3 ± 0.4[b]	118.7 ± 1.8[a]	0.0a
P	0.8694	0.0012	0.0001	0.0001	0.0001	0.0001

Means within a column followed by the same letter do not differ significantly from each other (P > 0.05; SAS, PROC GLM, SNK).

presenting the manuscript in this form. This study was supported by a collaborative project between CSIR-Crops Research Institute, Kwame Nkrumah University of Science and Technology and the Export Marketing and Quality Awareness Project (EMQAP) of the Ministry of Food and Agriculture (MoFA).

REFERENCES

Ahmed BI, Onu I, Mudi L (2009). Field bioefficacy of plant extracts for the control of post flowering insect pests of cowpea(Vigna unguiculata (L.) in Nigeria. J. Biopesticides,
2(1): 37- 43.

AVRDC (2003a). A farmer's guide to harmful and helpful insects in eggplant fields. AVRDC - (The World Vegetable Centre), Taiwan.

AVRDC (2003b). Effects of natural and synthetic pesticides in controlling Diamondback moth in cabbage. Progress report 2002. Shanhua, Taiwan: AVRDC-the World Vegetable Centre, pp. 157-158.

Bennett-Lartey SO, Oteng-Yeboah AA (2008). Ghana Country Report on the State of Plant and Genetic Resources for food and agriculture. CSIR-Plant Genetic Resources Research Institute, Bunso, Ghana.

Coulibaly O, Cherry AJ, Nouhoheflin T, Aitchedji CC, Al-Hassan R (2007). Vegetable producer perceptions and willingness to pay for biopesticides. Int. J. Vegetable Sci., 12(3): 27- 42.

Dively GP, Patton T, Miller A (2003). Evaluation of organic insecticides for control of cabbageworms and other insects on cole crops, in Field Efficacy Evaluation of Selected Conventional and Organic Insecticides for Control of Insect Pests in Maryland. Final report, University of Maryland, USA, pp. 6-7.

El Shafie HAF (2001). The use of neem products for sustainable management of homopterous key pests on potatoes and eggplants in the Sudan. Ph.D thesis. Institute of Phytopathology and Applied Zoology. Justus Liebig University of Giessen.

FAO (2000). Cabbage Ecological Guide. FAO Inter-Country Programme for IPM in Vegetables in South and SoutheastAsia. Rome, Italy.

FAO (2003). Eggplant Ecological Guide. FAO Inter-Country Programme for IPM in Vegetables in South and Southeast

Asia. Rome, Italy.

FAO (2008). Small farmer participation in export production: The case of Kenya. FAO, Rome, Italy.

Fiscian P (1999). Assessment of damage due to feeding of Leucinodes orbonalis Quen. (Lepidoptera: Pyralidae) on *Solanum* sp. J. Ghana. Sci. Assoc., 1(2): 21 - 25.

Gerken A, Suglo JV, Braun M (2001). Crop protection policy in Ghana. Pesticide Policy Project produced by Plant Protection and Regulatory Services Directorate of Ministry of Food and Agriculture, Accra.

Navon A (2000). Bacillus thuringiensis insecticides in crop protection - reality and prospects. Crop. Prot., 19: 669 - 675.

Obeng-Ofori D, Ankrah DA (2002). Effectiveness of aqueous neem extracts for the control of insect pest of cabbage (*Brassica oleracea* var *capitata* L.) in the Accra plains of Ghana. Agric. Food. Sci. J. Ghana, 1: 83 - 94.

Obeng-Ofori D, Owusu EO, Kaiwa ET (2002). Variation in the level of carboxylesterase activity as an indicator of insecticide resistance in populations of the diamondback moth, *Plutella xylostella* (L.) attacking cabbage in Ghana. J. Ghana. Sci. Assoc., 4(2): 52-62.

Obopile M, Munthali DC, Matilo B (2008). Farmers' knowledge, perceptions and management of vegetable pests and diseases in Botswana. Crop. Prot., 27: 1220 -1224.

Owusu-Ansah F, Afreh-Nuamah K, Obeng-Ofori D, Ofosu-Budu KG (2001). Managing infestation levels of major insect pests of garden eggs (*Solanum integrifolium* L.) with aqueous neem seed extracts. J. Ghana. Sci. Assoc., 3(3): 70-84.

22

The theoretical approach of ecoplexivity focusing on mass outbreaks of phytophagous insects and altering forest functions

Anne le Mellec[1]*, Jerzy Karg[2], Jolanta Slowik[3], Ignaczy Korczynski[4], Andrzej Mazur[4], Timo Krummel[1], Zdzislaw Bernacki[3], Holger Vogt-Altena[1], Gerhard Gerold[1] and Annett Reinhardt[1]

[1]Landscape Ecology Section, University of Göttingen, Goldschmidtstr 5, D-37077 Göttingen, Germany.
[2]Research Centre for Agricultural and Forest Environment, Polish Academy of Sciences, Field Station Turew, Szkolna 4, Pl-4-000 Kościan, Poland.
[3]Centre for Nature Conservation (CNC), University of Göttingen, von Sieboldstrasse 2, D-7075 Göttingen, Germany.
[4]Department of Forest Entomology, Poznan University of Life Sciences, Ul Wojska Polskiego 71c, PL-60-637 Poznan, Poland.

Epidemics of forest insects can have deep impacts on ecosystem functioning and dynamics, with consequences for forest economics and forest carbon feedback to climate change. Despite the many roles that insects fulfil in terrestrial ecosystems, their importance in nutrient cycling is not well known (Kosola et al., 2001). The only instances where herbivores are recognized to have a large effect on ecosystem function are mass outbreaks of particular species like herbivores. However, the climate change induced alterations in precipitation and temperature patterns will undoubtedly affect occurrence, intensity, frequency, magnitude and timing of these phenomena and thus, provoke an increasing susceptibility of hosts and a significantly larger habitat presence of pests. Records show that, in an increasing number of cases severe outbreaks can even cause the complete devastation of vast areas and thus, imply considerable economic losses at a large scale. Down to the present day, it remains uncertain how forest ecosystems will respond to the changing environmental conditions in the long run. This work reports on the possible alterations of forest functions due to mass outbreaks of phytophagous insects with respect to the changing ecosystem service of carbon sequestration ability of forests on the northern hemisphere.

Key words: Forest disturbances, insect mass outbreaks, forest functioning, carbon sequestration.

INTRODUCTION

The growing interest in the impacts of global climate change on forest ecosystems is not surprising as forests cover about 43% of the world's surface and account for some 70% of terrestrial net primary production (NPP) (Melillo et al., 1993), which transforms into 359 Gt of carbon in biomass and 787 Gt of carbon deposited in soils (WBGU, 1998). Forests are being bartered on world markets for carbon mitigation purposes since they represent 46% of the globally sequestered terrestrial carbon (WBGU, 1998; Nilsson, 1995).

According to the latest IPCC (2007) assessment report, disturbances in the form of insect infestations which will most certainly increase in areas featuring sub-continental to continental climatic conditions as a result of a decline in precipitation and rising temperatures, however, depending on the modelled warming scenario (IPCC, 2007; Dale et al., 2000). Yet till the present day, it remains uncertain how European forest ecosystems will respond to the changing environmental conditions in the long run. It can be assumed that, in specific regions, forests will not feature sufficient resilience capabilities to

---
*Corresponding author. E-mail: amellec@gwdg.de.

adapt to these fast changing climatic conditions characterized by the increased occurrence of forest disturbances like wild fires and mass outbreaks of insects. Until this very day, the conception and opinion has prevailed that forests on the northern hemisphere serve as gigantic carbon sinks. However, feedback effects indicate that, the assumption that forests represent one of the key mitigation factors for reducing the emission of greenhouse gases (IPCC, 2007) has become debatable (Stern et al., 2006), meaning that, forests are likely to become carbon sources.

Forest disturbances are inherent to natural forest dynamics (Ayres and Lombardero, 2000). On the other hand, disturbances such as wild fires and wind throw as well as mass outbreaks of insects often cause devastating ecological effects and substantial economic losses. In the United States of America, more than 20 million hectares of forested land are annually affected by insect mass outbreaks, resulting in drastic economic losses in the magnitude of about 20 billion US $ (Ayres and Lombardero, 2000). However, studies suggest that insect herbivore can also act as a regulator of forest primary production, leading to increasing above ground biodiversity after defoliation (Hartley and Jones, 2004).

Apparently, canopy herbivore plays an important role with respect to carbon (C) and N inputs by altering timing and quality of organic material reaching forest floor and affecting below ground processes (herbivore mediated production of organic matter (OM)), thus, directly as well as indirectly affecting ecosystem functioning. However, despite the many purposes insects serve in terrestrial ecosystems, their importance in nutrient cycling and advanced forest functions are not well understood (Rouault et al., 2006).

The assessment report of the IPPC predicts that, forest perturbations such as mass outbreaks of insects will increase dramatically in the future due to changes in precipitation and temperature patterns, thus, leading to the depletion of water resources and subsequently, increase the susceptibility of trees for diseases and insects attacks (IPCC, 2007).

Other than an expected drop in net primary production (NPP) (Ciais et al., 2005) the resultant diminishing C storage in wood biomass as well as alterations in carbon cycling characteristics (Baldocchi, 2005) are likely to have fatal consequences for the overall stability of forest stands resulting in a curtailed adaptability and resilience of forest ecosystems (Bale et al., 2002). Climate change induced alterations in precipitation and temperature patterns will undoubtedly affect the occurrence patterns of these phenomena (Dale et al., 2001) by changing the natural cycles of mass outbreaks through modifying magnitude, frequency, intensity and duration of the reproduction characteristics of pest insects (Begon et al., 1991; Cambell and Madden, 1990; Dale et al., 2001). This is based on indirect (accelerated reproduction turnover, lower mortality, decreased defense potential of trees) and direct processes (change in population density

of antagonists, parasitoids, predators) (Dale et al., 2001; Hunter, 2001). All in all, it can be postulated with confidence that global warming provokes an increasing susceptibility of hosts and a significantly larger habitat presence of pests (Hunter, 2001). Records show that, in some cases severe outbreaks can even cause the complete devastation of large areas and thus, imply considerable economic losses at a large scale.

Enhanced nutrient input to the soil-transformation from recalcitrant to easily degradable matter

During mass outbreaks of phytophagous insects, high amounts of organic matter (OM) enter the forest floor dry as insect modified fragments (pellets, damaged needles and leaves) or in solution by throughfall. Under these conditions, the matter transfer from the canopy source to the soil is enlarged. Studies have shown that, input of organic carbon into the forest floor can reach about 180 kg C ha^{-1} 6 months^{-1} (by throughfall) and up to 520 kg C ha^{-1} 6 months^{-1} (by frass pellets) during the vegetation season (le Mellec and Michalzik, 2008). Peak fluxes occurred within a short period of 5 weeks during June/July, where about 80% of the overall inputs enter the ground under herbivore activity, thus, stressing the relevance of altered timing of organic matter cycling (le Mellec and Michalzik, 2008). Guggenberger and Zech (1994) could demonstrate that, 40 to 50% of the organic carbon in throughfall consists of easily degradable carbohydrates and therefore, can be considered as a promoting co-substrate for decomposition processes. In this context, various studies have demonstrated that herbivorous insects are able to transform needle biomass or leaf constituents into more easily degradable secondary products (insect pellets, honeydew) (Chapman et al., 2003; Hunter, 2001; Schowalter et al., 1991; Stadler et al., 2001). Through the transformation into more easily decomposable organic matter, herbivorous insects are likely to promote decomposition activity and lead to an accelerated and quantitatively increased decomposition rate and release of nutrients (Ritchie et al., 1998; Hollinger, 1986; Chapman et al., 2003; Lovett and Ruesink, 1995; Stalder et al., 2001). In this context, studies suggest that mass outbreaks of insects affect above and below ground carbon sequestration by their defoliating activity: (1) due to a limited above ground C fixation as a result of frass induced tree mortality (limited net primer production (NPP)) (Langstrom et al., 2001; Amour et al., 2003; Cedervind and Långström, 2003); (2) due to the quantitatively increased and qualitatively changed matter input that causes altered soil processes (Lovett et al., 1995). This in consequence might lead to a limitation of below ground C sequestration and the reduction of soil organic carbon (SOC) storage caused by an enhanced release of DOC (dissolved organic carbon per soil percolates), CO_2 as well as nitrous oxide (N_2O) (Madritch and Hunter, 2003).

However, it remains still unclear to what extent those nutrient entries have the potential to change biogeochemical processes and therefore, the above and below ground functioning in forest ecosystems. From this point of view, forest stands with an enhanced vulnerability to mass outbreaks (due to limited above and below ground C sequestration ability and an enhanced production and release of CO_2 and N_2O) might expose themselves as forest stands with an increasing global warming potential. However, this scenario appears to be in stark contrast with the general perception that, forests of the northern hemisphere are supposed to act as C sinks. If climate change does in fact affect frequency, magnitude and interannual activity of forest disturbances like wild fires and insect outbreaks, forest ecosystems will certainly lose their ability to sequestrate C very significantly.

Here, we present a theoretical approach that deals with this phenomenon in application of new terms. This approach provides information of possible altered forest functions due to mass outbreaks on small (compartment), meso (ecosystem) and large scale (ecosystems).

METHODS

The model of ecoplexivity

Ecosystems can be considered as self-organized open systems Ellenberg (1972), in which various biological, chemical and physical processes determine processes through energy and matter fluxes. Within and between ecosystems, ecoparments (compartments in ecosystems) are linked by an exchange of energy and matter transfer under participation of different trophic levels (connectivity). The more diverse ecoparments are linked to each other, the more complex ecosystems are featuring a broad range of diversity (structural and biological). The degree of complexity within an ecosystem, is governed by the quantity and quality of trophic levels (feeding hierarchy in a food web such as primary producers, herbivores and primary carnivores), the quality and amount of information in cross linking (level of connectivity) as well as ecoparment involvement (multi-various ecoparments with overlapping schemes, particularly key ecoparments like the soil). Based on the fact that forest functioning is critically determined by connective and complex structures, we created the new term of "ecoplexivity" (connectivity + complexity).

The structure of this theoretical ecosystem approach is based on three scales: Small scale (process scale of ecoparments); meso scale (energy and matter transfer between ecoparments within an ecosystem); large scale (energy and matter exchange between (eco) systems).

The small scale (solid line ellipse) is the key level of ecosystem functioning comprising all chemical, physical and biological processes within ecoparments (canopy and soil). All functions within and between ecosystems are governed by these processes. The resulting compartment function (CF, canopy function and SF, soil function)) represents the sink or source capability for energy, water and matter characterised by processes (triangle) of either release (mobilisation) and/or sorption (immobilisation) processes (Figure 1).

The meso scale (dashed/dotted ellipse) is based on the energy, water and matter transfer between ecoparments like the canopy-to-soil transfer (Figure 1). Functions resulting from this level are intra-

ecosystem functions, such as above and below ground production (C sequestration) and biodiversity. The interaction between (eco) systems (forests-atmosphere interactions) is based on the largest scale through the exchange of energy, water and matter fluxes. Large scale functions can be reflected in source and sinks abilities of these systems like forests on the north hemisphere (dashed/dotted ellipse) acting as enormous sinks for C. The exchange of energy, water and matter between (eco) systems is called ecosystem-cross-linking (ECL) (double arrows).

Mass outbreaks of phytophagous insects and altering forest functions

Insect outbreaks are temporal and spatial dynamic disturbances which take place in ecosystems as a result of process cascades affecting different ecosystem compartments, time scales and trophic levels. According to Hollinger (1986), these disturbances (associated with the "Figure 8 model"), although being destructive, can become a driving force for constructive rebuilding processes for ecosystems in cases when ecosystem development (succession) is inhibited. The ecoplexivity approach has been devised to reflect possible alterations in ecosystem functioning due to mass outbreaks of phytophagous insects and the related elevated input of mediated organic matter.

Alteration of function on a small scale

Defoliating processes in the canopy compartment can provoke an alteration in the functionality by decreasing photosynthetic rates and by transferring large amounts of organic matter from one compartment (canopy) to another (soil). Studies suggest that, these elevated and chemically transformed matter inputs can trigger processes in the soil compartment due to increased soil microbial activity (mineralization and decomposition). Subsequently, the source functionality of both compartments (canopy and soil) is increased (Figure 2).

Alteration of function on a meso scale

The big black arrow in the graph indicates the alteration within the canopy-to-soil transfer due to increasing matter inputs by phytophagous insect frass from the canopy compartment during mass outbreak situations. The alterations of small scale processes (Canopy: reduced photosynthetic activity, soil: accelerated decomposition) also have effect on the meso scale by reducing carbon assimilation due to frass attack. In consequence, the above ground carbon fixation is limited by reduced photosynthesis resulting in reduced biomass production (enhanced tree mortality, limited timber growth and reduced fine root production). Therefore, NPP (net primary production) is reduced. NEP (net ecosystem production) is additionally reduced, particularly due to below ground carbon release. According to the acceleration hypothesis (Ritchie et al., 1998), SOM is mineralized at higher rates resulting in enhanced production of CO_2. Secondly, there is more vertical water percolation which includes DOC transport (loss) to the groundwater. This is due to reduced transpiration activity of infested or dead trees. As a possible consequence, it can be presumed that forest stands experience a shift from carbon sinks to carbon sources.

Alteration of function on a large scale

The largest scale which is represented by the ECL reflects the relation between the forest ecosystem and external systems such

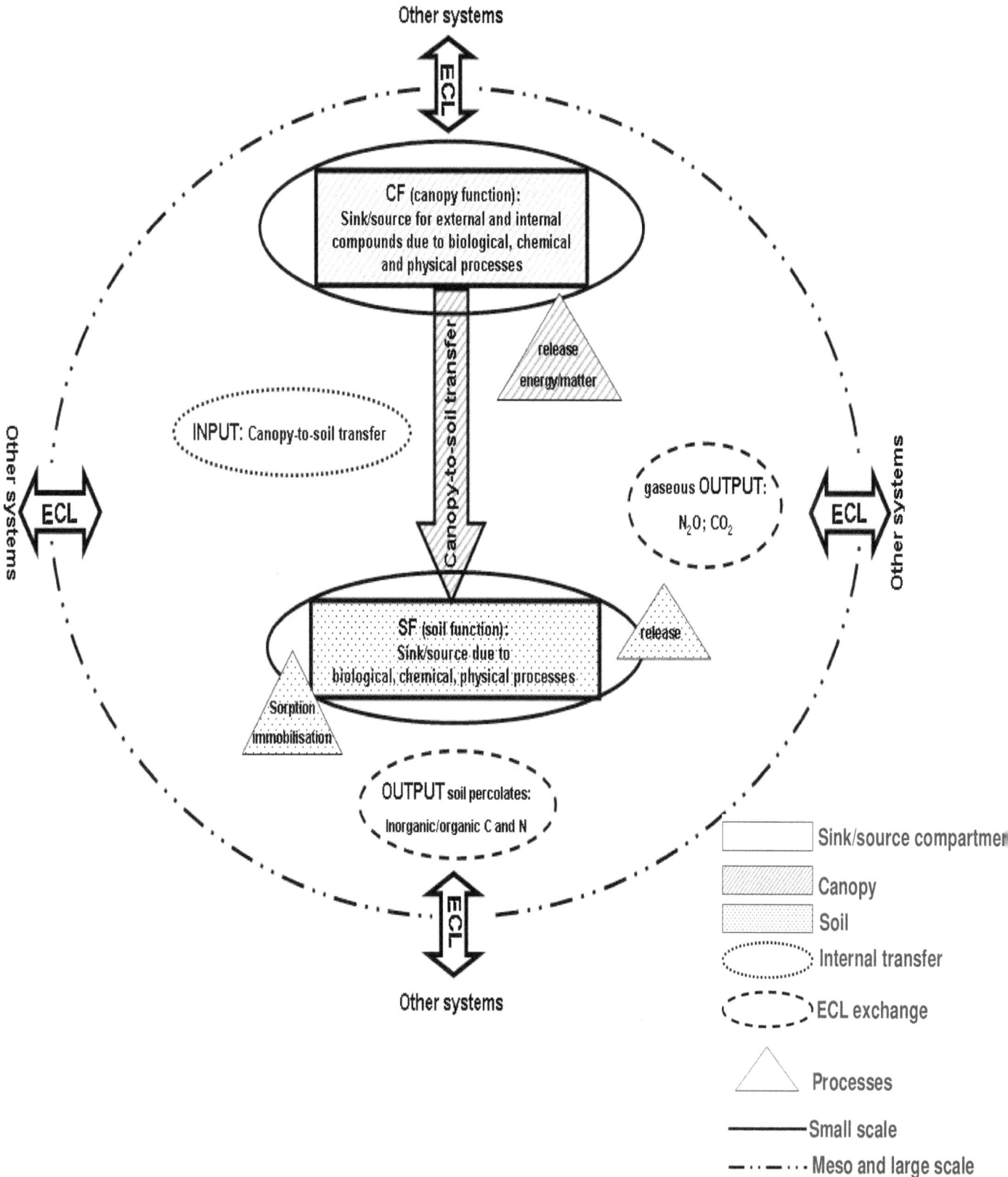

Figure 1. The model of ecoplexivity without mass outbreaks of phytophagous insects. The theoretical approach of ecoplexivity: Sink/source functionality of the canopy and the soil compartment due to abiotic and biotic processes (small scale: solid line ellipse). The meso scale (dash dotted line) shows the boundary between ecopartments through the canopy-to-soil transfer (under participation of various trophical levels and processes). There exist exchanges to other ecosystems (four double arrows) on meso and large scale (dashed dotted line) due to ecosystem cross-linking (ECL).

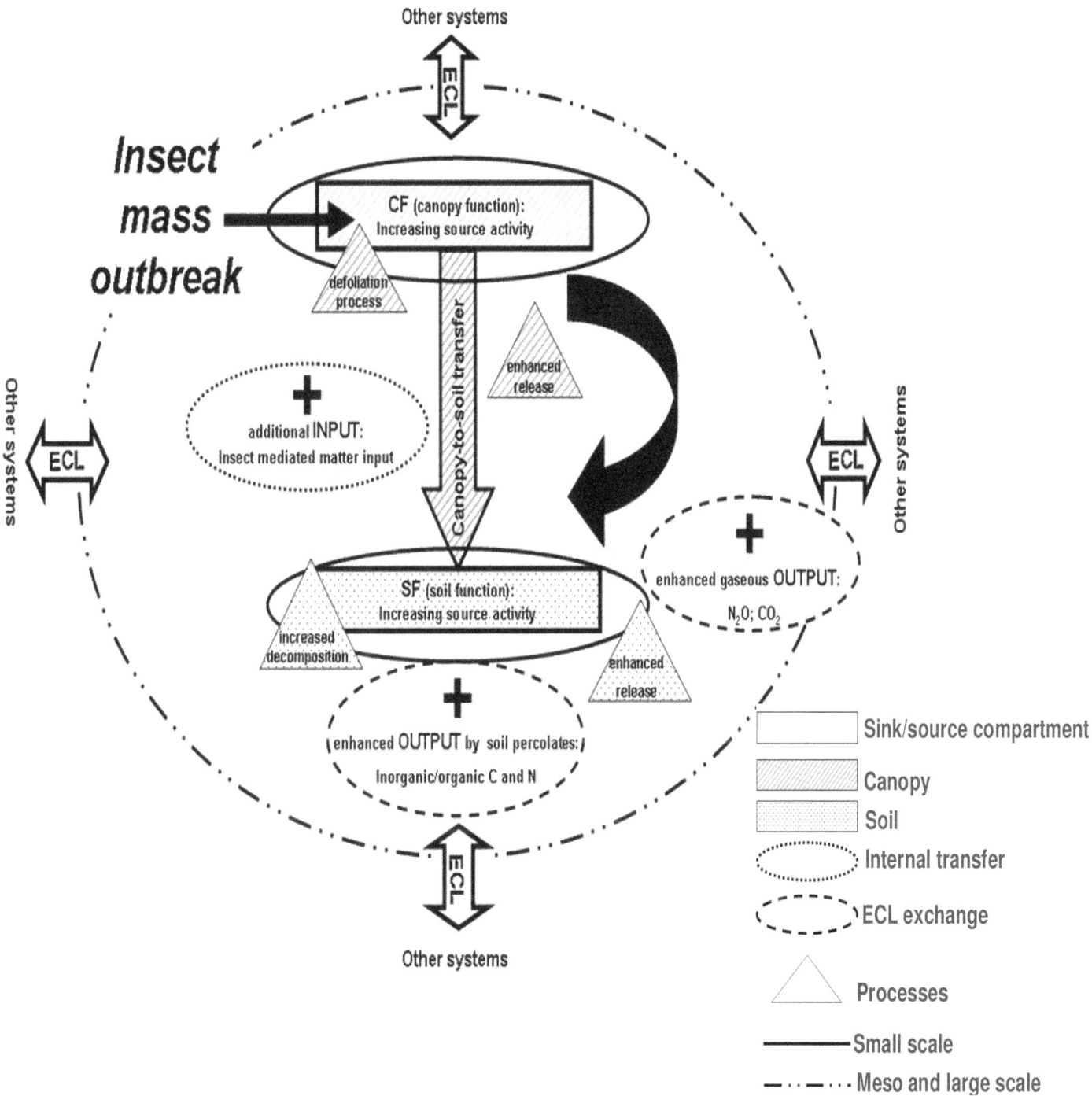

Figure 2. The model of ecoplexity with mass outbreaks of phytophagous insects. Due to mass outbreaks of phyotophagous, insects the canopy compartment becomes more the character of a source for energy and organic matter compounds. This leads to an increase of the canopy-to-soil transfer (big black arrow). Consequently, the soil processes alter due to these enhanced and qualitatively changed inputs. The soil compartment also becomes the character of a source due to the enhanced release of CO_2 and DOC. On meso as well as large scale, mass outbreaks lead to altered forest functions by reducing the forest abilities of carbon sequestration.

as the atmosphere; no matter which scenario will prove to be the 'true' one (increase of mean temperature by 1.4 to 4°C (WBGU, 1998), it can be foreseen that, heat waves like the ones in 2003 will increase in frequency and intensity in Europe and thus, affect forest

ecosystems. Furthermore, it is expected that not only summer temperatures will increase, but additionally, the amount of precipitation during the summer period will drop substantially for vast areas in Europe. This baleful scenario featuring a decline in precipitation and the reduction of soil water reservoirs will certainly induce severe limitations of water being available to plants (drought stress), which subsequently, become more susceptible to insect attacks. In the near future, climate change will undoubtedly impinge on the occurrence, intensity, frequency, magnitude and timing of mass outbreaks of pests (Dale et al., 2001). In the wake of this trend, tree hosts are likely to become extremely predisposed and will have to face an ever increasing habitat presence of pests. The earlier mentioned alterations on the small and meso-scale also bear the potential to make an impact on the large scale functionality through an increase in the soil-atmosphere transfer (increased release of CO_2 due to enhanced microbial decomposition).

DISCUSSION

To what magnitude those temporally limited available matter inputs will affect forest ecosystems remains worth discussing. However, studies have shown that those matter inputs can lead to an improvement in the nutrition scheme due to matter transformation from recalcitrant to more degradable compounds (quality aspect), as well as resulting in an increase in matter quantity provided that defoliation rate is below 70% (light frass). Defoliations under light frass conditions might appear as "first aid scenario" in monoculture coniferous stands of continental/sub-continental regions, where a large percentage of biomass (essential elements and nutrients) is accumulated in the canopy compartment, and where nutrient and water availability is limited. Here, outbreaks of canopy phytophagous insects might refill the internal matter cycling due to an enhanced matter release (frass activity) and canopy-to-soil transfer during insect outbreaks.

On the other hand, severe defoliation can cause a serious dieback of the host trees. This situation can be denominated as a "worst case scenario". Due to the absence of nutrient uptake by the trees, essential elements will simply disappear through runoff and with seepage output and thus, resulting in a nutrient loss for the ecosystem. Furthermore, forest stands which are marked by a negative NEP due to limited above and below ground C sequestration ability (and enhanced production and release of CO_2 and N_2O), might easily contribute to the increasing global warming potential. This aspect seems to contrast the general perception of forests of the northern hemisphere serving as vast C sinks. If climate change has the potential to modify frequency, magnitude and inter-annual activity of forest disturbances such as insect outbreaks (as the IPCC report predicts), then, forest ecosystems will certainly loose their ability to sequestrate C to a great extend and subsequently, might even take a shift from carbon sinks to carbon sources, as one of the most dramatic examples of an ECL.

REFERENCES

Armour H, Straw N, Day K (2003). Interactions between growth,

herbivory and long-term foliar dynamics of Scots pine. Trees: Struc. Funct., 17: 70-80.

Ayres MP, Lombardero MJ (2000). Assessing the consequences of global change for forest disturbances from herbivores and pathogens. Science Total Environ., 262: 263-286.

Baldocchi D (2005). The carbon cycle under stress. Nature, 437: 483-484.

Bale JS, Masters GJ, Hodkinson ID, Awmack C, Bezemer TM, Brown VK, Butterfield J, Buse A, Coulson JC, Farrar J, Good JEG, Harrington R, Hartley S, Jones TH, Lindroth RL, Press MC, Symrnioudis I, Watt AD, Whittaker JB (2002). Herbivory in global climate change research: Direct effects of rising temperature on insect herbivores. Global Change Biology 8: 1-16.

Begon M, Harper JL, Townsend CR (1991). Ökologie. Birkhäuser Verlag, Basel.

Cambell CL, Madden LV (1990). Introduction to plant disease epidemiology. New York, Wiley.

Cedervind J, Långström B (2003). Tree mortality, foliage recovery and top-kill in stands of Scots pine (Pinus sylvestris) subsequent to defoliation by the pine looper (Bupalus piniaria). Scand. J. For. Res., 18: 505-513.

Chapman SK, Hart SC, Cobb NS, Whitham TG, Koch GW (2003). Insect herbivory increases litter quality and decomposition: An extension of the acceleration hypothesis. Ecology, 84: 2867-2876.

Ciais P, Reichstein M, Viovy N, Granier A, Ogée J, Allard V, Aubinet M, Buchmann N, Bernhofer Chr, Carrara A, Chevallier F, De Noblet N, Friend AD, Friedlinstein P, Grünwald T, Heinesch B, Keronen P, Knohl A, Krinner G, Loustau D, Manca G, Metteucci G, Migletta F, Ourcival JM, Papale D, Pilegaard K, Rambal S, Seufert G, Soussana JF, Sanz MJ, Schulze ED, Vesala T, Valentini R (2005). European-wide reduction in primary productivity caused by heat and drought in 2003. Nature, 437: 529-533.

Dale VH, Joyce LA, McNulty S, Neilson RP, Ayres MP, Flannigan MD, Hanson PJ, Ireland LC, Lugo AE, Peterson CJ, Simberloff D, Swanson FJ, Stocks BJ, Wotton BM (2001). Climate Change and Forest Disturbances. BioScience, 51: 723-734.

Dale VH, Joyce LA, McNulty SG, Neilson RP (2000). The interplay between climate change, forests, and disturbances. Sci. Total Environ., 262: 210-204.

Ellenberg H (1972). Integrated Experimental Ecology. Methods and Results of Ecosystem Research in the German Solling Projekt. Berlin, Germany. New York, USA.

Guggenberger G, Zech W (1994). Composition and dynamics of dissolved organic carbohydrates and lignin-degradation products in two coniferous forests, N.E. Bavaria, Germany. Soil Biol. Biochem., 26: 19-27.

Hollinger DY (1986). Herbivory and the cycling of nitrogen and phosphorus in isolated California oak trees. Oecologia, 70: 291-297.

Hunter MD (2001). Insect population dynamics meets ecosystem ecology: Effects of herbivory on soil nutrient dynamics. Agric. For. Entomol., 3: 77-84.

IPCC (2007). Climate Change 2007: The Physical Science Basis Contribution of Working Group 1 to the Fourth Assessment Report of the Intergovernmental Panel on Climate Change [Solomon S., Qin D. Manning M., Chen Z., Marquis M., Averyt K.B., Tignor M. and Miller H.L. (eds.)]. Cambridge University Press, Cambridge, United Kingdom and New York, NY, USA, p. 996.

Kosola KR, Dickmann DI, Paul EA, Parry D (2001). Repeated insect defoliation effects on growth, nitrogen acquisition, carbohydrates, and root demography of poplars. Oecologia, 129: 65–74.

Långström B, Annila E, Hellqvist C, Varama M, Niemelä P (2001). Tree mortality, needle biomass recovery and growth losses in Scots pine following defoliation by Diprion pini and subsequent attack by Tomicus piniperda. Scandinavian J. For. Res., 16: 342-353.

le Mellec A, Michalzik B (2008). Impact of a pine lappet (Dendrolimus pini) mass outbreak on C and N fluxes to the forest floor and soil microbial properties in a Scots pine forest in Germany. CJFR, 38: 1829-1849.

Lovett GM, Ruesink AE (1995). Carbon and nitrogen mineralization from decomposing gypsy moth frass. Oecologica, 104: 133-138.

Madritch MD, Hunter MD (2003). Intraspecific litter diversity and nitrogen deposition affect nutrient dynamics and soil respiration.

Oecologia, 136: 124-128.

Melillo JM, McGuire DA, Kicklighter DW, Moore B, Vorosmarty CJ, Schloss AL (1993). Global climate change and terrestrial net primary production. Nature, 363: 234-240.

Nilsson S (1995). Valuation of global afforestation programs for carbon mitigation. Clim. Change, 30: 249–257.

Ritchie ME, Tilman D, Knops JMH (1998). Herbivore effects on plant and nitrogen dynamics in oak savanna. Ecology, 79: 165-177.

Rouault G, Candau JN, Lieutier F, Nageleisen LM, Martin JC, Warzée N (2006). Effects of drought and heat on forest insect populations in relation to the 2003 drought in Western Europe. Ann. For. Sci., 63: 613-624.

Schowalter TD, Sabin TE, Stafford SG, Sexton JM (1991). Phytophage effects on primary production, nutrient turnover, and litter decomposition of young Douglas fir in western Oregon. For. Ecol. Manage., 42: 229-243.

Stadler B, Solinger S, Michalzik B (2001). Insect herbivores and the nutrient flow from the canopy to the soil in coniferous and deciduous forests. Oecologia, 126: 104-113.

Stern N, Peters S, Bakhshi V, Bowen A, Cameron C, Catovsky S, Crane D, Cruickshank S, Dietz S, Edmonson N, Garbett S-L, Hamid L, Hoffman G, Ingram D, Jones B, Patmore N, Radcliffe H, Sathiyarajah R, Stock M, Taylor C, Vernon T, Wanjie H, Zenghelis D (2006). Stern Review: The Economics of Climate Change. London, UK.

WBGU (1998). The accounting of biological sources and sinks in the Kyoto Protocol: Progress or setback for the global environment? Bremerhaven, Germany.

23

Quantitative survey of stored products mites infesting wheat flour in Jeddah Governorate

Abir Sulaiman Al-Nasser

Faculty of Applied Sciences for Girls, Umm Al-Qura University, Makkah, Kingdom of Saudi Arabia.
E-mail: asnasser@uqu.edu.sa, al-nasser.abir@hotmail.com.

In this study, a faunistic survey of mites was conducted in different product stores such as bakeries, warehouses, mills, department stores and houses during a nine month study period (October 2009 to June 2010), in Jeddah governorate. The survey was carried out on wheat flour as different samples 150 gm each were collected monthly from three regions; North, Middle and South. The results of the current study identified five species of mites belonging to four families: *Dermatophagoides farinae* (Pyroglyphidae), *Acarophenax tribolii* (Acarophenacidae), *Cheyletus malaccensis* (Cheyletidae), *Blattisocius tarsalis* and *Blattisocius keegani* (Ascidae). The mites' counts ranged from 0.15 to 20.44 mites/150 gm wheat flour. The highest counts for the majority of mites species collected from the three regions were recorded during the months of November, December, January and February, according to the decreasing degrees of temperature and relative humidity during these months. *D. farinae* was the most dominant species in all samples collected from the three regions, followed by *A. tribolii* in samples from North and Middle. The predatory species *C. malaccensis was* detected in three regions but in few numbers, whereas, *B. tarsalis* and *B. keegani* were detected only in samples from the South. These results indicate the effect of temperature and relative humidity on the distribution density of stored products mites.

Key words: Stored products mites, quantitative survey, wheat flour, temperature, relative humidity.

INTRODUCTION

Stored products mites are found in different product stores and sometimes, in high concentration. Mites flourish in warm and damp environments where they feed on protein rich substances such as grain, fungi and other micro-organisms. Both house dust mites and storage mites occur in damp dwelling (Harju et al., 2006). Mites cause significant grain weight losses and decrease of germinability (Zdarkova and Reska, 1976). Their activities cause heating of grain mass and moisture translocation which permits the development of molds and germination of the grain, contamination by alive and dead mites, different stages as well as exuviae and feaces resulting in being harmful for human consumption (Hughes, 1976). Mites are vectors of toxicogenic fungi (Hubert et al, 2003). which contribute to contamination of food and feed with mycotoxins (Franzolin et al., 1999) and Hubert et al, 2004). It is possible for the workers and even the customers to be exposed to mites and their allergens (Harju et al., 2006), either by handling, inhaling or/and ingestion of mites- contaminated food.

Respiratory allergy, bronchial provocation, wheezing,

rhinitis, contact dermatitis, urticaria and asthma have been associated with storage mites (Cuthbert and Jeffrey, 1993) and Olsson, Van Hage- Hamsten, 2000), Korunic, 2001), Borghetti et al., 2002), Musken et al., 2003), Wang et al., 2003), Matsumoto and Satoh, 2004), Walusiak et al., 2004), Stejskal and Hubert, 2008), also, there is increasing evidence that ingestion of mites contaminated food can affect human health, causing several intestinal allergic reactions such as vomiting, gastrointestinal disorders, abdominal cramps, wheezing, cough, difficult breathing, throat discomfort, dyspnea, urticaria and angioedema (Castillo et al., 1995), Scala, 1995), Blanco et al, 1997), Sanchez-Borge et al., 1997), Antunes et al., 1999), Sanchez-Borges et al., 2001), Guerra Bernd et al., 2001), Dutau, 2002), Wen et al, 2005), Sánchez-Borges et al., 2008), Sánchez-Borges et al., 2009), Yan et al., 2008). Prolonged ingestion of mite contaminated foods causes various alterations in the small intestinal tissues of rats (Saleh and Al-Nasser, 2007).

The present study aims to estimate the distribution of

stored products mites in wheat flour in different regions of Jeddah Governorate.

MATERIALS AND METHODS

Collection of mites

Wheat flour samples of 150 gm each were collected monthly from different product stores such as bakeries, warehouses, mills, department stores and houses. Samples were collected from three different regions in Jeddah (North-Middle and South); three replicates were taken from each region during a nine month period from October 2009 June 2010.

Extraction and identification of mites

Samples were collected in paper bags and transported to the laboratory with labels including the region and date of collection. Mites were extracted using modified Tulgren funnel apparatus (Aspaly et al., 2007) and left for a period of 48 h. Extracted mites were received in Petri dishes containing 70% ethyl alcohol, and then mounted in Hoyer's medium on glass slides. Mites counts were assessed as the number of individuals in each replicate for the three regions monthly with the aid of a dissecting binocular microscope, specimens were identified microscopically and classified into their taxonomical rank by using different specific keys Hughes,1976), Krantz, 1978).

Statistical analysis

Data were analyzed using the 2 ways ANOVA test to compare the mite species numbers monthly in the three regions of Jeddah.

RESULTS

Results of the present study identified five species of store products mites belonging to four families: *Dermatophagoides farinae* (Pyroglyphidae), *Acarophenax tribolii* (Acarophenacidae), *Cheyletus malaccensis* (Cheyletidae), *Blattisocius tarsalis* and *Blattisocius keegani* (Ascidae). Wheat flour samples collected from three regions of Jeddah (North-Middle and South) varied in the mites numbers during the sampling period from October 2009 to June 2010.

Results in Table 1 represent the mean numbers of mites for each species per 150 gm of wheat flour collected monthly from the three regions. According to the mean numbers of mites from the nine months, *D. farinae* was the most dominant species in the three regions, recording a highest rate in the North and the Middle (13.73 and 13.53), respectively. These increasing rates were significant in comparison with the other species. The highest number of *D. farinae* was in the month of January, with 20.89, 20.89 and 19.56 in the North, Middle and South, respectively. We observed a significant difference in May in all the three regions (9.33, 7.11 and 4.89).

The mean number of *A. tribolii* during the nine months was 10.96 and 11.01 in the North and the Middle with

significant differences in comparison with the other species. *A. tribolii* was not detected in the South, and it reached the highest number in February (89.20 and 22.22) and the lowest number in June (2.67 and 0.00) in the North and the Middle, respectively (Table1).

C. mallaccensis was detected in all samples collected during the sampling period but in few numbers. Table1 shows that *C. mallaccensis* recorded significant increase in the Middle region (4.25), in comparison to the North and South regions (3.11 and 3.06). The distribution of *C. mallaccensis* varied all through the nine months in the three regions recording the highest number during November in the Middle (7.11), a non significant decrease was observed during the month of June (0.00, 0.44 and 0.44) in the North, South and Middle regions, respectively.

On the other hand, *B. tarsalis* and *B. keegani* were not detected in the North and the Middle during the nine months of study; these two species were recorded just in the South region with a mean number of 6.87 and 5.43 for *B. tarsalis* and *B. keegani* respectively (Table 1).

Distribution of the five species of mites throughout the nine months of study is represented in Table 2. It is obvious that *D. farinae* (12.84) was the most dominant species in Jeddah followed by *A.tribolii* (7.32), *C. mallaccensis* (3.47), *B. tarsalis* (2.29) and *B. keegani* (1.81) with significant differences.

According to the numbers of mites in each region, regardless of the species, it was observed that the distribution of mites slightly varied in the three regions (Table 3). The differences between the mean numbers of studied species were not significant except for the month of October where a significant difference appeared between the Middle region (3.02) and South region (1.78).

Increasing rate of temperature and relative humidity was observed during the season of sampling except for the months of December, January and February which recorded a slight decrease in the temperature degrees (Table 4).

DISCUSSION

The results of the current study showed that wheat flour samples collected during nine months, from three regions in Jeddah contained small numbers of stored product mites. We noticed that *D. farinae* was the most prevalent species in all the collected samples. *D. farinae* is a free living mite belonging to the family Pyroglyphidae which is meanly found in house dust, feeding on organic debris, fungi, saprophages and detretivores, but it could be found in stored products too. Several studies referred to the presence of *D. farinae* in different stored food stuff such as wheat flour (Zaher, 1986), Matsumoto et al., 2001), Al-Nasser, 2007; Yi et al., 2009), bran and maize (Saleh et al., 1985), Bran (Baker, 2000), wheat and grain residue (Mahgoob et al., 2006), broad bean and rough rice (El-Sayed and Ghallab, 2007).

Table 1. Distribution of five mites' species in three locations (North, Middle and South) in Jeddah during the nine months from October 2009 to June 2010.

Insect	Location	October	November	December	January	February	March	April	May	June	Total	Mean
Dermatophagoides farinae	North	9.78 [a]	18.67 [a]	20.89 [a]	20.89 [a]	18.67 [b]	14.22 [a]	9.78 [a]	7.11 [b]	3.56 [a]	123.56 [a]	13.73 [a]
	Middle	9.33 [a]	18.67 [a]	19.56 [a]	20.89 [a]	18.67 [b]	13.33 [a]	8.44 [a]	9.33 [a]	3.56 [a]	121.78 [a]	13.53 [a]
	South	3.56 [bc]	17.33 [a]	15.56 [b]	19.56 [a]	16.00 [c]	12.00 [a]	9.78 [a]	4.89 [c]	2.67 [a]	101.33 [b]	11.26 [b]
Acarophenax tribolii	North	1.33 [d]	18.22 [a]	20.44 [a]	20.44 [a]	20.89 [ab]	13.33 [a]	4.00 [b]	0.00 [f]	0.00 [a]	98.67 [b]	10.96 [b]
	Middle	1.78 [cd]	19.56 [a]	19.11 [a]	20.00 [a]	22.22 [a]	13.33 [a]	3.11 [bc]	0.00 [f]	0.00 [a]	99.11 [b]	11.01 [b]
	South	0.00 [d]	0.00 [f]	0.00 [e]	0.00 [d]	0.00 [f]	0.00 [d]	0.00 [d]	0.00 [f]	0.00 [a]	0.00 [g]	0.00 [g]
Blattisocius tarsalis	North	0.00 [d]	0.00 [f]	0.00 [e]	0.00 [d]	0.00 [f]	0.00 [d]	0.00 [d]	0.00 [f]	0.00 [a]	0.00 [g]	0.00 [g]
	Middle	0.00 [d]	0.00 [f]	0.00 [e]	0.00 [d]	0.00 [f]	0.00 [d]	0.00 [d]	0.00 [f]	0.00 [a]	0.00 [g]	0.00 [g]
	South	1.78 [cd]	12.00 [b]	12.44 [c]	12.44 [b]	15.11 [c]	5.33 [b]	2.22 [bc]	0.44 [ef]	0.00 [a]	61.78 [c]	6.87 [c]
Blattisocius keegani	North	0.00 [d]	0.00 [f]	0.00 [e]	0.00 [d]	0.00 [f]	0.00 [d]	0.00 [d]	0.00 [f]	0.00 [a]	0.00 [g]	0.00 [g]
	Middle	0.00 [d]	0.00 [f]	0.00 [e]	0.00 [d]	0.00 [f]	0.00 [d]	0.00 [d]	0.00 [f]	0.00 [a]	0.00 [g]	0.00 [g]
	South	1.78 [cd]	9.78 [bc]	10.22 [c]	9.78 [b]	11.56 [d]	4.44 [bc]	1.33 [cd]	0.00 [f]	0.00 [a]	48.89 [d]	5.43 [d]
Cheyletus malaccensis	North	1.33 [d]	6.22 [de]	5.78 [d]	5.33 [c]	3.56 [e]	2.22 [cd]	1.33 [cd]	1.78 [de]	0.00 [a]	27.56f	3.06f
	Middle	4.00 [b]	7.11 [cd]	6.22 [d]	6.67 [c]	5.78 [e]	3.56 [bc]	2.67 [bc]	1.78d [e]	0.44 [a]	38.22 [e]	4.25 [e]
	South	1.78 [cd]	4.00 [e]	5.78 [d]	4.44 [c]	5.33 [e]	2.22d	1.78 [cd]	2.22d	0.44 [a]	28.00 [f]	3.11 [f]
F 0.05		*	*	*	*	*	*	*	*	n.s	*	*
LSD 0.05		2.10	3.03	2.59	2.70	2.63	2.74	2.08	1.76	-	9.60	1.07

Values within columns not sharing common superscript letters are differ significantly at p < 0.05.

Favorable conditions for high populations of house dust mites are 25 to 32°C at relative humidity and 5 to 87% (Saleh et al., 1985). Dust mites populations flourish best in wheat flour compared to the other varieties of flour and at ambient temperature with high humidity instead of the air conditioned environment (Yi et al., 2009). High relative humidity in combination with temperatures at about 25°C leads to high density of mites in whole wheat (Danielsen et al., 2004).

In the present study, D. farinae was present in the wheat flour samples collected all over the nine months despite the increasing rate of temperature and relative humidity in Jeddah which is in agreement with previous study (Edrees and Saleh, 2008), reporting that D. farinae was the only species collected during summer season from dust samples collected in Jeddah, as it can undergo dryness and low humidity in comparison with other species of dust mites.

A. tribolii was detected in the present study in samples collected from the North and Middle regions, belonging to the family Acarophenacidae which is mealy parasitic on Tribolium spp. (Lopez, 2005), sucking their haemolymph. A. tribolii was recorded previously in flour and rice

(Saleh,1980); wheat, corn, lentil and bean (Wafa et al.,1996); bran (Baker, 2000); wheat flour and grainresidue (Al-Nasser, 2007), Mahgoob et al., 2006). Results of the present study revealed the presence of C. malaccensis in all samples collected during the nine months from the three regions but it was detected in small numbers. C. malascensis belongs to the family Cheyletidae, this predator is found in stored products and/or in house dust predating on Pyroglyphids and Acarids mites (Kandil,1974; Hafez, 1977) C. malaccensis is a major predatory mite in grain storage systems (Athanassiou et al., 2002; Putatunda, 2005;

Table 2. Distribution of five mites' species in Jeddah during the nine months from October 2009 to June 2010.

Insect	October	November	December	January	February	March	April	May	June	Total	Mean
Dermatophagoides farinae	7.56 [a]	18.22 [a]	18.67 [a]	20.44 [a]	17.78 [a]	13.19 [a]	9.33 [a]	7.11 [a]	3.26 [a]	115.56 [a]	12.84 [a]
Acarophenax tribolii	1.04 [c]	12.59 [b]	13.19 [b]	13.48 [b]	14.37 [b]	8.89 [b]	2.37 [b]	0.00 [c]	0.00 [b]	65.93 [b]	7.32 [b]
Blattisocius tarsalis	0.59 [c]	4.00 [d]	4.15 [d]	4.15 [cd]	5.04 [c]	1.78 [c]	0.74 [cd]	0.15 [c]	0.00 [b]	20.59 [d]	2.29 [d]
Blattisocius keegani	0.59 [c]	3.26 [d]	3.41 [d]	3.26 [d]	3.85 [c]	1.48 [c]	0.44 [d]	0.00 [c]	0.00 [b]	16.30 [d]	1.81 [d]
Cheyletus malaccensis	2.37 [b]	5.78 [c]	5.93 [c]	5.48 [c]	4.89 [c]	2.67 [c]	1.93 [bc]	1.93 [b]	0.30 [b]	31.26 [c]	3.47 [c]
F 0.05	*	*	*	*	*	*	*	*	*	*	*
LSD 0.05	1.21	1.75	1.49	1.56	1.52	1.58	1.2	1.02	0.78	5.54	0.62

Values within columns not sharing common superscript letters are differ significantly at p< 0.05.

Table 3. Distribution of mites in three locations (North, Middle and South) in Jeddah during the nine months from October 2009 to June 2009.

Location	October	November	December	January	February	March	April	May	June	Total	Mean
North	2.49 [ab]	8.62 [a]	9.42 [a]	9.33 [a]	8.62 [a]	5.96 [a]	3.02 [a]	1.78 [a]	0.71 [a]	49.96 [a]	5.55 [a]
Middle	3.02 [a]	9.07 [a]	8.98 [a]	9.51 [a]	9.33 [a]	6.04 [a]	2.84 [a]	2.22 [a]	0.80 [a]	51.82 [a]	5.76 [a]
South	1.78 [b]	8.62 [a]	8.80 [a]	9.24 [a]	9.60 [a]	4.80 [a]	3.02 [a]	1.51 [a]	0.62 [a]	48.00 [a]	5.33 [a]
F 0.05	*	n.s.	n.s.	n.s.	n.s.	n.s.	n.s.	n.s.	n.s.	n.s.	n.s.
LSD 0.05	0.94	-	-	-	-	-	-	-	-	-	-

Values within columns not sharing common superscript letters are differ significantly at p< 0.05.

Table 4. Temperature degrees and relative humidity in Jeddah during the nine months from October 2009 to June 2010.

Month	Temperature (°C)			Relative humidity (%)		
	Maximum	Minimum	Mean	Maximum	Minimum	Mean
October	37.9	24.4	30.3	98	8	62
November	33.0	22.8	27.6	98	13	66
December	30.7	20.7	25.0	97	16	66
January	32.0	19.6	25.0	96	6	59
February	31.2	19.3	24.7	97	18	66
March	33.9	20.2	26.5	96	11	61
April	35.8	23.8	29.3	100	14	58
May	37.0	24.1	30.0	91	4	57
June	40.2	25.8	32.2	97	3	53

Palyvos et al., 2008).

It has been detected previously in wheat (El-Desouky, 1991; Athanassiou et al., 2005), bran (Al-Nasser 2007, Saleh et al., 1985; Baker, 2000; El-Desouky, 1991) wheat flour (Al-Nasser, 2007; Saleh et al.,1985; El-Sayed and Ghallab, 2007; Putatunda, 2005) rice (Al-Nasser AS, 2007), Saleh et al.,1985), maize, broad bean and rough rice (El-Sayed and Ghallab, 2007), grain residue (Putatunda, 2005; Palyvos et al., 2008; Hubert et al., 2006) and, different grains (Mahgoob et al., 2006; Habibpour et al., 2002), in stores dust (Zheltikova et al., 1997).

A previous study reported that there is a significant positive correlation between the occurrence and population density of the predatory and pest mites, indicating that the occurrence of Cheyletus predators and their prey are either density dependent or regulated by the same physical factors such as temperature, humidity and cleaning, and that decreasing temperatures decrease the development of *Cheyletus* spp. more rapidly than those of pest mites (Lukas et al., 2007).In the present study, it was obvious that increasing temperature and relative humidity were the reason for decreasing the development of the predatory mite because of the hot weather in Jeddah during the sampling period.

B. tarsalis and *B. keegani* were detected in small numbers during the present study and were not detected in the North nor the Middle; belonging to the family Ascidae, *Blattisocius* species were predators which feed on eggs of different grain beetles (*Tribolium confusium, T. castaneum* and *Oryzoephilus surinamensi*) (Barker, 1967), and on the eggs of Acarids mites (Thind and Ford, 2006).

B. keegani was reared on two stored grain mites Fawzy MMH, 1996), *Suidasia nesbetti* (Hughes) and *Grammolichus aegypticus* Shereef and Fawzy. The same species was reared on the larval stages of *Tyrophagus putrescentiae* (Schrank) and *Rhizoglyphus robini* Claparéde (El-Sanady, 2005).

Previous studies recorded the presence of *Blattisocius* species in different stored foodstuff: bean (Saleh, 1980), wheat bran (Saleh et al., 1985), bran (Baker, 2000), Athanassiou et al., 2001), rice and barley (Al-Nasser, 2007), maize, wheat flour, wheat, broad bean and rice (El-Sayed and Ghallab, 2007).

Decrease of mite numbers observed in the current study could be related to the climatic conditions during the seasons of sampling, which were inconvenient to the development and reproduction of the mites. We suggest that the components of wheat flour and its moisture content could be effective in decreasing the mites' numbers. Previous study mentioned that the environmental and storage conditions may influence food contamination and mite development (Brazis et al., 2008).

A study on dust mites in groceries related the presence of mites in high numbers to the moisture, mould damage or lack of cleaning and they reported that having no mites in a sample does not mean that there are no mites in the premises because mite densities can be low or the sampled location may not be ideal for mites to thrive.

Conclusion

The previous discussed results provide limited but important information concerning the presence of mites in stored foodstuff especially wheat flour and their possibility to affect human health if handled, inhaled and/or ingested. Despite the decreasing rates of mites during the sampling period of this study, nevertheless, mites and their allergen could increase and accumulate in the infested wheat flour during the prolonged periods of storage especially if the climatic conditions were convenient for their development and reproduction.

There is a need to undertake similar studies on a wide range of foodstuff and to investigate the levels of mites allergen in food and to establish the threshold for allergic response with consideration to the prolonged period of storage and to the climatic changes (temperature and relative humidity) throughout the year which might be somehow convenient for the development and reproduction of mites and cause an increase in their population and allergen in the foodstuff.

ACKNOWLEDGEMENTS

The author would like to thank Professor Dr. Mohamed Zaky Ali, Zoology Department, Faculty of Sciences at Qena, South Valley University, Egypt, for revision of the manuscript in addition to the support provided by Professor Dr. Amina Essawy Essawy, Zoology Department, Faculty of Sciences, Alexandria University, Egypt, in the statistical analysis.

Information about temperature degrees and relative humidity were provided by the Presidency of Meteorology and Environment Protection in King Abdul Aziz University, Jeddah.

REFERENCES

Al-Nasser AS (2007). Studies on some stored products mites in Jeddah Governorate of Saudi Arabia. Ph.D. Scie. Thesis. Girls College ,King Abdul Aziz Univ. Jeddah. K.S.A.

Antunes HBB, Aruda LK, Bernd LAG (1999). Anaphylaxis induced by mite-contaminated corn flour. J. Allergy, 54: 204.

Aspaly G, Stejskal V, Pekar S, Hubet J (2007). Temperature – dependent population growth of three species of stored product mites(Acari:Acaridida). J. Exp. App. Acarol., 42: 37-46.

Athanassiou CG, Kavallieratos NG, Palyvos NE, Sciarretta A, Trematerra P (2005). Spatiotemporal distribution of insects and mites in horizontally stored wheat. J. Econ. Entomol., 98(3): 1058-1069.

Athanassiou CG, Palyvos NE , Eliopoulos PA , Papadoulis GT (2001) Distribution and migration of insects and mites in flat storage containing wheat. J. Phytoparasitica, 29(5): 379-392 .

Athanassiou CG, Palyvos NE, Eliopoulos PA, Papadoulis GT (2002) Mites associated with stored seed cotton and related products in Greece. J. Phytoparasitica, 30: 387-394.

Baker AA (2000). Studies on some stored product mites. M. Scie Thesis. Faculty of Agric., Alexandria Univ. Egypt.

Barker PS (1967). Bionomic of *Blattisocius keegani* Fox (Acarina: Ascidae), a predator on eggs of pests of stored grain. Can. J. Zool., 45(1): 93-99.
Between toxigenic *Aspergillus flavas* Link and mites (*Tyrophagus*
Blanco C, Quiralte J, Castillo R, Delgado J, Ateaga C, Barber D, Carrillo T (1997). Anaphylaxis after ingestion of wheat flour contaminated with mites. J. Allergy Clin. Immunol., 99: 308-313.
Borghetti C, Magarolas R, Badorrey I, Radon K, Morera J, Monso E (2002). Sensitization and occupational asthma in poultry workers (Barc). J. Med. Clin., 118(7): 251-255.
Brazis P, Serra M, Selles A, Dethioux F, Biourge V, Puigdemont A (2008). Evaluation of storage mite contamination of commercial dry dog food. J. Vet. Dermatol., 19(4): 209-214.
Capriles-Hulett MA, Caballero-Fonseca F (2008). Additional information on the pancake syndrome. J. Ann. Allergy Asthma Immunol., 101: 2-221.
Castillo S, Sanchez-Borges M, Capriles A, Suarez-Chacon R, Caballero F, Fernandez-Caldas E (1995). Systemic anaphylaxis after ingestion of mite-contaminated flour. J. Allergy. Clin. Immunol., 95:304.
Cuthbert OD, Jeffrey IG (1993). Barn Allergy: An allergic respiratory disease of farmers. J. Sem. Respir. Med., 14: 73-82
Danielsen C, Hansen LS, Herling C (2004). The influence of temperature and relative humidity on the development of *Lepidoglyphus destructor* (Acari: Glycyphagidae) and its production of allergens :a laboratory experiment. Exp. Appl. Acarol., 32(3): 151-170.
Dutau G (2002). Les acariens, de nouveaux allergènes alimentaires masqués. J. Allergol. Immunol. Clin., 42 : 171-177 .
Edrees NO, Saleh SM (2008). Population dynamics of house dust mites in Jeddah Governerate. Egypt. J. Exp. Biol., 4: 139-146.
El-Desouky TM (1991).Studies on certain stored product pests in Menoufia Governerate (Egypt). M. Sci. Thesis. Faculty of Agriculture, Menoufia Univ. Shebin El-Kom, Egypt.
El-Sanady MA (2005).Studies on some stored product mites and their predators. Ph.D. Thesis, Faculty of Science (Girls) Al-Azhar Univ. Egypt, p. 193.
El-Sayed FMA, Ghallab MMA (2007). Survey on mites associated with major insect pests infesting stored grains in Middle Delta. J. Egypt. Soc. Acarol., 1: 29-38.
Fawzy MMH (1996). Biological studies on house dust mites in Egypt. Ph.D. Thesis, Fac. Agric. Cairo Univ. Egypt, p. 237.
Franzolin MR, Gambale W, Cuero RG, Correa B (1999). Interaction *putrescientiae* Schrank) on maize grains effects on fungal growth and aflatoxin production. J. Stored Prod. Res., 35: 215-224.
Guerra Bernd LA, Arruda LK, Antunes HBB (2001). Oral anaphylaxis to mites. J. Allergy., 56: 88-89.
Habibpour B, Kamali K, Meidani J, Adler C, Navarro S, Scholler M, Stengard-Hansen L (2002). Insects and mites associated with stored products and their Arthropoda parasites and predators in Khuzestan province (Iran). Proc. of the IOBC-WPRS working group "Integrated protection in stored products", Lisbon, Portugal. Bulletin-OILB-SROP., 25(3): 89-91.
Hafez SM (1977). Studies on predacious and parasitic mites of stored food. Ph.D. Thesis. Faculty of Agriculture Ain Shams University.
Harju A, Husman T, Merikoski R, Pennanen S (2006). Exposure of workers to mites in Finnish groceries. J. Ann. Agric. Environ. Med., 13: 341-344.
Hubert J, Stejskal V , Kubatova A , Munzbergova Z, Vanova M, Zdarkova E (2003). Mites as selective fungal carriers in stored grain habitats. J. Exp. Appl. Acarol., 29(1-2): 69-87.
Hubert J, Munzbergova Z, Kucerova Z, Stejskal V (2006). Comparison of communities of stored product mites in grain mass and grain residues in the Czech Republic . J. Exp. Appl. Acarol., 39(2): 149-158.
Hubert J, Stejskal V, Munzbergová Z, Kubátová A, Váòová M, ïárková E (2004). Mites and fungi in heavily infestedstores in the Czech Republic. J. Econ. Entomol., 97: 2144-2153.
Hughes AM (1976). The mites of stored food and houses. Min. Agric. Fish Food (London) H.M.S.O., pp. 1-2.
Kandil MM (1974). Effect of food on the biology of Family Cheyletidae. M.SC. Thesis. Faculty of Agriculture Cairo University.
Korunic Z (2001). Allergenic components of stored agro products. Arh.

Hig. Toksikol., 52(1): 43-48.
Krantz GW (1978). A manual of Acarology. Oregon State Univ. Book Stores Ltd. Corvallis, Oregon.
Lopez JE (2005).Parasite prevalence and the size of host population: an experimental test. J. Parasitol., 91(1): 32-37.
Lukas J, Stejskal V, Jarosik V, Hubert J, Zdarkova E (2007). Differential natural performance of four *Cheyletus* predatory mite species in Czech grain stores. J. Stored. Prod. Res., 43: 97-102.
Mahgoob AEA, Badawy AI, Badoor IM (2006). Survey of mites associated with grain residues and mixed flour in warehouses and mills in great Cairo . Arab- Universities. J. Agric. Sci., 14(1): 509-529.
Matsumoto T, Gotto Y, Miike T (2001). Anaphylaxis to mite-contaminated flour. J. Allergy, 56: 247-265.
Matsumoto T, Satoh A (2004). The occurrence of mite-containing wheat flour. J. Ped. Allergy Immunol., 15(5): 469-471
Musken H, Franz JT, Wahl R, Paap A, Cromwell O, Masuch G, Bergmann KC (2003). Sensitization to different mite species in German farmers *in vitro* analyses. J. Investig. Allergol. Clin. Immunol., 13(1): 26-35.
Olsson S, Van Hage- Hamsten M (2000). Allergens from house dust and storage mites: similarities and differences, with emphasis on the storage mite *lepidoglyphus destructor*. J. Clin. Exp. Allergy, 30: 912-919.
Palyvos NE, Emmanouel NG, Saitanis CJ (2008). Mites associated with stored products in Greece. J. Exp. Appl. Acarol., 44(3): 213-226.
Putatunda BN (2005). Mites (Acarina) associated with stored food products in Himachal Pradesh, India: a taxonomic study. J. Entomol. Res., 29(1): 79-82.
Saleh SM (1980). Studies on some mite species Ph.D. Sci. Thesis. Faculty of Agriculture. Alexandria University, Egypt.
Saleh SM, Al-Nasser AS (2007). Histopathological effects of the flour mites, *Acarus siro* on the small intestine of Wister rats. Egyptian J. Exp. Biol. Zool., 3: 179-184.
Saleh SM, El-Helaly MS, El Gayar FH (1985). Survey on stored product mites of Alexandria (Egypt). J. Acarologia, 26 (1): 87-93 .
Sanchez-Borge M, Capriles-Hulett A, Fernandez-Caldas E, Suarez-Chacon R, Caballero F, Castillo S, Sotillo E (1997). Mite-contaminated foods as a cause of anaphylaxis. [25]Sanchez-Borges. J. Allergy. Clin. Immunol., 99(6): 738-743
Sanchez-Borges M, Capriles- Hulett A, Suarez-Chacon R, Fernandez-Caldas E (2001). Oral anaphylaxis from mite ingestion. J. Allergy. Clin. Immunol. Int., 13: 33-35.
Sánchez-Borges M, Suárez-Chacon RI, Capriles-Hulett A, Caballero-Fonseca F, Iraola V, Fernández-Caldas E (2009). Pancake Syndrome (Oral Mite Anaphylaxis). J. World. Allergy Organ., 2(5): 91-96.
Sánchez-Borges M, Suárez-Chacon RI, Capriles-Hulett A, Caballero-Fonseca F (2008).additional information on the pancake syndrome. J. Ann. Allergy Asthma Immunol., 101: 2-221.
Scala G (1995). House dust mite ingestion can induce allergic intestinal syndrome. J. Allergy, 50: 517-519.
Stejskal V, Hubert J (2008). Risk of occupational allergy to stored grain arthropods and false pest-risk perception in Gzech grain stores. Ann. Agric. Environ. Med. J., 15(1): 29-35.
Thind BB, Ford HL (2006). Laboratory studies on the use of two new arenas to evaluate the impact of the predatory mites *Blattisocius tarsalis* and *Cheyletus eruditus* on residual populations of the stored product mite *Acarus siro*. J. Exp. Appl. Acarol., 38(2/3): 167-180.
Wafa AK, El Kifl MA, Hegazy AH (1996). Survey of stored grain and seed mites. Bull. Soc. Entomol. Egypt, 50: 225-232.
Walusiak J, Krawczyk - Adamus P, Hanke W , Wittczak T, Palczynski C (2004). Small non-specialized farming as a protective factor against immediate – type occupational respiratory allergy. J. Allergy., 59(12): 1294-1300.
Wang DY, Goh DYT, Ho AKL, Chew FT, Yeoh KH, Lee BW (2003). The upper and lower airway responses to nasal challenge with house-dust mite *Blomia tropicalis*. J. Allergy, 58(1): 78-82.
Wen D, Shyur SH, Ho C (2005). Systemic anaphylaxis after the ingestion of pancake contaminated with the storage mite *Blomia freemani*. J. Ann. Allergy, Asthma Immunol., 95(6): 612-614 .
Yan TS, Tham E, Tzien YC, Cheng YF, Jial C, Nge C, Yan CK , Wah LB (2008). Anaphylaxis following the ingestion of flour contaminated

by house dust mites- a report of two cases from Singapore. Asian Pac. J. Allergy Immunol., 26: 165-170.

Yi FC,Chen JY, Chua KY, Lee BW (2009). Dust mite infestation of flour samples. J. Allergy, 64(12): 1788-1789.

Zaher MA (1986). Predacious mites and non-phytophagous mites in Egypt (Nile valley and Delta) Pl. programme. USA project, p. 557.

Zdarkova E, Reska M (1976). Weight losses of ground nuts *Arachis hypogaea* L. from infestation by the mites *Acarus siro* L. and *Tyrophagus putrescentiae* (Schrank). J. Stored Prod. Res., 12: 101-104.

Zheltikova TM, Gervazieva VB, Zhirova SN, Mokronosova MA, Sveranovskaia VV (1997). Storage mites as the source of household allergens . J. Zh. Mikrobiol. Epidemiol. Immunobiol., 6: 73-76.

24

The role of cockroaches and flies in mechanical transmission of medical important parasites

Gehad T. El-Sherbini[1]* **and Eman T. El-Sherbini**[2]

[1]Department of Parasitology, Faculty of Pharmacy, October 6 University Cairo, Egypt.
[2]Department of Zoology, El Nahda University, Beni Sweif, Egypt.

Arthropods can be found on human or animals as ecto-parasites. Vectors can contaminate stored foods and transmit illness, or introduce diseases in new areas. Pet species facilitating infestations, and different risk factors related to infestation were identified. To determine the possible role of insects in dissemination of medically important parasites, study was carried out in residential areas of Khaldyia Village, El–Fayoum governorate, Egypt In 2009 - 2010. Parasites of medical importance were isolated and identified. The cockroaches collected from residential areas were 45 as the control group. A total of 178 cockroaches were collected, over a period of one year. Flies collected from human feces were also observed. Flies were abundant in defecation areas and around houses. This study aimed to isolating parasite objects from the exteriors of the bodies of flies, and Isolation and identification of parasites from internal and external surface of cockroaches, and to investigate parasite transmission rates among flies in an unsanitary community.

Key words: Mechanical transmission, flies, cockroaches.

INTRODUCTION

Arthropods are probably the most successful of all animals because of one or other reasons. They are found in every type of habitat and in all regions of the world. They feed on a wide variety of plant or animal material and have been known as major cause diseases for centuries.

Without the vector, the parasite life cycle would be broken and the pathogen cannot survive. Vectors can cause harm in different ways. They may cause illness, and this may happen through the consumption of food contain human entero pathogens, mechanically transmitted by flies or cockroaches. Muscoids dipterans have always been associated with human and domestic animals due to the abundance of food resources found in stables and domestic garbages. These flies are of major concern for veterinary medicine due to their capacity to act as vectors of several pathogenic organisms such as protozoa cysts, helminth parasites, enteropathogenic bacteria, and enterovirus (Graczyk et al., 2001).

Stored food products may be damaged or contaminated

by live or dead insects, faeces from them, odours, webbing or cast skins. Furthermore, vectors such as mosquitoes maybe introduced, and established in areas in which they have not previously been found (Chandler and Read, 1962), and where vector borne diseases can spread. Closed living accommodation favour the spread of ecto-parasites. Overcrowding, bad hygiene and lack of ventilation made the place an ideal environment for infection.

Soil transmitted helminths (STH) are relatively common parasites in the slum and rural area of many countries (Che Ghani et al., 1993; Sornmani et al., 1983; 2004), this high prevalence is closely related to poor environmental hygiene, and impoverished health services (Montresor et al., 1998). The main source of transmission is defecation outside latrines by heavily infected persons (Mott, 1989). While contaminated water might be the major transmission mode, indirect transmission by non biting flies cannot be excluded (Chandler and Read, 1962; Getachew et al., 2007). Many authors have indicated that the primary school children are an ideal target group for (STH) (Bundy et al., 1992), as children frequently defecate indiscriminately around their houses, particularly in the courtyards, sitting room, drains, even houses

*Corresponding author. E-mail: gody_55@yahoo.com.

has latrines (Lai and Ow Yang, 1993; Yu et al., 1993).

Over 50 species of synanthropic flies have been reported to be associated with unsanitary conditions and are involved in dissemination of human pathogens in the environment(Olsen, 1998).

In unsanitary communities, garbage, dead animal carcasses, and piles of feces, are often scattered around the houses. Flies are commonly found both indoor and outdoors. They persist on excrement, dead animal bodies, and contaminated areas where faecal matter, large amounts of organic waste, and piles of garbage are left exposed and unattended (Sualiman et al., 1989), theoretically, flies can transmit helminths through mechanical or biological means (Harwood and James, 1979). Human pathogens can also be transmitted as airborne particles for short distances from fly- electrocuting traps, as electrocuting traps do not alter the infectivity of pathogens transported by flies (Olsen, 1998). There are size limitations regarding the transmittal from the communicated sites. Bigger particles such as helminth eggs are transported by flies on their external surface, that is, exoskeletons, while small cystic stages of human infectious intestinal protozoa can be ingested as well as transported on the exoskeleton.

When infected persons excrete in open areas, there is an increased risk of contact between flies and pathogen-positive faecal matter (Getachew et al., 2007). Several studies have shown that eggs of Ascaris lumbericoides, Trichuris trichiura, hook worm, Entrobious vermicularis, Taenis sp., Hymenolips nana, Toxocara canis, hook worm larvae, and Strongyloides larvae are carried by many species of house flies (Getachew et al., 2007; Sualiman et al., 1989).

Cockroaches are among the most notorious pests of premises, they frequently feed on human faeces, and therefore they can disseminate cysts of enteric protozoans in the environment if such faeces are contaminated (19). They can not only contaminate food by leaving droppings and bacteria that can cause food poisoning (Che Ghani et al., 1993) but also they can transmit bacteria, fungi, and other pathogenic microorganisms (Czajka et al., 2003; Kopanic, 1994). Cockroaches feed on garbage and sewage and so have copious opportunities to disseminate human pathogen (Cotton et al., 2000; Pai et al., 2005). Also their nocturnal and filthy habits make them ideal carriers of various pathogenic microorganisms (Allen, 1987). Some parasites have been found in external surface or internal parts of body of cockroaches and some studies have shown that exposure to cockroach antigens may play an important role in asthma- related health problems (Montresor et al., 1998; Mott, 1989).

The study was carried out in an unsanitary community where parasitic infections and soil contamination with helminth ova were high and where flies were abundant in defecation areas and house hold environment.

This study aimed to isolate parasite objects from the exteriors of the bodies of flies, and Isolation and identification of parasites from internal and external surface of cockroaches, and to investigate parasite transmission rates among flies in an unsanitary community. The Ethics Committee for this study protocol has been approved by the governorate.

MATERIALS AND METHODS

Study area

The study area was Khalidiya village. It is located north of the Fayoum city. The village presents severe lack of hygiene services, although it occupies a unique tourist site that is far less than half a kilometer from Qarun lake.

This area was chosen because when we conducted a study of parasite infections in this area, we observed many piles of faecal matter in the nearby mangrove swamp.

Flies were seen everywhere in the area on food during meals, around children eating confectionery, flies were above sleeping children, dropped food, baby toys, and garbage. The abundance of flies around the house hold was attributable to the stock of rotten fish, and could also be caused because of the product of the cuttlefish harvest.

Fly collection

Flies were collected from the mangrove swamp and the nearby community, since many flies were seen on every pile of faecal matter in the swamp. The best method for collecting them was by stool bait trap.

Field-expedient bottle traps

Fly traps can be fashioned from disposable plastic water bottles. The simplest of these were constructed by cutting off the top and inverting it to form a cone leading into the body of the bottle, where bait is placed. Flies attracted to the bait are trapped inside the bottle and disposed of when the bottle becomes too full to be effective (Figure 1). Baits may consist of spoiling fruit or meat, food residue, and similar fragrant items. Once flies are attracted into the bottle, their natural pheromones increase attractiveness of the trap to other flies. These traps can be hung (no higher than 2.5 to 3 m) or placed on the ground out of traffic areas.

Under adverse environmental conditions, such as constant high wind, rain, or dust storms which cause fly baits less effectiveness, it may becomes necessary to employ alternatives for dispensing baits. One of it is to add poison in the bait of the trap which is illustrated earlier, or fashion a trap filled to a depth of 5 cm with poison fly bait and in which four 6 mm holes are cut near the top of the bottle to allow the flies access (Figure 2). The trap should be hung between 1 and 3 m above the ground. These traps also work well indoors. The contents must be shaken periodically so that dead flies do not accumulate on the surface of the bait, inhibiting contact between newly attracted flies and the poison. Another technique is to place the bait in a box to keep it from blowing away or becoming soaked or dust-coated. A simple example is to put granular fly bait in a flat box constructed from scrap wood, clearly labeled with the appropriate warning, and place the box on the ground where flies can access it. Such boxes should be checked periodically to dump dead flies and recharge them with bait. Dead flies should be disposed with waste material, ideally with medical waste when possible. An added advantage to this method is that it prevents troops from collecting and misusing the bait. These bait stations work well when placed near latrines, showers, and waste disposal sites

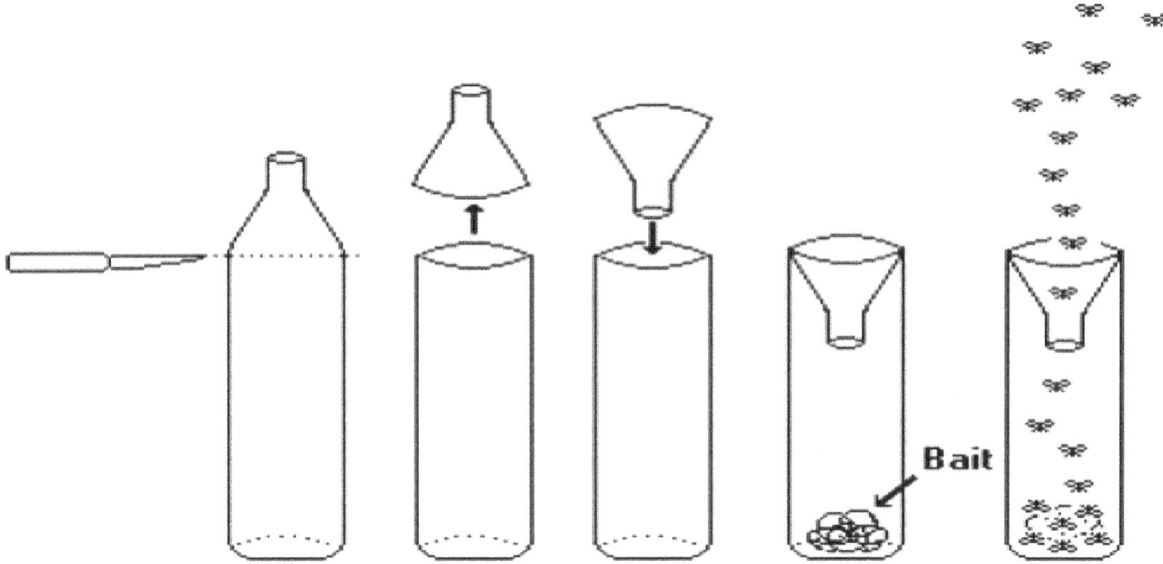

Figure 1. Plastic water bottle fly trap (inverted cone model).

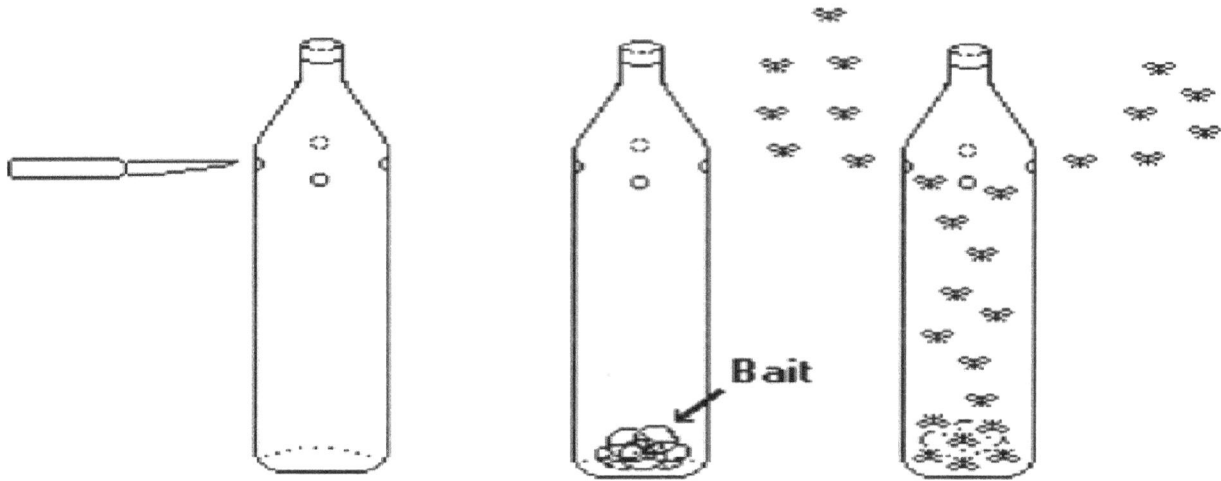

Figure 2. Plastic water bottle fly trap (multi-hole model).

(burn locations, dump sites, etc.

A stool sample with a mass of flies on its surface was chosen as the bait was placed in a fly trap left in the area and left for 1 h. To attract the attention of the flies, other piles of faecal matter were covered with sand, soil, leaves, or a wood during fly collection. Due to the strong sunlight and high temperatures during the day time summer (July), most of the flies become somnolent and were easily collected in the collecting chamber and transferred to collecting tubes. In the community rotten fish was used to lure and trap the flies.

Fly examination

The flies were divided into 2 groups, as follow:

1. Flies, from mangrove swamp, represented a homogeneous contamination of the fly population since they were exposed to the same source of infection and community group.
2. Helminths ova on the body surfaces were examined using manual shaking technique. 10flies were pooled and stored in a test tube with formalin detergent solution (FDS), while 200 flies were processed per test.

Specimens were brought to the laboratory for processing.

Laboratory processing

The collected flies were washed by manual shaking for 1 min. The flies were removed by clean forceps and kept for identification.

The remaining preparation was centrifuged at 2000 rpmfor2min. All sediments were examined under a light microscope for parasite objects.

Cockroach collection

One hundred and seventy-eight cockroaches were collected, over a period of one year, 133 from different wards of houses of the village as the test group and 45 from residential areas, situated within 2 km premises from the study areas as the control group.

The test group of insects captured (mostly at night time or early morning) from the floor of wards and kitchens, basements or bathrooms of residential areas.

Each cockroach was collected in a sterile test tube transported to the laboratory and anaesthetised by outing at 0°C for 5 min examined under the dissecting microscope and identified using standard taxonomic keys. For comparing control and test group, Chi –square test was applied.

Isolation and identification of parasite from external surface

After identification, 2 ml of sterile normal saline (0.9%) was added to the test tube and the cockroaches were thoroughly shaken for 2 min.

Isolation of parasitic cyst was carried out by using 1 ml of the washing result which was centrifuged at 2000 rpm for 5 min. The deposit was then examined after staining with 1% lugols iodine under light microscopy and identified (Beaver et al., 1984).

Isolation and identification of parasites from internal surfaces

After external washings, cockroaches were placed in flasks rinsed with 70% alcohol for 5 min. (to decontaminate external surfaces as 70% alcohol is bactericidal). They were transferred to other flasks and allowed to dry at room temperature. Cockroaches were then washed with normal saline for 2t o 3 min to remove traces of alcohol. Only whole and live captured cockroaches were utilised for the study. After being immobilised at 0°C the gut of the cockroach was dissected out and macerated in 2 ml of normal saline. The resulting macerate was then processed in a similar way as described previously and the results recorded. For parasites ova/cysts, about 1 ml of the washing result was centrifuged at 2000 rpm for 5 min. and the resulting deposit examined after staining with 10% Lugol iodine under light microscope and identified (Beaver et al., 1984).

RESULTS

Flies with helminthic objects on the body surface

A total of 576 house flies (195 male and 381 female) were studied, all were identified as *Musca domestica*, and *Chrysomya megacephala*.

108 flies from the swamp area and 68 house flies in the community were studied. The results are shown in Table 1 and 2.

Among 28 flies from the swamp area, 17 Hook worm eggs, 17 *T. trichura* and 1 *Ascaris* were detected (25.9%). The average egg count per positive fly was 1:3. Most flies carried only one egg on the body surface, while 17.9 and 7.1 % had 2 and 3 eggs respectively.

In the community 8 house flies (11.8%) were

contaminated with Hook worm and *T. trichura* eggs, the average egg count per positive fly was 1.0.

Faecal dots attached to the inner surface of the collection chamber were washed with FD solution and examined for helminth objects, this contamination occurred when the flies were trapped in the chamber. In the laboratory, the washing was processed by sedimentation method.

0.5 ml of sediment was obtained containing 27 *T. trichura* and 27 hook worms; the total number of eggs carried by the 508 flies in the mangrove swamp was derived from the pooled eggs from the sediment, egg in the two trials. The average number of eggs on the body surface of a fly was 0.4. Mites, the ectoparasite of the flies were also isolated from the body surfaces of the house flies, the number of mites per fly ranged 1 to 40.

This study revealed that cockroaches trapped from different sites (toilets, parlours, kitchens and bedrooms) in the houses with pit latrines and water system shared the same parasites. The parasites included: cysts of *Entomoeba hystolitica*, oocysts of *C. parvum*, *C. cayetenensis* and *Isospora belli*, cysts of *Balantidium coli*, ova of *A. lumbricoides*, *Anchylostoma deodunale*, *Enterobius. vermicularis*, ovae *Trichuris. trichura* and larvae of *Strongyloides stercoralis*. (Table 3)

Cockroaches trapped in the toilets of houses with pit latrines had a mean parasites count of 98 parasites/ml while those trapped in the houses with water system had a mean parasitic count of 31 parasites/ml. On the other hand a mean parasitic count of 19 parasites/ml were recorded from kitchens of houses with water system (Table 4).

Medically important parasites were isolated from external and internal surface of 98% of test cockroach and 8.9% of control cockroach.

Human parasites isolated from test group of cockroaches showed adult *E. vermicularis* and 8 *Ascaris* egg in two cockroaches but observation of control group did not show any parasites.

DISCUSSION

In the village, people did not consider parasites infection a serious problem, most did not submit stool for examination. Some infected cases refused to take antiparasitic drugs.

House flies, bush flies, and blow flies were common around the house holds, in garbage and in human and animal excreta (Getachew et al., 2007: 30; Sualiman et al., 1989; Monzon et al., 1991)

House flies are a proven mechanical transmitter of pathogens to human food (Sulaiman S,Sohadi AR, Yunus H, Iberahim R, 1988). Ten intestinal helminth eggs and larvae has been isolated from flies collected around house hold, in un urban slum area, on an open defecation area, garbage heap, a small open air market, larvae has been isolated from flies collected around household

Table 1. Percentage of contaminated flies in study area.

Area	No. examined	Positive flies [No. (%)]	Helminths eggs and larvae			
			Hook worm	T. trichura	Ascaris	H. nana
Swamp	108	28 (25.9)	17	17	1	4
Community	68	8(11.8)	5	5	0	1

Table 2. Number of parasite eggs on the body surfaces of flies in the swamp area.

Trails	Number of flies	Helminths eggs and larvae				Total
		Hook worm	T. trichura	Ascaris	H. nana	
Manual	400		86	34	-	121
Contamination rate	108	17	17	1	-	35
Washed sediment	-	27	27	-	-	54
Total	508	130	78	2	-	210

Table 3. Distribution of medical important parasites by sites and toilet facilities.

Source	No. of cockroaches studied	
	Pit toilet	Water system
Toilets	35	22
Kitchens	27	22
Living-rooms	16	10
Bed-rooms	14	8

Parasites identified—Cysts of E. histolytica, cysts of Balantidium coli, ova of Ascaris lumbricoides, Anchylostoma deodunalae, Enterobius vermicularis, ova of Trichuris trichura and larvae of Strongyloides stercoralis.

Table 4. Mean parasite count of cockroaches by site and toilet facilities.

Source	Pit toilet	Water system
	Mean parasitic count (parasite/ml)	Mean parasitic count (parasite/ml)
Toilets	98	31
Kitchens	50	19
Living-rooms	47	11
Bed-rooms	38	11

Parasites identified—Cysts of E. histolytica, cysts of Balantidium coli, ova of Ascaris lumbricoides, Anchylostoma deodunalae, Enterobius vermicularis, ova of Trichuris trichura and larvae of Strongyloides stercoralis.

in un urban slum area, on an open defecation area, garbage heap, a small open air market, and meat butchers near human dwellings (Getachew et al., 2007;30; Sualiman et al., 1989; Beaver et al., 1984). Because the mangrove swamp was the post- tsunami defecated area for the villagers, piles of faecal matter attracted flies. After feeding and resting, the flies travelled into the community, about 100 m from the feeding site. Flies that had direct contact with parasite positive faeces were efficient carriers, because at least 25.9% were contaminated with pathogens. In the defecation area every 2 to 3 flies carried at least 1 parasite objects on the body surface.

After feeding, they rested in the area and contaminated the environment with the pathogens on their footpads, hairs, bristles, and external mouth parts.

An almost invisible dot of faeces in the environment might contain eggs or larvae that develop further and then transmit to humans.

In this community 11.8% of flies had eggs on their body surface and could transmit them to human food, and household surroundings. 25.9% of infected flies had 2 to 3 eggs adhering to their body surface. The study found that 508 flies could leave 0.5 ml of faecal sediment in the collection chamber. This was considered to be the amount of pathogenic faecal matter distributed into the environment by 508 flies, thus, a fly carried 0.001 g faecal mass on the body surface after feeding on human waste.

We did not investigate the presence of parasite objects in flies' guts. Nevertheless many researchers have reported higher parasite detection rates in the gastrointestinal lumen than on body surface (Getachew et al., 2007; Monzon et al., 1991; Khan and Huq, 1978).

Sulaiman et al. (1989) found hook worm eggs and larvae in the gut of lies, but found more on external surfaces. From this investigation it may be concluded that only one person with a light soil transmitted infection can contaminate both defecation areas and disease vectors. Over 25.9% of the fly population was contaminated. After resting and contaminating the environment with infective matter carried on the body surface, they transmitted the infection to the community, at the rate of 11.8%. The discovery of A. lumbercoides eggs on the flies supported the supposition that housefly was a potential STH transmitter, which could carry and spread pathogens to other places, since they are able to travel up to 20 miles to unsanitary sites (Umeche and Mandah, 1989).

The current study showed that the house fly is a potential mechanical vector for parasite infection, and therefore its role in disease transmission should be not being underrated. In high risk areas, health education targeting the elderly should emphasize personal and environmental hygiene. In areas where open air defection is common, food must be strictly protected from house flies, since in this study are in every 11 flies around the house was found positive for helminth eggs. Other microorganisms causing bacterial infection have been reported (Getachew et al., 2007).

The control or eradication of house flies should be attempted, to stop intestinal parasite transmission in the community, in addition to drug administration.

The results of the present study revealed contamination of almost all cockroaches collected from homes with different parasites which are significantly higher in comparison to control group. The importance of cockroaches as carrier of parasitic worm, cysts, or eggs, is because there are some reports about the presence of parasitic forms on or in cockroaches (Greenberg, 1973). The finding of the present study showed the parasitic contamination in high numbers. The presence of E. vermicularis infestation indicates that the cockroaches had opportunity to get in touch with infested patients or contaminated clothes which emphasises their vectorial potential for parasitic diseases (Chan et al., 2004).

Conclusions

Synanthropic insects such as flies and cockroaches can significantly contribute to the spread of food- borne parasites diseases in both developing and developed countries.

REFERENCES

Allen BW (1987). Excretion of viable tubercle bacilli by Blatta orientalis (the oriental cockroach) following ingestion of heat-fixed sputum smears: a laboratory investigation. Trans. R. Soc. Trop. Med. Hyg., 81: 98-99.

Beaver PC, Jung RC, Cupp EW (1984). Clinical Parasitology. 9 edn. Philadelphia: Lea and Febiger.

Bundy DAP, Hall A, Medley GF, Savioli L (1992). Evaluation measures to control intestinal parasitic infections. World Health Stat. Q., 45: 168-79.

Chan OT, Lee EK, Hardman JM, Navin JJ (2004). The cockroach as a host for Trichinella and Enterobius vermicularis: implications for public health. Hawaii. Med. J., 63: 74-77.

Chandler AC, Read CP (1962). Introduction to parasitology with special references to the parasites of man. 10th ed. New York: John Wiley & Sons.

Chandler AC, Read CP (1962). Introduction to parasitology with special references to the parasites of man. 10th ed. New York: John Wiley & Sons.

Che Ghani BM, Oothuman P, Hashim BB, Rusli BI (1993). Patterns of hookworm infections in traditional Malay villages with and without JOICFP Integrated Project in Peninsular Malaysia-1989. In: Yokogawa M, Editors. Collected papers on the control of soil transmitted helminthiases, Tokyo: APCO, 5: 14-21.

Cotton MF, Wasserman E, Pieper CH, Van Tubbergh D, Campbell G, Fang FC, Barnes J (2000). Invasive disease due to extended spectrum beta-lactamase-producing Klebsiella pneumoniae in a neonatal unit: the possible role of cockroaches. J. Hosp. Infect., 44: 13-17.

Czajka E, Pancer K, Kochman M, Gliniewicz A, Sawicka B, Rabczenko D, Stypulkowska- Misiurewicz H (2003). Characteristics of bacteria isolated from body surface of German cockroaches caught in hospitals. Przegl. Epidemiol., 57: 655-662.

Getachew S, Gebre-Michael T, Erko B, Balkew M, Medhin G (2007). Non-biting cyclorrhaphan flies (Diptera) as carriers of intestinal human parasites in slum areas of Addis Ababa, Ethiopia. Acta Trop., 103: 186-194.

Graczyk TK, Cranfield MR, Fayer R, Bixler H (1999). House flies (Musca domestica) as transport hosts of Cryptosoridium parvum. Am. J. Trop. Med. Hyg., 61: 500-504.

Graczyk TK, Knight R, Gilman RH, Cranfield MR (2001). The role of non-biting flies in the epidemiology of human infectious diseases. Microb. Infect., 3: 231-235.

Greenberg B, Flies, disease I (1973). Ecology, classification and biotic association. New Jersey: Princeton University Press, Vol. 1.

Harwood RF, James MT (1979). Entomology in human health, 7th ed. New York: Macmillan Publishing.

Khan AR, Huq F (1978). Disease agents carried by flies in Dacca city. Bangladesh. Med. Res. Counc. Bull., 4: 86-93.

Kopanic RJ (1994). Cockroches as vectors of Salmonella: laboratory and field trials. J. Food. Prot., 57: 125-132.

Lai KPF, Ow Yang CK (1993). Soil-transmitted helminthiasis in a rubber and oil-palm estate in Selangor, Peninsular Malaysia. In: Yokogawa M, Editors. Collected papers on the control of soil-transmitted helminthiases, Tokyo: APCO, 5: 72-77.

Montresor A, Crompton DWT, Hall A, Bundy DAP, Savioli L (1998). Guidelines for the evaluation of soil-transmitted helminthiasis and schistosomiasis at community level. WHO/CTD/SIP/98.1

Monzon RB, Sanchez AR, Tadiaman BM, Najos OA, Valencia EG, de Rueda RR (1991). A comparison of the role of Musca domestica (Linnaeus) and Chrysomya megacephala (Fabricius) as mechanical vectors of helminthic parasites in a typical slum area of Metropolitan Manila. Southeast Asian. J. Trop. Med. Public Health, 22: 222-228.

Mott KE (1989). The World Health Organization and the control of intestinal helminths. In: Yokogawa M, Editors. Collected papers on the control of soil-transmitted helminthiases, Tokyo: APCO, 4: 189-200.

Olsen AR (1998). Regulatory action criteria for filth and other extraneous materials. 3. Review of flies and foodborne enteric disease. Reg. Toxicol. Pharmacol., 28: 199-211.

Pai HH, Chen WC, Peng CF (2005). Isolation of bacteria with antibiotic resistance from household cockroaches (Periplaneta americana and

Blattella germanica). Acta Trop., 93: 259-265.

Sornmani S, Vivatanasesth P, Harinasuta C, Potha U, Thirachantra S (1983). The control of Ascariasis in a slum community of Bangkok. In: Yokogawa M, Editors. Collected papers on the control of soil-transmitted helminthiases, Tokyo: APCO, 290(35): 260-266.

Sualiman S, Mohammod CG, Marwi MA, Oothuman P (1989). Study on the role of flies in transmitting helminths in a community. In: Yokogawa M, Editors. Collected papers on the control of soil-transmitted helminthiases, Tokyo: APCO, 4: 59-62.

Sulaiman S, Sohadi AR, Yunus H, Iberahim R (1988). The role of some cyclorrhaphan flies as carriers of human helminths in Malaysia. Med. Vet. Entomol., 2: 1-6.

Umeche N, Mandah LE (1989). *Musca domestica* as a carrier of intestinal helminths in Calabar, Nigeria. East. Afr. Med. J., 66: 349-352.

Yu S, Xu L, Jiang Z, Chai Q, Zhou C, Fang Y (1993). Environmental and human behavioral factors in propagation of soil-transmitted helminth infections. In: Yokogawa M, Editors. Collected papers on the control of soil-transmitted helminthiases, Tokyo: APCO, 5: 83-88.

25

Bio-efficacy of synthetic chemicals, botanicals and microbial derivatives against scale insect *Coccus hesperidum* Linn. in arecanut

B. K. Shivanna[1]*, S. Gayathridevi[2], R. Krishna Naik[3], B. Gangadhara Naik[4], H. Shruthi[5] and R. Nagaraja[6]

[1]Department of Agricultural Entomology, UAS, AINRP (T), ZARS, Shimoga -577204, Karnataka, India.
[2]Department of Agricultural Entomology, UAS, College of Agriculture, Shimoga -577204 Karnataka, India.
[3]Department of Computer Science, UAS, College of Agriculture, Shimoga -577204 Karnataka. India
[4]Department of Plant Pathology, UAS, College of Agriculture, Shimoga -577204, Karnatak, India
[5]Department of Agricultural Microbiology, UAS, College of Agriculture, Shimoga -577204
[6]Department of Agricultural Microbiology, UAS, KVK, Shimoga -577204, Karnataka, India

Ignorance of plant protection in areca palms at early stages can cause considerable loss from the sucking pests, particularly, *Coccus hesperidum* Linn. (Hemiptera: Coccidae) during unfavorable weather conditions. In order to overcome this, replicated field trials at five different locations were conducted during 2008/2009 and 2009/2010. Synthetic chemicals (chlorpyriphos 20 EC at 2.5 ml/l, endosulfan 35 EC at 2 ml/l, bupfrofezin 25 SC at 1 ml/l and methomyl 40 SP at 2 g/l), Aazadirachtin 0.03% at 3 ml/ (botanical group) and spinosad 45 SC at 0.5 ml/l (microbial derivative) including an untreated check were imposed twice at an interval of 15 days. Treatmental effects were assessed five days after each spray from 2 cm^2 leaf area. Pooled results indicated that all the insecticide treatments were found to be significantly superior over untreated check control by recording the lowest population of scales. Spinosad and bupfrofezin were found to be significantly superior and were on par with methomyl by registering lowest number of scales. Methomyl was on par with ruling insecticide endosulfan and was significantly different from standard check chlorpyriphos with lesser population of scales. Azadirachtin recorded higher scale population than other insecticide treatments. Microbial derivative spinosad and bupfrofezin were found to be effective against arecanut scales than other treatments and can be used in managing arecanut scales.

Key words: Bio-efficacy, synthetic chemicals, botanicals, microbial derivative, *Coccus hesperidum,* arecanut.

INTRODUCTION

Arecanut is largely cultivated in the plains and foothills of Western Ghats and north eastern regions of India. Area and production in different states indicate that Karnataka, Kerala and Assam account for over 90%. The arecanut palm, *Areca catechu* L. (Aracaceae) has been an important commercial crop and is the source of arecanut commonly referred to as betelnut or supari in India. Since time memorial, it is being used in masticatory (chewing), religious and social ceremonies (Murthy, 1968). Due to lack of scientific knowledge and ignorance by the cultivators on agronomic aspects, pest and diseases considerable crop losses were encountered in fields. An array of insect and non insect-pests infests all parts of the palm, such as stem, leaves, inflorescence, roots and nuts in one or other stage of the crop growth. As many as 102 insect and non-insect pests have been reported to be associated with arecanut palm (Nair and Daniel, 1982).

*Corresponding author. E-mail: bkshivanna@gmail.com.

Table 1. Effect of different insecticides on control of arecanut scales *Coccus hesperidum* Linn.(Coccidae: Hemimptera).

| Treatment | PTC | | | 5 DAT | | | | | |
| | | | | First spray | | | Second spray | | |
	2008	2009	pooled	2008	2009	pooled	2008	2009	pooled
Chlorpyriphos 20 EC 2.5 ml/l	15.21 (4.92)*	16.20(4.55)	15.90(4.05)	4.13 (2.53)	4.20 (2.48)	4.13 (2.16)	4.1 (20.9)	4.2 (2.12)	4.12(2.16)
Spinosad 45SC 0.5ml/l	15.11 (4.80)	16.20(4.59)	15.60(4.00)	1.53 (1.40)	1.2 (1.40)	1.31 (1.34)	1.6 (1.24)	1.6 (1.41)	1.6 (1.44)
Endosulfan 35 EC 2ml/l	15.43 (4.55)	16.12(4.32)	15.71(4.05)	4.4 (2.39)	4.2 (2.36)	4.3 (2.18)	4.2 (2.04)	1.8 (1.73)	3.0 (1.73)
Azadiarachtin 0.03% 4 ml/l	14.9 (4.33)	16.21(4.52)	15.21(4.03)	4.30 (2.50)	4.30 (2.52)	4.3 (2.19)	4.3 (2.37)	4.2 (2.17)	4.28(2.17)
Buprofezin 25SC 1ml/l	14.8 (4.27)	16.21(4.41)	15.50(4.03)	2.0 (1.58)	1.80 (1.92)	1.9 (1.54)	1.52 (1.46)	1.8 (1.84)	1.53(1.48)
Methomyl 40 SP 2g/l	16.1 (4.25)	16.31(4.38)	16.3 (4.09)	4.16 (2.34)	4.20 (2.39)	4.15 (2.16)	4.10 (2.10)	1.8 (1.85)	2.86(1.72)
Control Untreated check	16.2 (4.65)	16.40(4.49)	16.30(4.10)	16.20 (4.20)	10.40 (3.48)	13.10 (3.56)	16.21(4.71)	16.21(4.71)	16.21(4.14)
CV %	6.30	5.17	0.88	15.70	17.26	7.68	5.66	17.18	8.16
CD @ 5%	0.43	0.41	0.06	1.11	1.14	0.29	0.24	0.69	0.33

PTC= Pretreatment count, DAT= days after treatment * Figures in parenthesis are √x+0.5 transformed values.

Many species of scale insects infests the areca leaves. Among them, *Coccus hesperidum* Linn (coccidae: Hemiptera), a scale insect, is severe on undersurface of the leaves. Colonized feeding on under surface of the leaves by both nymphs and adults results in the production of yellow patches on the leaves, which under severe infestation, cover the entire leaf (Rao and Bavappa, 1961).

The honeydew secreted by this insect invites the sooty mould, which interfered with the photosynthesis of the palm. Heavy colonization in young seedlings results in severe blotching and drying of leaves (Daniel, 2003). Suggested neem formulations against foliage feeding *C. hesperidum*, such as nimbicidine and mulineem (Daniel, 2003) are in vogue and needs efficient molecules for the management of scales in arecanut plantation.

MATERIALS AND METHODS

A multi location field trial in three districts (five locations) was conducted for two consecutive seasons during 2008/2009 to 2009/2010 in randomized block design with seven treatments and three replications. The treatments which were replicated thrice are as follows: 1) chlorpyriphos 20 EC 2.5 ml/l, 2) spinosad 45SC 0.5 ml/l (microbial group), 3) endosulfan 35 EC 2 ml/l, 4) azadirachtin 0.03% 3 ml/l (botanical group), 5) buprofezin 25 SC 1ml/l, 6) methomyl 40 SP 2g/l and untreated check 7) control. Two insecticidal sprays were given at an interval of 15 days. The spray fluid was applied to the lower surface of leaves at the rate of 500 l/ha with a knapsack sprayer.

Ten plants were randomly selected in each plot by tying with luggage labels. A day before spraying, that is, pretreatment count (PTC) and five 5 days after spraying treatment, observations on number of scales per 2 cm² leaf area on top, bottom and middle leaves of selected plants were recorded. The efficacy was computed as reduction in number of scales compared to untreated check control.

The data on the (average of top, bottom and middle leaf of each plant) mean of three replications were considered for statistical analysis. Data were square root transformed and analyzed statistically.

RESULTS AND DISCUSSION

The results with respect to Table 1 were significant, indicating differential efficacy of the treatments imposed. Pooled data of two years in all the locations showed significant treatmental differences for scales population in areca leaves. Number of scales/2 cm² leaf/plant, and least number of scales (1.48 and 1.44 scales/ 2 cm² leaf area/plant) were observed in the second spray on the areca palm treated with buprofezin and spinosad respectively and are were found to be significantly superior over rest of the treatments. The level of scales population in standard check methomyl (1.72 scales/ 2 cm^2 leaf area/plant) was on par with spinosad, bupfrofezin and endosulfan. However, the plant based azadirachtin displayed moderate level of control (2.17 scales/2 cm^2 leaf area/plant) and was significantly different from the unsprayed control which recorded the highest population of 4.14scales/ 2 cm^2 leaf area/plant.

The reduction in scales population was due to the efficacy of newer molecules, such as bupfrofezin and spinosad. Literature on these molecules (bupfrofezin and spinosad) against scales was meager. However, minimum population of scales observed in azadirachtin treated plots was in confirmation with the results reported earlier by Daniel (2003) and Nair and Menon (1963).

REFERENCES

Daniel M (2003). NATP Final report on Development of IPM packages for plantation crops. CPCRI, Kasargod, p.184

Murthy KN (1968). Areca nut growing in north east India. Indian Farming, 18: 21.

Nair CPR, Daniel M (1982). Pests. In Bavappa et al. (eds): The Arecanut palm. (Bavappa KVA, Nair MK, Prem Kumar T, editors). CPCRI. Kasaragod, pp. 151-184.

Nair RB, Menon R (1963). Major and minor pests of arecanut crop. Areca catechu Linn. Arecanut J., 14: 139-147

Rao KSN, Bavappa KVA (1961). Nursery diseases and pests of arecanut and their control. Arecanut J., 12: 136.

26

Chromosome study in few species of Acridids (Acrididae: Tryxalinae): Karyotypic analysis and distribution patterns of constitutive heterochromatin

Pooja Chadha* and Anupam Mehta

Department of Zoology, Guru Nanak Dev University, Amritsar, Punjab India.

Chromosomes with detailed karyotypic information (nature, number, size, relative length, length of X-chromosome, nature of X-chromosome) and C-banding patterns of few species of grasshoppers belonging to sub-familiy Tryxalinae are discussed. The karyotypes comprises of acrocentric chromosomes with complement number 2n = 23 (male). Constitutive heterochromatin distribution was found at centomeric, interstitial, terminal sites along with thick and thin bands among all the species except in *Acrida turrita* which possessed only centromeric C-bands. The number and location of C-bands in Acridids exhibit intra specific variations. In the present communication the karyotypic analysis and C-banding patterns are analyzed for further differences between genera belonging to the same sub-family.

Key words: Acrididae, orthoptera, C-banding.

INTRODUCTION

Orthoptera has been considered as a classical material for karyological investigations. The size and number of their chromosomes are such that both qualitative and quantitative studies on chromosomal anomalies can be detected easily (Turkoglu and Koca, 2002). The karyotype is found to have a cytotaxonomic value. Acridoid group is known for its karyotypic uniformity or conservatism (Aswathanarayana and Ashwath, 2006).

The introduction of C-banding technique offers a simple mean of defining constitutively heterochromatic regions. C-banding technique has made it easier to assess the changes in constitutive heterochromatin and have revealed the existence of remarkable degree of C-band variations within species (King and John, 1980; Lopez-Fernandez and Gosalvez, 1981). Hsu (1974) hypothesized that the heterochromatin has passive role of body guard i.e. it is used by the cell as a body guard to protect the vital euchromatin by forming a layer of dispensable shield on the outer surface of nucleus. This heterochromatin is subjected to both qualitative and quantitative variations. Many studies on the C-bands

have been done in Acridoids which are known to possess high level of chromosomal variations. The number and location of C-bands in Acridids exhibit both intra- and interspecific variations (Yadav and Yadav, 1993). The present communication deals with the chromosome complement and distribution of constitutive heterochromatin along with the differences between genera belonging to same sub-family are discussed.

MATERIALS AND METHODS

The males of few species of Acrididae were collected in and around Guru Nanak Dev University campus, Amritsar (Punjab). The testis was excised following standard Colchicine-Hypotonic-Cell suspension-Flame dry technique (Yadav and Yadav, 1983). The flame dried slides were treated for C-banding following method of Sumner (1990) with slight modifications. The chromosomes were classified after Levan et al. (1964).

RESULTS

The perusal of Table 1 revealed that among the four species studied, all belong to sub-family Tryxalinae and these are *Acrida turrita* Linn., Acrida *exaltata* Walk.,

*Corresponding author. E-mail: poojachadha77@yahoo.co.in.

Table 1. Nature of chromosomes and morphometric characters in some species studied.

S/No.	Species	Male 2n number	Chromosome size			*Range of relative length of autosomes	Length of X- chromosomes	Nature of chromosomes			Nature of X-chromosome
			L	M	S			m	sm	a	
	Sub-family: Tryxalinae										
1	*Acrida turrita*	23	3	6	2	(19.9 ± 0.39) - (144.2 ± 0.48)	199.0 ± 2.26	-	-	All	Largest
2	*Acrida exaltata*	23	3	6	2	(35.1 ± 0.36) - (128.9 ± 1.32)	148.4 ± 0.42	-	-	All	Largest
3	*Phlaeoba infumata*	23	3	6	2	(38.0 ± 0.72) - (128.2 ± 0.46)	141.0 ± 1.18	-	—	All	Largest
4	*Phlaeoba antennata*	23	3	6	2	(23.2 ± 0.32) - (139.0 ± 1.02)	151.1 ± 1.05	-	-	All	Largest

L: Long, M: Medium, S: Small, m: metacentric, sm : submetacentric, a: acrocentric   * relative length of shortest and longest autosomes.

Phlaeoba infumata Brunn., *Phlaeoba antennata* Brunn. The diploid chromosome number was found to be 23 and sex-determining mechanism was found to be XO: XX type among all the species investigated. Table I shows the various morphometric characters of all the species investigated. It was ascertained that the chromosome morphology is acrocentric for all the species. X-chromosome is the marker as it is the largest element among all the species studied. Figures 1 to 4 showing spermatogonial metaphase of four species respectively.

Table 2 is showing the position of constitutive heterochromatin among male grasshoppers studied. It was observed that the centromeric bands of constitutive heterochromatin were found among all the species under study. The interstitial bands were seen in *A. exaltata*, *P. infumata*, *P. antennata* only. Terminal bands were seen in all the species investigated except in *A. turrita*. Thick bands were present in all the species while thin bands were seen in three species of hoppers excluding *A. turrita*. Figures 5 to 8 showing C-banded spermatogonial metaphase of the species.

DISCUSSION

The karyology of every species is unique in itself and provides an identity to species (Channaveerappa and Ranganath, 1997). The short horned grasshoppers are characterized by possessing acrocentric chromosomes. Due to great cytogenetic uniformity the short horned hoppers are considered as an example of "Karyotypic conservatism" (Aswathanarayana and Ashwath, 2006).

In the present study, 4 Acridids have been investigated which belong to sub-family Tryxalinae. It is revealed that hoppers belonging to family Acrididae have 23 chromosome numbers. The sex-determining mechanism is found to be XO/XX type among all the studied species. Yadav and Yadav (1986) reported similar results in relation to chromosome number and sex-mechanism among Haryana population of Acridoideans. While studying the chromosomes of 11 species of grasshoppers from Simla (H.P), Sharma and Gautam (2002) also revealed similar results. So, the short horned grasshoppers of different regions are showing cytogenetic uniformity regarding chromosome number and sex-determining mechanism. During the present investigation, the chromosomes are found to be acrocentric in nature. Upto six metacentrics through fusions have been reported in Tryxalines *Myrmeleotettix maculates* (John and Hewitt, 1966) and *Stauroderus scalaris* (John and Hewitt, 1968). Meanwhile, Aswathanarayana and Ashwath (2006) observed a series of structural changes involving 6th, 7th and 9th pair exhibiting hetero and homomorphism in *Gastrimargus africanus orientalis*. Mayya et al. (2004) reported short arms in chromosomes of *Aiolopus thalessimus tumulus* and *Acrotylus humbertianus*. Whereas, no such change have been reported in present study. The X-chromosome is found to be largest of all the other chromosomes among the 6 species

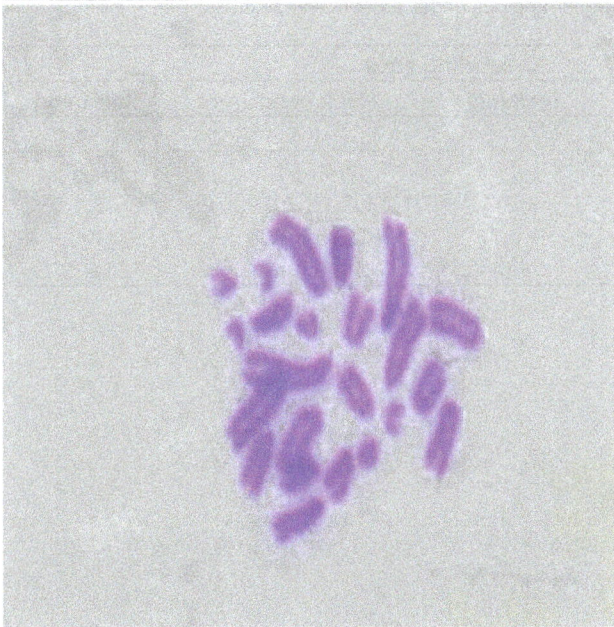

Figure 1. Spermatogonial metaphase of *Acrida turrita*

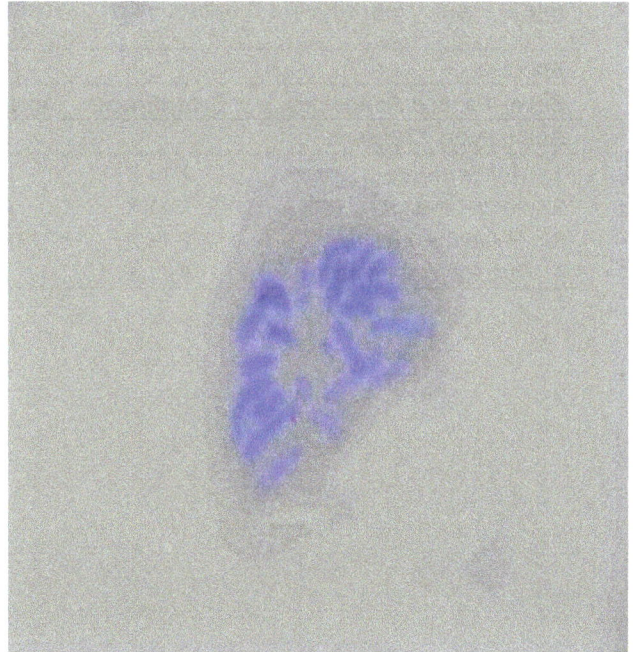

Figure 3. Spermatogonial metaphase of *Phlaeoba infumata*

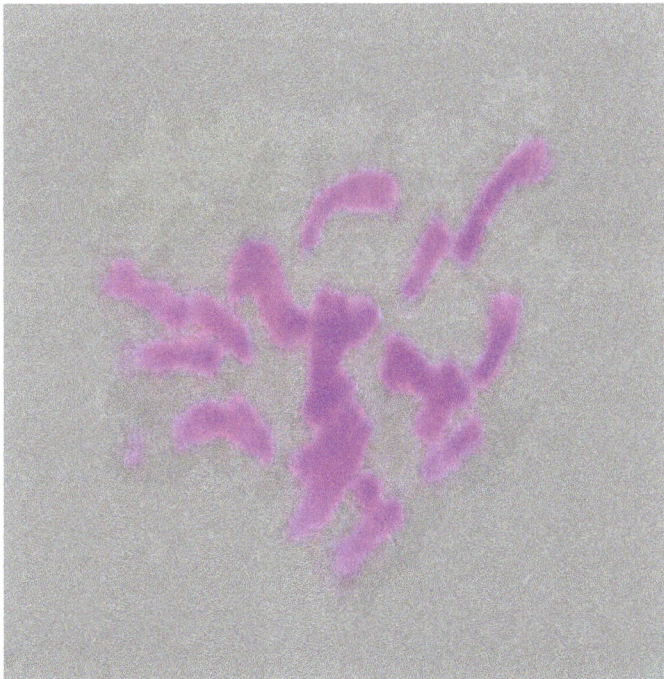

Figure 2. Spermatogonial metaphase of *A. exaltata.*

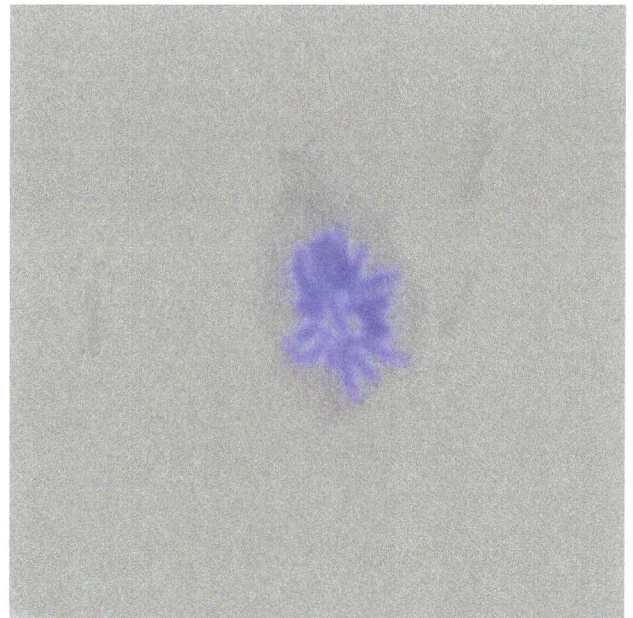

Figure 4. Spermatogonial metaphase of *P. anteenata.*

investigated. Mayya et al. (2004) also reported the X-chromosome to be largest in all the species except in *A. thalessimus tumulus* and *Spathosternum prasiniferum*. The C-band represents the constitutive heterochromatin in the homologous chromosome of a Yasminieh, 1971). C- banding pattern in various species

of grasshoppers provide important clues that have karyotype. This type of DNA consist of short repeated chromatin and said to be genetically inert (Yunis and polynucleotide sequences. C-bands exhibit centromeric, interstitial and terminal sites. It is a variant state of occurred during the course of evolution. Many studies have shown a remarkable degree of C-band variation.

Table 2. Showing the position of C-heterochromatin in male grasshopper species under present investigation.

Species	C-banding sites				
Super-Family: Acridoidea	**Centromeric**	**Interstitial**	**Terminal**	**Thick band**	**Thin band**
Sub family: Tryxalinae					
1. Acrida turrita	1-11,X	____	____	1,4,6,7	____
2. Acrida exaltata	1-11,X	1,X	1,2,4	1,2,3,4,5,6,7,8,9,X	3,4
3. Phlaeoba infumata	1-11,X	3,5,7,X	1,6,7	1,2,3,5,6,7,X	1,4
4. Phlaeoba antennata	1-11,X	2	4,5	1,2,3,4,X	1,3,6

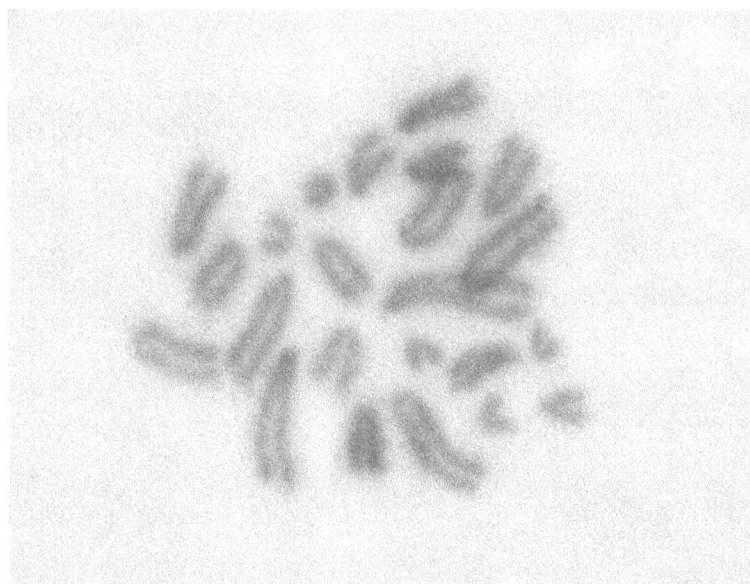

Figure 5. C-banded spermatogonial metaphase of *A. turrita*.

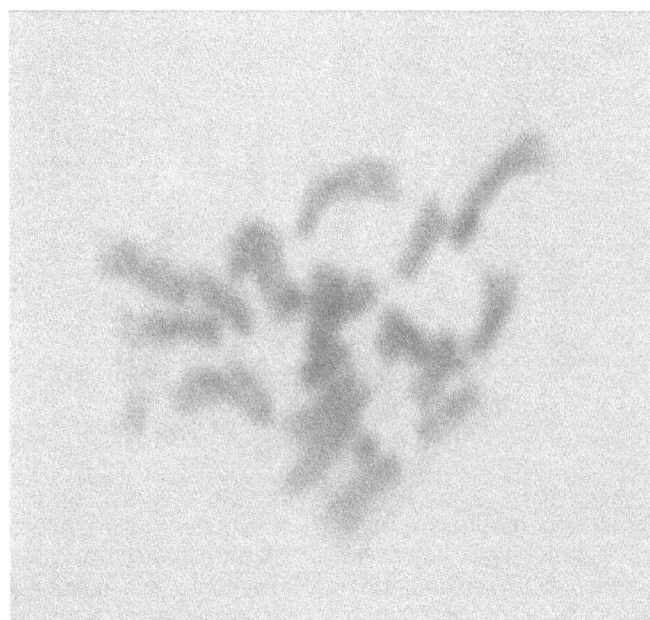

Figure 6. C-banded spermatogonial metaphase of *A. exaltata*

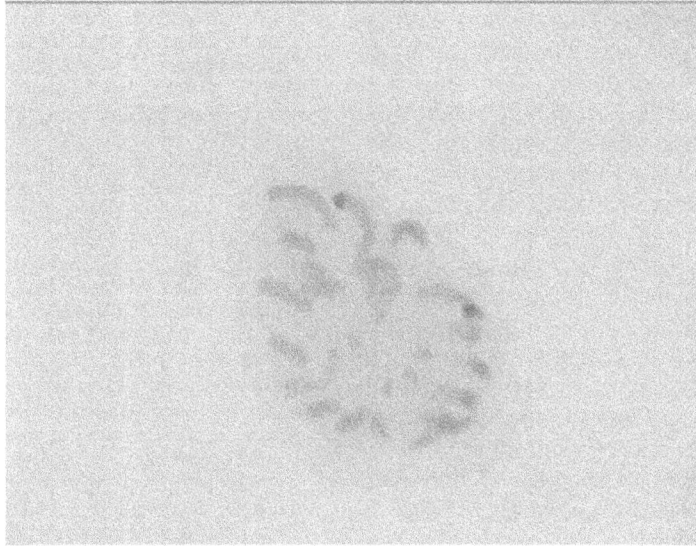

Figure 7. C-banded spermatogonial metaphase of *P. infumata.*

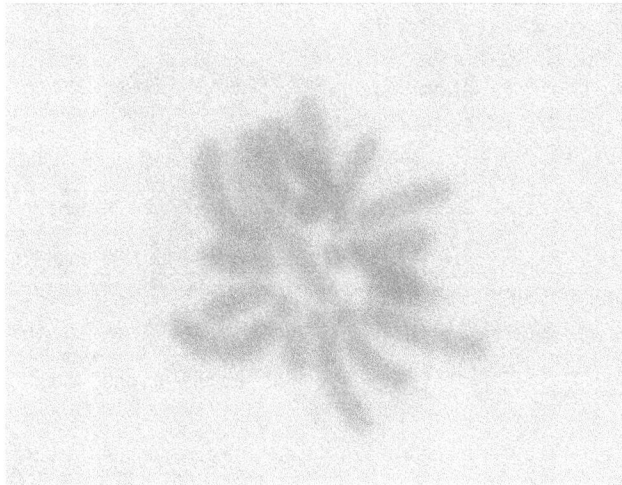

Figure 8. C-banded spermatogonial metaphase of *P. anteenata.*

In the present study, C-bands are also found to be present in three locations that is centromeric, interstitial and terminal. But their extent is found to vary among the species studied. All the four species of Acridids studied possess centromeric C-bands. Yadav and Yadav (1993) and Mayya et al. (2004) reported the prevalence of centromeric bands in short horned grasshoppers as a very common feature. Kumaraswamy and Rajasekarasetty (1976) reported centromeric C-bands in *A. turrita*. Aswathanarayana and Ashwath (2006) also revealed centromeric bands in *A. turrita*. According to Yadav and Yadav (1993), restriction of C-heterochromatin to centromeric regions is considered to facilitate whole arm translocation.

C-bands found within the centre of the body of chromosome is termed as interstitial C-band. In the present study, interstitial C-bands are seen in all the species of grasshoppers except in *A. turrita*. The interstitial C-bands were also encountered in 10 species of their study by Yadav and Yadav (1993). Mayya et al. (2004) revealed the presence of interstitial C-bands at different locations on chromosomes among Acridoid species. These interstitial bands are found to be inactivated centromere in some species of *Hieroglyphus nigrorepletus* (Yadav and Yadav, 1993). These interstitial C-bands might have an effect on the expression of the flanking euchromatic segment (Aswathanarayana and Ashwath, 2006). Terminal bands were exhibited by all the species except *A. turrita*. Similarly, the absence of terminal bands was also reported in *A. turrita* by Yadav and Yadav (1993). On the other hand, our studies revealed the presence of centromeric, interstitial, terminal bands along with thick and thin bands.

The comparison of interspecific C-banding patterns of the same sub-family has no clear correlation. The species from the same genus have not shown uniformity in their C-banding patterns (John and King, 1977; Santos and Giraldez, 1982) which has attributed to dynamic nature of heterochromatin (Yadav and Yadav, 1993). Same situation has been seen in present study the species from same sub-family differ in their C-band distribution (*A. turrita, A. exaltata,*). Likely, such comparisons are such that one cannot be sure that chromosomes of similar relative lengths are necessarily homologous in all genomes (King and John, 1980). Perhaps the only exceptions are the X and the megameric chromosomes which presumably have a common origin within the Acrididae (White, 1973). The immediate tendency for C-heterochromatin to vary in grasshoppers has been considered by many reports (Santos et al., 1983; Yadav and Yadav, 1983) and present report.

The pattern and distribution of C-heterochromatin distribution varies among Acridoid taxa, especially karyologically conservative ones. These variations are to be governed by some hidden mechanism of change, other than gross chromosome rearrangements operating in the process of speciation.

ACKNOWLEDGEMENTS

Our sincere thanks go to Head, Department of Zoology for providing laboratory facilities and a heartful thanks to Prof. A. S. Yadav, Kurukshetra University for identification of grasshoppers.

REFERENCES

Aswathanarayana NV, Ashwath SK (2006). Structural polymorphism and C-banding pattern in a few Acridid grasshoppers. Cytology, 71(3): 223-228.

Channaveerappa H, Ranganath H (1997). Karyology of few species of south Indian acridids. II Male germ line karyotypic instability in Gastrimargus. J. Biosci., 22(3): 367-374.

Hsu TC (1974). Procedures for inducing C-bands and G-bands in mammalian chromosomes. Mammalian Chromosomes Newslett., pp. 151: 88.

John B, Hewitt GM (1966). Karyotype stability and DNA variability in the Acrididae. Chromosoma (Berl.), 20 : 155-172.

John B, Hewitt GM (1968). Patterns and pathways of chromosome evolution within the Orthoptera. Chromosoma (Berl.), 25: 40-47.

John B, King M (1977). Heterochromatin variations in *Cryptobothrus chrysophorus* II Patterns of C-banding. Chromosoma, 65: 59-79.

Kumaraswamy KR, Rajasekarasetty MR (1976). Pattern of C-Banding in *Acrida turrita* (Acrididae: Orthoptera). Curr. Sci., 45(21): 762-763.

King M, John B (1980). Regularities and restrictions governing C-band variation in Acridid grasshoppers. Chromosoma, 76: 123-150.

Levan A, Fredga K, Sandberg AA (1964). Nomenclature for centromeric position of chromosomes. Hereditas, 52: 201-220.

Lopez-Fernandez C, Gosalvez J (1981). Differential staining of a heterochromatic zone in *Arcyptera fusca*. Experentia, 37: 240.

Mayya S, Sreepada KS, Hegde MJ (2004). Non-banded and C-banded Karyotypes of ten species of short- horned grasshoppers (Acrididae) from South India. Cytologia, 69(2): 167-174.

Santos JL, Giraldez R (1982). C-heterochromatin polymorphism and variation in chiasma localization in *Euchorthippus pulvinatus gallicus* (Acrididae: Orthoptera). Chromosoma, 85: 507-518.

Santos JL, Arana P, Giraldez R (1983). Chromosomal C-banding pattern in Spanish Acridoidea. Genetica, 61: 65-74.

Sumner AT (1990). Chromosome banding. U. Hymen ed. London, Boston, Sydney, Wellington, pp. 39-104.

Sharma T, Gautam DC (2002). Karyotypic studies of eleven species of grasshoppers from north-western Himalayas. Nucleus, 45(1-2): 27-35.

Turkoglu S, Koca S (2002). Karyotype, C- and G- band patterns and DNA content of *Callimenus* (= *Bradyporus*) *macrogaster macrogaster*. J. Insect Sci., 2(24): 1-4.

White MJD (1973). Animal cytology and evolution. Third edition, Cambridge University Press.

Yunis JJ, Yasminieh WG (1971). Heterochromatin, Sat- DNA and cell function. Science, 174: 1200-1209.

Yadav JS, Yadav AS (1983). Analysis of hopper chromosomes with banding techniques. I. *Phlaeoba infumata* (Brunn.) (Tryxalinae: Acrididae) and *Atractomorpha crenulata* (F.) (Pyrgomorphinae: Acrididae). Biology, 29(1): 47-53.

Yadav JS, Yadav AS (1986). Chromosome number and sex-determining mechanisms in thirty species of Indian Orthoptera. Folia Biol., 34(3): 277-284.

Yadav JS, Yadav AS (1993). Distribution of C-heterochromatin in seventeen species of grasshoppers (Acrididae: Orthoptera). Nucleus., 36(1-2): 51-56.

27

Habitat association and movement patterns of the violet copper (*Lycaena helle*) in the natural landscape of West Khentey in Northern Mongolia

Gantigmaa Chuluunbaatar[1], Kamini Kusum Barua[2]* and Michael Muehlenberg[2]

[1]Institute of Biology, Mongolian Academy of Science,Ulan Bataar, Mongolia
[2]Centre for Nature Conservation, Georg-August University,Gottingen, Germany.

We studied the habitat association and movement patterns of the Lycaenidae species, *Lycaena helle* using the mark-release-recapture method in a heterogenous natural habitat of West Khentey in Northern Mongolia. Butterfly individuals were collected using nets during a standardized one hour sampling in different biotopes. *L. helle* was found to predominantly inhabit the wet mesophile grasslands and herb meadows, although moderate occurrences were also recorded in the riparian woodlands, birch forests of the river valley and in the mixed forests of *Larix sibirica* and *Betula platyphylla*. We investigated the movement patterns of the Violet Copper within ecologically open landscape and recorded individual occurrences across the different habitat types of West Khentey. The mean distances between first and subsequent captures were found to be greater for both the sexes (107 ± 76 m for females and 44 ± 41 m for males). The single greatest movement between recaptures was 386 m for females and 163 m for males. We could conclude from our studies that *L. helle* had a more closed distribution range within the natural landscape of our study area as the high plant diversity could be considered to be an important factor restricting their movement patterns as unlike that of their counterparts in fragmented landscapes of Central Europe.

Key words: Lycaena helle, habitat occupancy, mobility, natural landscape

INTRODUCTION

Lycaena helle (Denis and Schiffermuller, 1775) is a very rare butterfly species in Central Europe (Fischer et al., 1999) and has been listed as endangered in Germany (Bundesamt fuer Naturschutz, 1998). This species is living in small populations on fragmented and isolated habitat islands in Central Europe (Fischer et al., 1999, van Swaay and Warren, 1999).

Anthropogenic activities like afforestation, peat extraction and management methods for improving cattle grazing like drainage, burning and chemical treatments have largely contributed to habitat loss and degradation throughout Central Europe (Kudrna, 1986). The local

extinction and decline of many butterfly species are related to changes in habitat quality (van Swaay and Warren 1999, Summerville et al., 2002, Rodriguez et al., 1994).

Many authors documented the influence of landscape patterns on butterfly community (Schneider et al., 2003; Natuhara et al., 1999; Saarinen, 2002; Dover et al., 1997; Schneider and Fry, 2001; Pullin, 1997; Rodriguez et al., 1994; Summerville and Crist, 2003; Summerville and Crist, 2004). Sparks and Carey (1995) found an influence of the floral composition on butterfly diversity. Soederstroem et al. (2001) showed that tree species diversity and cover had a positive effect on butterfly species, but high proportion of large trees had a negative effect on butterfly species richness. Dover et al. (1997) discussed the importance of shelter in the open country-side for butterflies. Features of landscape are the most important predictors that influence the population and community ecology of species (Hunter, 2002; Tews et al.,

*Corresponding author. E-mail: kbarua@gwdg.de, kaminikusum@gmail.com.

2004; Rodriguez et al., 1994; Pullin, 1997; Root, 1972; Ehrlich and Murphy, 1987; Dennis and Eales, 1997).

In contrast to the central European landscape, Mongolia which is a landlocked country in Central Asia but biogeographically located in a similar Palaearctic zone as Europe has more pristine habitats and biotopes. There is also a close similarity in the Palaearctic faunal composition, specifically in case of the butterfly fauna, where some species which are listed as critical and endangered in Central Europe or even reported to be extinct have been found to be common in Mongolia (Muehlenberg et al., 2000). The diverse biomes in Mongolia range from taiga forest, mountain forest steppe, meadow steppe and desert. The forest zone covers approximately 5% of the country's total geographical area and are mainly concentrated in the northern part and also forming the southernmost edge of the largest continuous forest system on earth, the Siberian taiga. Unlike other countries, Mongolia with its rich cultural heritage and unique ecosystems has integrated sustainable development into nature conservation (MNE, 1996).

The main aim of this study was to investigate the occurrence of Violet Copper in different habitat types within the natural landscape of West Khentey. In contrast to the human modified landscape in Europe, the Khentej habitats represent natural conditions least altered by human activities. Our specific objectives were: (1) to characterise the influence of landscape structure and vegetation on Violet Copper population by comparing their habitat occupancy and; (2) to determine the mobility of Violet Copper in natural landscape.

METHODS

Study area

The southern areas of Siberian region are located in Mongolia and comprises of high mountainous areas of northern Mongolia, the basin of Lake Khubsgul, Orkhon-Selenge, and Khentey Mountains. The Khentey and Khubsgul belong to high mountain region and are still covered in large parts with primary boreal forest. This region is the coldest area in Mongolia and is known to have continuous and isolated regions of permafrost (Gantsetseg and Sharkhuu, 2002). The Khan Khentey is a strictly protected area having global importance for biodiversity conservation due to its largely natural landscape (Muehlenberg and Samiya, 2000). The West Khentey which is a part of the Khan Khentej mountain range belongs to the Euroasiatic-Boreal-Forest region and is located in the transition zone between the closed forest of the Siberian mountain taiga in the north and the Central Asian steppe in the south (Dulamsuren, 2004). Our field work was carried out near Khonin Nuga Research Station (49°4' 48" N, 107°17' 15" E). The zonal vegetation of the study area primarily consists of dark taiga forests with Pinus sibirica, Abies sibirica and Picea obovata in the upper montane belt, light taiga forests with Larix and Betula platyphylla or Pinus sylvestris in the northern and eastern slopes of the lower montane belt as well as meadow and mountain steppes on the southern and western slopes of the lower montane belt (Dulamsuren et al., 2004; 2005a, b).

The vegetation pattern in the Khentej region strongly depends on altitude and exposition and is known to represent a separate floristic bioregion. In the transition forest steppe zone, forests are found only on the northern slopes while on the southern slopes, high insulation which dries out the soil moisture has facilitated the dominance of steppe vegetation (Valendik et al., 1998). Muehlenberg et al. (2000) described eight different types of vegetation in the West Khentey: mountain taiga, mountain forest, meadow steppe, mountain dry steppe, shrublands, riparian woodland, herb meadows and wet grasslands. The forest area in West Khentey region has only some patches of climax coniferous forests, because a history of frequent fires has been known to cause mixed forest of variable successional stages, such that boreal coniferous forests are of high structural diversity and spatial heterogeneity, due to the natural disturbances (Gunin et al., 1999; Goldammer and Furyaev 1996).

The mountain taiga belt in the Khentey region ranges from about 1200 to 1600 m above sea level with extensive P. sibirica forest covering the northern, northwestern and western slopes. The mountain forest (about 800-1200 m above sea level) consists of Larix sibirica and B. platyphylla forests on the northern and western slopes. The meadow steppe is not covered by trees although trees were known to exist in former times. The species composition in these grasslands mainly comprised of Aster alpinus, Campanula glomerulata, Trisetum sibiricum and Lilium pumilum. The mountain dry steppe occur on the southern slopes where trees are not known to exist and in these patches elements of the Daurican, Central Asiatic and Manchurian elements have immigrated. Shrubland which borders the lower mountain stratum in the valley has only a few species like Betula fructicosa, Busseola fusca, Crataegus sanguinea and Salix sp. forming a dense growth. The riparian woodland is dominated by such tree species like Populus laurifolia, Betula platyphylla or Populus obovata. The herb meadows in the river valley have such dominant species like Filipendula palmata, Filipendula ulmaria, Heracleum dissectum, Achillea alpine, Gerum alleppicum, Sanguisorba officinalis, Lilium dahuricum and Elymus dahuricus. The wet grassland is characterised by Carex meyeriana, Carex dichroa, Carex enervis, Carex caespitose, Carex schmidtii, Ligularia sibirica, Caltha palustris, Halenia corniculata and Comarum palustre.

Climate

The climate of Khentey region is characterised by cold and dry winters with mean January temperatures as low as -23 to -28°C (Tsendedash, 1995; Tsegmid, 1989). Frost occurs from end of August to early June on 280-300 days/year (Tsedendash, 1995). Mean maximum monthly temperatures range from -22.1°C in January to 19°C in July. Temperature extremes are 36.4°C in June and -40.1°C in January. Mean annual precipitation in the Khentey region is higher than in other parts of Mongolia, ranging from 380 to 450 mm. Most of the rainfall occurs in summer between June and August. Dewfall is frequent from spring to autumn (Velsen-Zerweck, 2002; Gantigmaa, 2005).

Study site

The mobility of adult Violet Copper was recorded in open areas of herbaceous plant meadows with shrub layers. These natural habitats were heterogeneous with shrubs and herb meadows (HM) on the terrace in the river valley and mountain dry steppe (MDS) on southern slopes (Figure 1). The size of this habitat is less than 10 ha, but it includes grassland with two different plant communities of herb meadow (e.g. Iris sanguinea and Alopecurus arundinaceus community) and Carex-rich wet grassland (bog area). The shrub layer had such species like Salix sp. and Padus asiatica (Dulamsuren, 2004).

Mountain dry steppe (MDS) had a sparse vegetation cover

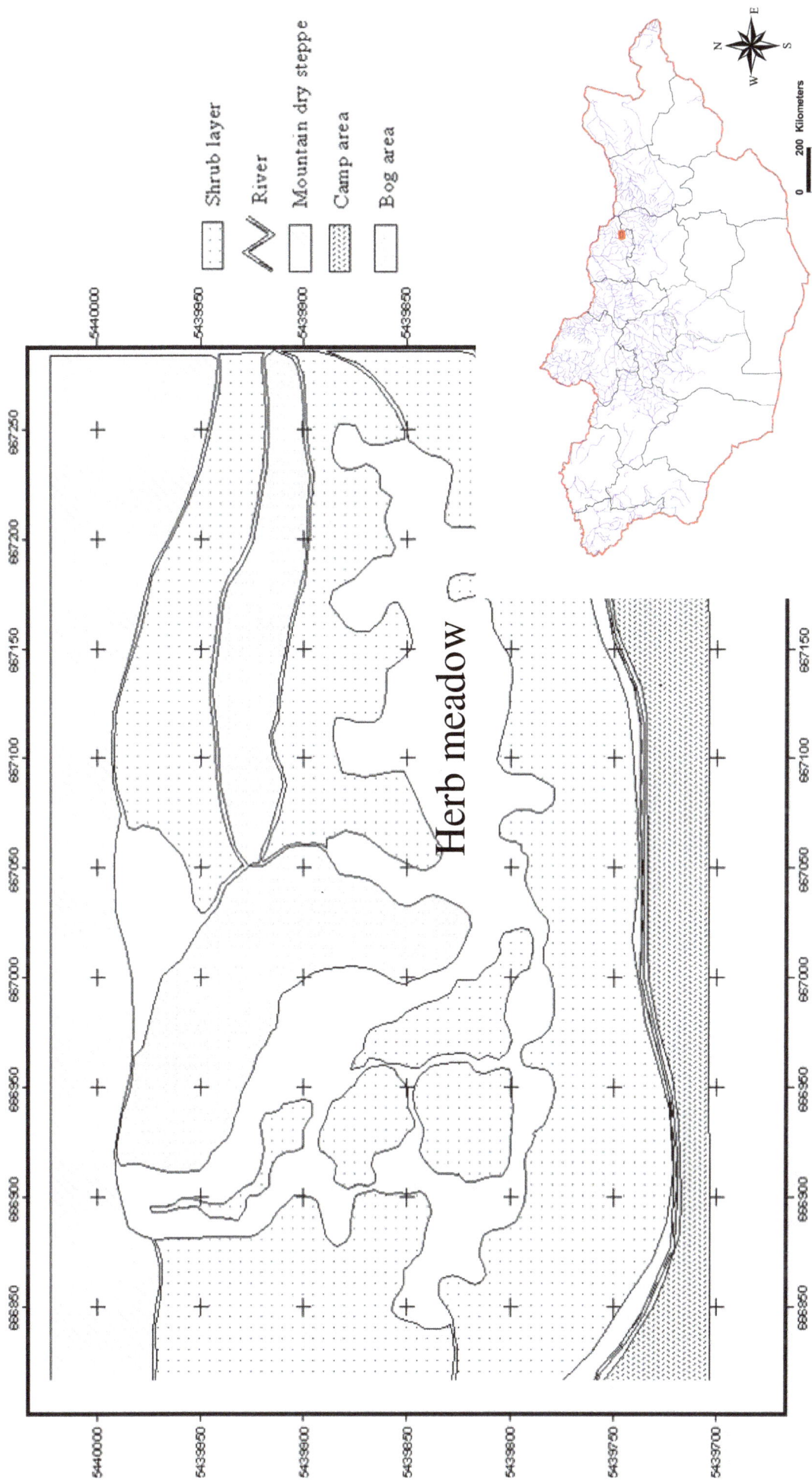

Figure 1. Study area showing the mosaic of natural habitats which are heterogeneous with shrubs and herb meadow on the terrace in the river valley.

Table 1. Number of encountered individuals of *L. helle* in different habitat types of West Khentey.

Habitat types	Standardised 1 h sample	
	Number of individuals	Percentage of individuals (%)
Birch forests of the valley	8.66 ± 13.27	11
Mesophile grassland	24.75 ± 17.19	42
Mixed Forest	4.0 ± 1.4	3
Riparian woodland	9.5 ± 9.19	11
Moist clearings in forest	8.33 ± 7.76	8
Herb meadow with shrubs	28.50 ± 30.4	24
Mountain dry steppe	1.0 ± 0.00	0.01

Table 2. Recapture rate of *L. helle* in the study sites of West Khentey.

Capture-mark-recapture results	*L. helle*		
	Male	Female	Total number of individuals
Marked individuals	187	205	392
Recaptured individuals	49	35	84
Recapture rate (%)	26%	17%	21%

dominated by *Potentilla - Carex, Potentilla acaulis, Potentilla viscosa, Artemisia* sp., *Koeleria macrantha, Poa, Thymus, Pulsatilla, Oxytropis sp,* and *Lilium pumilium*. In contrast, the herb meadow (HM) showed a predominantly *Carex-Artemisia sp.* association, including other important genera of larval food plants, such as *Bromus, Galium, Achillea, Poa, Equisetum, Dianthus, Polygonium, Sanguisorba, Vicia, Spiraea, Scutellaria, Potentilla,* and *Carum sp. Carex sp* were found to be widely distributed in both habitats and were utilised as foodplants by many species of butterflies.

Habitat association and movement patterns of adult Violet Copper

Field data on habitat association of Violet Copper were collected from different habitat types. During a two-year study period (2004-2005) all observations on the movement patterns of L. helle were made during the month of June, covering a 21 days sampling regime. We investigated the vegetation structure in each habitat type and during a standardised one hour sampling in each habitat, we counted the total number of individuals encountered by the netting method. This amounted to three hours of sampling per day in all the habitat types. During sampling an approximate area of 50 m² in each habitat type was covered on foot by two individuals. For the mark-recapture method, one observer marked the individuals and released them at the point of capture. The exact GPS position of all individuals recorded during the mark-release-recapture studies was plotted on a map, in order to get measurement estimates of straight distances moved between the subsequent captures.

RESULTS

Habitat association of adult Violet Copper

Our findings indicated that the Violet Copper (*L. helle*) was found in almost all habitat types (ANOVA; $F_{(6, 10)}$ = 1, 04; $p<0.4$) of West Khentey (Table 1).

This species was predominantly found in wet mesophile grasslands and herb meadows. Its occurrence in the riparian woodlands, birch forests of the valley and mixed forests of *L. sibirica* und *B. platyphylla* were comparatively lower, while occurrence in the mountain dry steppes was almost negligible

Mark-release-recapture

Overall 392 individuals of *L. helle* were marked during the mark-recapture study, of which 205 were females and 187 were males. Of these, 84 individuals (21%) were recaptured at least once within a period of 20 days. 26% of males and 17% of females were recaptured at least once (Table 2). The maximum time interval between mark and recapture was 18 days for males and 13 days for females.

The recapture rate of males was comparatively higher than that of the females. 24.5% of total recaptured individuals were encountered three times, 4.6% more than 4 times.

Movement patterns of adult Violet copper

Recapture results showed great differences between mean distances moved by males and females (ANOVA: $F_{(1.89)}$ = 13.50; $p<0.001$) (Table 3). The mean distance between first and subsequent recaptures were greater for both sexes (107 ± 76 and 44 ± 41 meters for females and males respectively) than reported in other studies (Fischer et.al., 1999). 42.5% of the *L. helle* individuals moved linear distances which were less than 40 m, 22% of the individuals moved less than 100 m and 35.5% of the

Table 3. The mean distance moved by Lycaena helle in different habitat types of West Khentey.

Mobility parameters	Male	Female	Total
Mean distance between first and subsequent recapture	44 ± 41	107 ± 70	79 ± 711
Mean distance ± SD (m)	59 ± 41	181 ± 74	09 ± 83
Mean distance ± SD (m) between two marginal points for multiple recaptures	63 ± 33	211 ± 54	116 ± 83
Mean distance between first and last recaptures ± SD (m)	53 ± 32	144 ± 69	92 ± 68
Maximum distance (m)	163	386	386

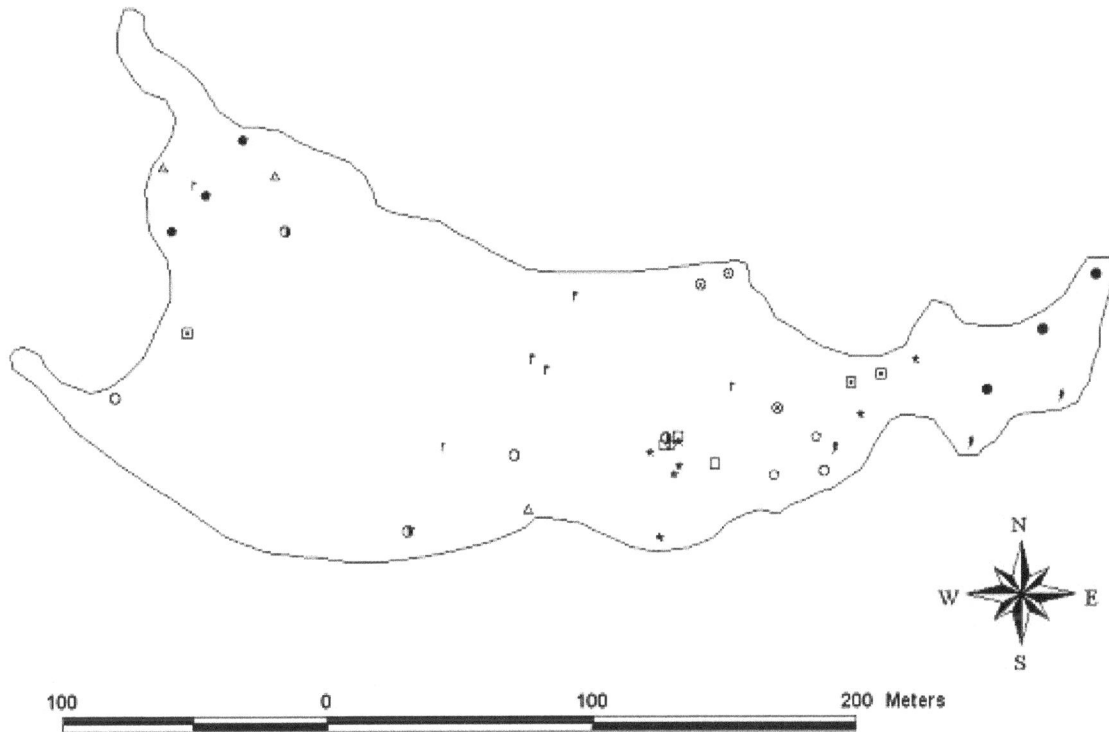

Figure 2. Movement patterns of Violet Copper (*L. helle)*, using data collected from individuals observed at least three times. Symbols indicate the position of each individual captured and same symbols show the movement positions of an individual.

individuals moved less than 200 m. The single largest movement between recaptures was 386 m for females and 163 m for males. Mean distances moved between recaptures were significantly different for both sexes (for male 44 ± 41 m, female 181 ± 74 m; ANOVA, $F_{(1, 82)}$ =92.63; p<0.001) (Table 3). In the case of multiple recaptures, if we connected the two marginal points covered by each individual, the average restricted movement was significantly different for both sexes (ANOVA; $F_{(1.26)}$ = 83.16; p<0.001). Most of the marked adults were recaptured from very near to their points of release (Figure 2).

Out of the 84 individuals recaptured at least once, six were recaptured maximum number of times within 0.9 ha area and 12 were recaptured at least three times within a 0.2 ha area in which they were marked. For example, the individuals marked by the symbol (•) were recaptured

two times after the first release, but the area covered by this individual was 0.19 ha.

Habitat characteristics

The observations on the movement patterns of *L. helle* were made during June 2004 at study sites which comprised of grassland with two different plant communities of herb meadow (e.g. *Iris sanguinea* and *Alopecurus arundinaceus* community) and *Carex*-rich wet grassland (boggy area). Shrub layer contained *Salix* and *Padus asiatica* shrubs (Dulamsuren, 2004). This site was dominated by such species like *I. sanguinea* , *Carex arnellii*, *Equisetum arvense*, *Galium* sp. The site also had such plant species which were blooming earlier in the spring, such as, *Anemone crinita*, *Ranunculus japonicus*,

Table 4. Percentage of visits of the Lycaena helle to different plant species in the study sites of West Khentey.

Plants	Female (%)	Male (%)	Total %
Anemone crinita	20	14	18
Polygonum viviparum	16	14	16
Callianthemum isophoides	13	8	12
Filipendula palmata	10	11	11
Ranunculus acer	8	13	9
Thalictrum sp.	6	6	6
Iris sanguinea	3	13	6
Carex	4	2	4
Spiraea salicifolia	4	2	3
Rosa acicularis	2	6	3
Galium boreale	4		3
Rumex acetosella	2	2	2
Spiraea media	1	5	2
Trollius asiaticus	2	2	2
Rumex acetosa	1		1
Galium verum	1		0.4
Phlomis tuberosa	1		0.4
Potentilla anserina	1		0.4
Rheum undulatum	1		0.4
Sanguisorba officinalis		2	0.4
Vicia unijuga		2	0.4

Trollius asiaticus, Callianthemum isophoides (Gantigmaa, 2005). The total number of observed visits of L. helle to flowers was 455 during the mark-release-recapture study. *L. helle* visited in total 21 plant species (Table 4). Our observations revealed that females and males visited as many as 19 and 15 plant species, respectively and also fed on the nectar of 7 plant species, most preferably two species with white flowers: *A. crinita* (18% of all visits), *C. isophoides* (12%) and *Thalictrum* sp. (6%) (Table 4). During the last part of the flight period, they also visited plants with yellow flowers (more frequently *R. japonicus*). Although, both the sexes of *L. helle* were observed to more frequently visit *P. viviparum* and *F. palmate*, both these plant species were less abundant in the study sites.

DISCUSSION

L. helle prefers wet mesophile grasslands, moist clearings in forest and along streams, springs and bogs with an abundance of its food plants (Van Swaay and Warren 1999). However in our study in West Khentey, we found this species predominantly in wet mesophile grasslands and herb meadows although it was also observed flying in the riparian woodland, birch forests of the valley and mixed forests of *L. sibirica* and *B. platyphylla*. Its occurrence in the mountain dry steppes and meadow steppes was relatively low. During our field studies, we made a close observation on the foraging behaviour of the adult butterflies and the plant species they visited. Although there is lack of sufficient knowledge about the foodplants of *L. helle* from Mongolia, *Polygonium bistorta* and *Polygonium viviparum* have been reported as larval foodplants in Europe (Van Swaay and Warren, 1999). We found the males to be more active than the females and they also showed a tendency to persist for a longer time period in a habitat patch as compared to the females. Our results also revealed a higher recapture rate for the male butterflies. Our results which also showed a higher number of individuals encountered in the wet grasslands and herb meadows clearly defined the strong habitat association of *L. helle* with open natural landscape.

However, in our study we found that the mean distances between the first and subsequent recaptures of *L. helle* was higher as compared to previous studies, in Central Europe (Fischer et al., 1999). When compared with the mean recapture distance of another species, scarce copper (*Lycaena virgaureae*) the mean distance between first and subsequent recapture values were found to be lower (Schneider et al., 2003; Gantigmaa, 2005). The mean distances between first and last re-captures were shorter than the mean average distances. These observations could be interpreted in defining the territorial behaviour of this species (Schurian and Fiedler, 1996; Fiedler, 1999). So while in open habitats, butterfly individuals could actually fly over large distances in search of food resources, in our study we found that in the natural habitats of West Khentej where the plant diversity was high and there was also an abundance of food resources, the butterflies did not actually have to fly over large distances as unlike in fragmented landscapes (Schneider et al., 2003).

The vast nature reserves and wildernesses in the West Khentey forest steppe ecosystem supports a full complement of native vegetation (Dulamsuren, 2004) and butterfly fauna (Gantigmaa, 2005) in North Mongolia. The forest steppe ecosystem of West Khentey region includes a natural forest-swamp mosaic and has the most ideal conditions for supporting native species. The mosaic structure of this natural landscape also fulfills the habitat requirements of *L. helle*. In previous studies this species has been reported to be restricted to sheltered locations in the vicinity of scrubs and trees (Fiedler, 1999) instead of moist meadows (Van Swaay and Warren, 1999). The West Khentej region offers an unique opportunity to study the role of abiotic factors such as climate and soils, with disturbances generated by occurrence of wildfires in driving the patterns of biodiversity within an otherwise "no human impact natural system". The natural vegetation patterns and prevailing abiotic conditions forming a very dynamic habitat mosaic have created the most suitable environment for the coexistence of many species of butterflies along with other native fauna within this system. More in-depth investigations are necessary to access the influence of the abiotic factors on the vegetation structure which could be crucial in determining the overall butterfly species diversity and richness.

ACKNOWLEDGEMENTS

This study was granted by DFG (German Research Foundation), within the Graduating Colleague Programme 'Biodiversity'. D. Myagmarsuren and A. Enkhmaa provided invaluable assistance during the mark-release-recapture field work done in 2004-2005. University of Goettingen provided field facilities. P. Tungalag and B. Oyuntsetseg provided help for identification of plant specimens. We are grateful for their kind support.

REFERENCES

Bundesamt FN (ed.) (1998). Rote Liste gefaehrdeter Tiere Deutschlands. – Schriftenreihe Naturschutz. Heft 55, BfN, Bonn-Bad Godesberg. 434 pp.

Dennis RLH, Eales T (1997). Patch ocupancy in *Coenonympha tullia* (Mueller, 1764) (Lepidoptera; Satyrinae): habitat quality matters as much as patch size and isolation. J. Insect Conserv., 1: 167-176.

Denis M, Schiffermüller I (1776). Systematisches Verzeichniß der Schmetterlinge der Wienergegend herausgegeben von einigen Lehrern am k. k. Theresianum. Names are dated to 1775.

Dover JW, Sparks TH, Greatorex-Davies JN (1997): The importance of shelter for butterflies in open landscapes. J. Insect Conserv., 1: 89-97.

Dulamsuren Ch (2004). Floristische Diversität, Vegetation und Standortbedingungen in der Gebirgstaiga des Westkhentej, Nordmongolei. PhD Thesis, Dissertation in Biology, University of Goettingen, 267pp.

Dulamsuren Ch, Kamelin RV, Cyelev NN, Hauck M, Muehlenberg M (2004). Additions to the flora of the Khentej, Mongolia. Part 2. – Willdenowia, 34: 505-510.

Dulamsuren Ch, Hauck M, Muehlenberg M (2005 a). Vegetation at the taiga forest-steppe borderline in the western Khentej Mountains, northern Mongolia. – Annales Botanici Fennici, 42: 411-426.

Dulamsuren Ch , Hauck M, Muehlenberg M (2005 b). Ecophysiological causes of the forest border moving towards the steppe - a case study in the Khentey Mountains, Northern Mongolia.- Proceedings of International Conference "Ecosystems of Mongolia and frontier areas of adjacent countries: natural resources. Biodivers. Ecolo. Prosp., 83-84.

Ehrlich PR, Murphy DD (1987). Conservation lessons from Long-Term Studies of Checkerspot Butterflies. Conserv. Biol. 1: 122-131.

Fiedler K (1999). Lycaenid-ant interactions of the Maculinea type: tracing their historical roots in a comparative framew Ecol. ork .J. insect Conserv. 2(1): 3-15.

Fischer K, Beinlich B, Plachter H (1999). Population structure, mobility and habitat preferences of the violet copper *Lycaena helle* (Lepidoptera; Lycaenidae) in West Germany: implications for conservation. J. Insect Conserv., 3: 43-52.

Gantsetseg B, Sharkhuu Kh (2002). Distribution and evolution permafrost in Mongolia of a part, is located in South-Siberian region of Russia. http: //www.ogbus.com/.

Gantigmaa Ch (2005). Butterfly communities in the natural landscape of West Khentej, northern Mongolia: diversity and conservation value. PhD Thesis, Dissertation in Biology, University of Göttingen.126pp.

Goldammer JG, Furyaev W (1996). Fire in ecosystems of boreal Eurasia. Kluwer Academic Publishers, Dordrecht, Boston, London, 528 p.

Grubov VI (1982). Key to the vascular plants of Mongolia (with an atlas). St. Peterburg.

Gunin DP, Vostokova AE, Dorofeyuk LN, Tarasov EP, Black CC (1999). Vegetation Dynamics of Mongolia. Kluwer Academic Publishers. Netherlands, 238 pp.

Hunter MD (2002) Landscape structure, habitat fragmentation, and the ecology of insects. Agricultural and Forest Entomology 4: 159-166.

Kudrna O (1986). Butterflies of Europe. AULA-Verlag, Darmstadt, 323 pp.

MNE (1996). Nature and Environment in Mongolia. Ministry of Nature and Environment, Ulaanbaatar, Mongolia (in Russian).

Muehlenberg M, Slowik J, Samja R, Dulamsuren Ch, Gantigmaa Ch, Woycechowski M (2000). The conservation value of West Khentej, North Mongolia. Evaluation of plant and butterfly communities. Fragmenta Floristica et Geobotanica 45: 63-90.

Muehlenberg M, Samiya R (2000). The forest steppe in North Mongolia: state, conservation value and threat. – Proceeding of Abstracts of the International Conference of Central Asian Ecosystems – 2000, Ulanbator, 101-102.

Muehlenberg M (2004). The inventory list of West Khentey region. Zentrum fuer Naturschutz. University of Goettingen. 120 p.

Natuhara Y, Imai C, Takahashi M (1999). Pattern of land mosaics affecting butterfly assemblage at Mt Ikoma, Osaka. Ecological Research 14: 105-118.

Pullin A (1997). Habitat requirements of Lycaena dispar batavus and implications for re-establishment in England. J. Insect Conserv. 1: 177-185.

Rodriguez J, Jordano D, Fernandez Haeger J (1994). Spatial heterogeneity in a butterfly-host plant interaction. J. Anim. Ecol. 63: 31-38.

Root RB (1972). Organization of a plant-arthropod association in simple and diverse habitats: The fauna of collards (*Brassica oleracea*). Ecol. Monographs, 43: 95-124

Saarinen K (2002). Butterfly communities in relation to changes in the management of agricultural environments. Dissertations in Biology, University of Joensuu, 94 pp.

Schneider C, Fry GLA (2001). The influence of landscape grain size on butterfly diversity in grasslands. J. Insect Conserv., 5: 163-171.

Schneider C, Dover J, Fry LA (2003). Movement of two grassland butterflies in the same habitat network: the role of adult resources and size of the study area. Ecol. Entomol., 28: 219-227.

Schurian KG, Fiedler K (1996). Adult behaviour and early stages of Lycaena ochimus (Herrich-Schaefer [1851] (Lepidoptera: Lycaenidae). Nachr. Entomol. Verein Apollo, N.F. 16: 329-343.

Sparks TH, Carey PD (1995). The responses of species to climate over two centuries: an analysis of the Marsham phenological record, 1736-1947. J. Ecol., 83: 321-329.

Summerville KS, Veech JA, Crist TO (2002). Does variation in patch use among butterfly species contribute to nestedness at fine spatial scales? Oikos 97: 195-204.

Summerville KS, Crist TO (2003). Determinants of Lepidopteran community composition and species diversity in eastern deciduous forests: roles of season, ecoregion, and patch size. Oikos 100:134-148.

Summerville KS, Crist TO (2004). Contrasting effects of habitat quantity and quality on moth communities in fragmented landscapes. Ecograph., 27: 3-12.

Soederstroem B, Svennsson B, Vessby K, Glimskaer A (2001). Plants, insects and birds in semi-natural pastures in relation to local habitat and landscape factors. Biodivers. Conserv., 10: 1839–1863.

Tews J, Brose U, Grimm V, Tielbörger K, Wichmann MC, Schwager M Jeltsch F (2004). Animal species diversity driven by habitat heterogeneity/diversity: the importance of keystone structures. J. Biogeogr., 31: 79-92.

Tsegmid C (1989). Some results of studies on microclimate and soil humidity of microassociations in mossy Larix forest of the eastern Khentey. Tesisi docl. nauchnoi konferentsii, posveshennie voprosam vozobnovleniya and resursi lesa MNR. 170–176.

Tsedendash G (1995). Khentein nuruunii oi-urgamalshil [Forest vegetation of the Khentej Mountains]. Dissertation, Ulaanbaatar MG (In Mongolian.).

Valendik EN, Ivanova GA, Chuluunbaatar ZO. Goldammer JG (1998). Fire in forest ecosystems of Mongolia. International Forest Fire News No. 19: 58-63.

Van Swaay C, Warren M (1999). Red Data Book of European Butterflies (Rhopalocera). Nature and environment, No. 99. Council of Europe Publishing.

Von Velsen-Zerweck M (2002). Socio-economic causes of forest loss in Mongolia. Doctoral Dissertation, Wissenschaftsverlag Vauk Kiel KG, Goettingen, 357 pp.

28

Study of using the bacterium *Bacillus thuringiernsis israelensis* in microbial control of *Musca domestica vicina*, Diptera Muscidae (Muscidae, Diptera)

Najlaa, Y. Abozinadah*, Faten, F. Abuldahb, Nawal and S. Al-Haiqi

Department of Biology, Faculty of Science, King Abdul Aziz University, Jeddah, Saudi Arabia.

The second instar larvae of *Musca domestica* were treated with *Bacillus thuringiensis israelensis*, (B.t.i.) at concentrations of 0.5, 1.0, 1.5 and 2.0% under laboratory conditions. Cumulative mortality percentage increased gradually with the increase in B.t.i concentrations and this is represented by straight regression lines indicating homogenity. The LC_{30} and LC_{50} were 0.87 and 1.305%, respectively. The tested concentrations indicate significant prolongation in larval duration compared to control. There was an inverse relation between the concentration and pupation percentage and an increase in pupal duration. An insignificant decrease in percentage of adult emergence was observed. Sex ratio was not affected. 2.0% Concentration of B.t.i. caused the greatest reduction in female fecundity (964.5 eggs/female), and the hatchability of eggs also decreased. The pupae and adults appeared malformed when treated with B.t.i. as a 2nd larval instar.

Key words: Housefly, larvicide, bioassay, mortality, LC_{30}, fecundity, fertility, malformation

INTRODUCTION

The house fly *Musca domestica vicina* (Muscidae: Diptera) is considered as one of the most major and medical pest in all tropical and subtropical parts of the world (Cohen et al., 1991). The chemical control methods by insecticides are currently used in spite of their power of contamination, votalization, and bioaccumulation (Moon, 2002). The different groups of insecticides are considered the main factors affecting the ecosystem. From this point of view, it is necessary to minimize the applications of insecticides which is considered as the main source of environmental pollution and affected the human health. There are many different insect species that have been successfully controlled by microbial agents (Hogsette, 1999). The bacterium, *Bacillus thuringienis israelensis* (B.t.i.) is considered highly beneficial for its specific activity. Several studies have shown the toxicity of endotoxins from various B.t. strains against *M. domestica* (Indrasith et al., 1992; Hodgman et

al., 1993; Lonc et al., 1997; Johnson et al., 1998; Zhong et al., 2000; Labib and Rady, 2001; Padmanabhan et al., 2005; Luga et al., 2008). The present investigation aimed to evaluate the efficiency of this biocide at different concentrations against 2nd larval instar of *M. domestica*, as well as to evaluate the effects of these concentrations on some biological parameters (Figure 1 to 10).

MATERIALS AND METHODS

Test insects

Sources of colony

Adults susceptible strain of house fly *M. domestica vicina* used in the present study were obtained from a well established colony originated from the Biology Department, Faculty of Science for girls, King Abdul Aziz University.

Rearing technique

Egg masses were used to maintain a colony in the laboratory under constant conditions of temperature and humidity (27 ± 2°C and 60

*Corresponding author. E-mail: mohed@live.com.

■ Observed ■ Expected

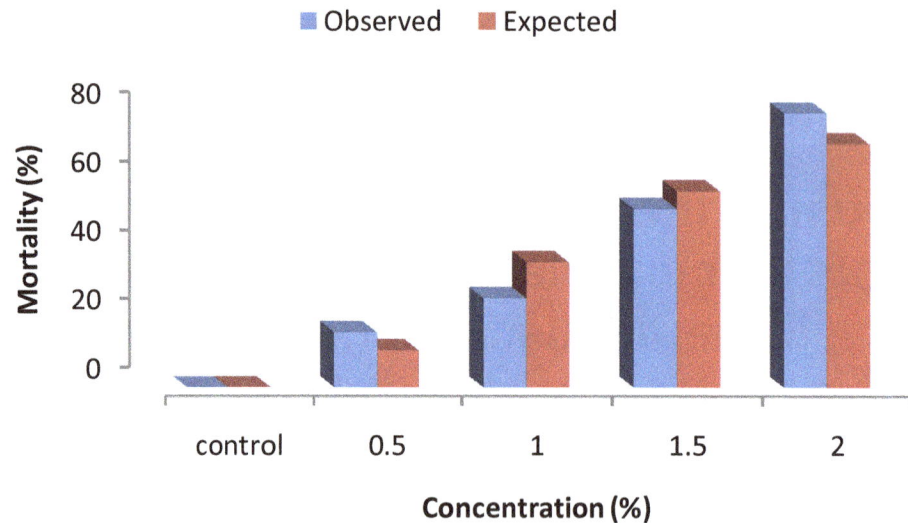

Figure 1. Susceptibility of 2nd instar larvae of *M. domestica* vicina to different concentration of *Bacillus thuringiensis israelensis* after 48 h of treatment.

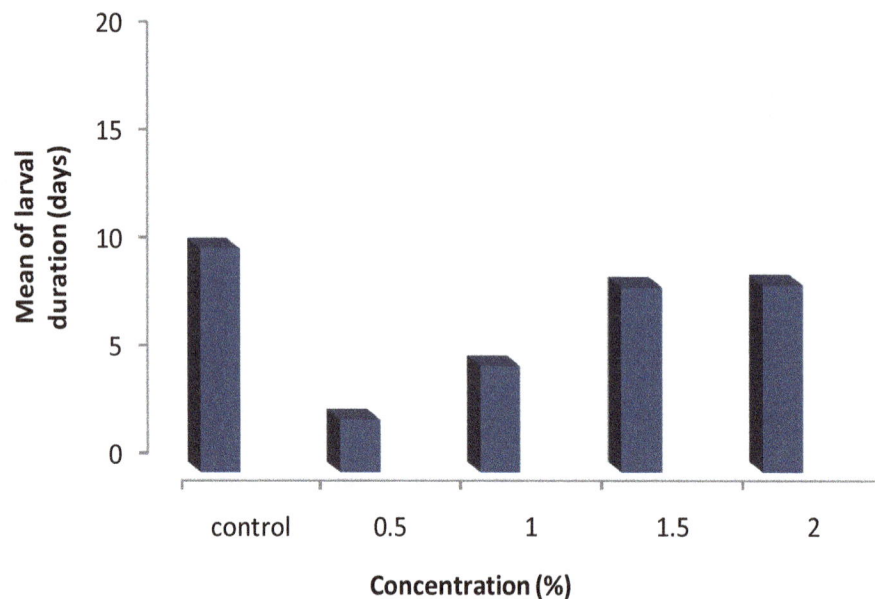

Figure 2. Effect of different concentrations of *Bacillus thuringiensis israelensis* on larval duration (days) of *M. domestica* treated as 2nd instar larvae.

±5% R.H.) Each egg mass was placed in a clean Petri dish (10 cm diameter), previously washed with 10% formalin solution to avoid any contamination according to a constant technique described by Lelwallen (1954). Full grown larvae were allowed to pupate in clean glass Petri dishes. Following emergence, the adults were provided with a piece of cotton soaked in 10% sugar, 2% milk solution as a source of food.

Source of the bacterial pathogen

The bacterium *B. thuringiernsis israelensis* was chosen as a

pathogen for this study because of its wide use in biological control. The powder was obtained from Valent Biosciences, U.S.A.

Susceptibility of house fly to the bacterial pathogen

Experimental larvae

Newly moulted 2nd instar larvae of house fly were segregated from the stock colony in clean glass Petri dishes (10 cm diameter) and starved for about 24 h.

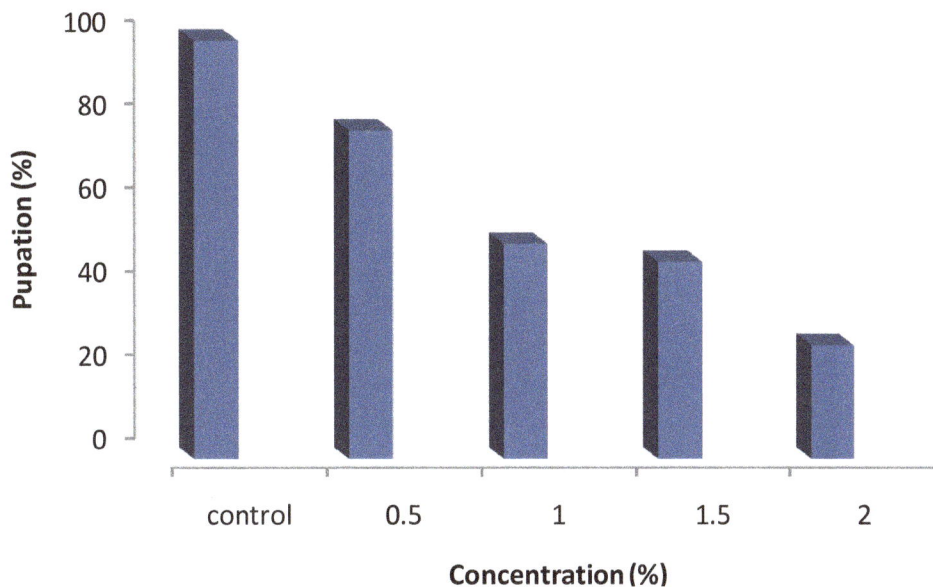

Figure 3. Effect of different concentrations of *Bacillus thuringiensis israelensis* on the percentage of pupation of *M. domestica* treated as 2nd instar larvae.

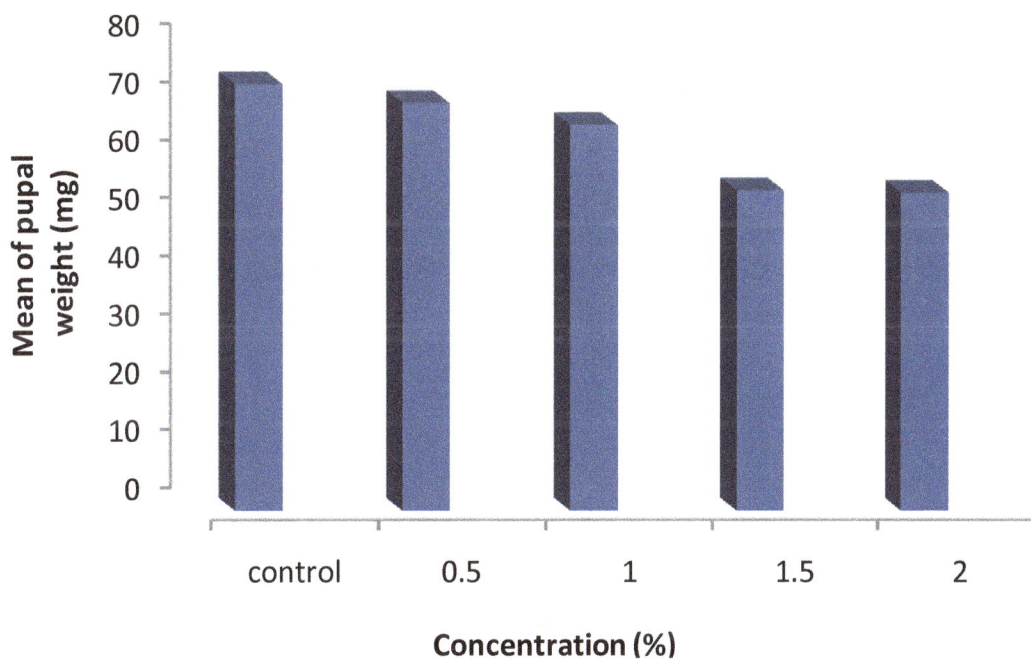

Figure 4. Effect of different concentrations of *Bacillus thuringiensis israelensis* on pupal weight (mg) of *M. domestica* treated as 2nd instar larvae.

Treatment technique

Four concentrations (0.5, 1.0, 1.5, and 2.0%) of the bacterial patho-gen were used. Fifty of the starved larvae were distributed in five replicates, were used for each concentration to feed for 48 h on treated larval media under constant laboratory conditions (27 ± 2% and 60 ± 5% R.H). The same technique described before was used except that the control larvae were allowed to feed on untreated media.

Biological studies

Final mortality percentage of treated and control larvae were recorded 48 h post treatment. Daily inspections were carried out until emergence occurred and the number of individuals (larvae, pupae, and adults) were recorded or each concentration. The larval

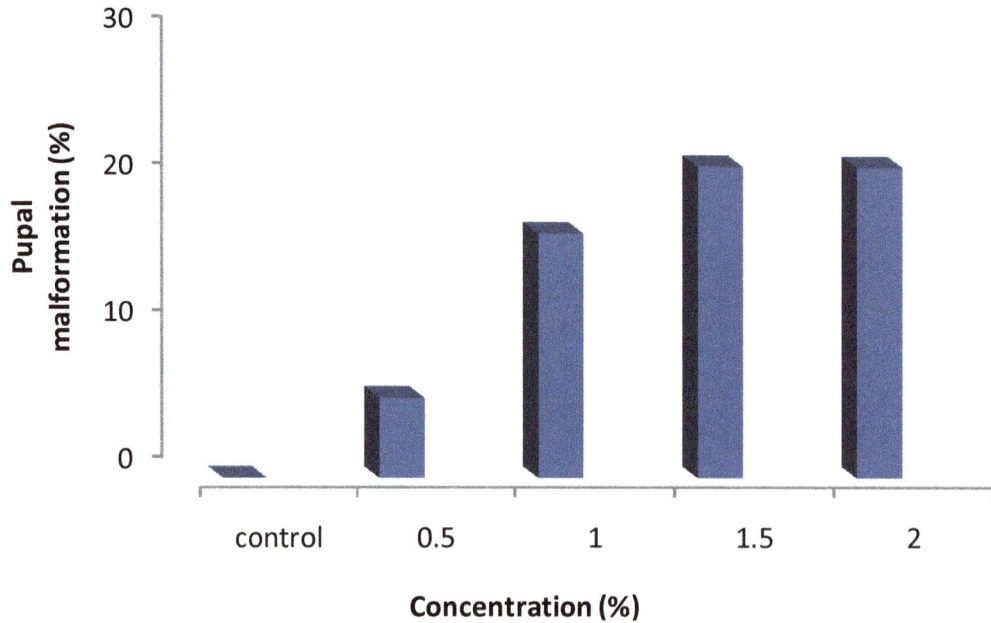

Figure 5. Effect of different concentrations of *Bacillus thuringiensis israelensis* on the percentage of pupal malformation of *M. domestica* treated as 2nd instar larvae.

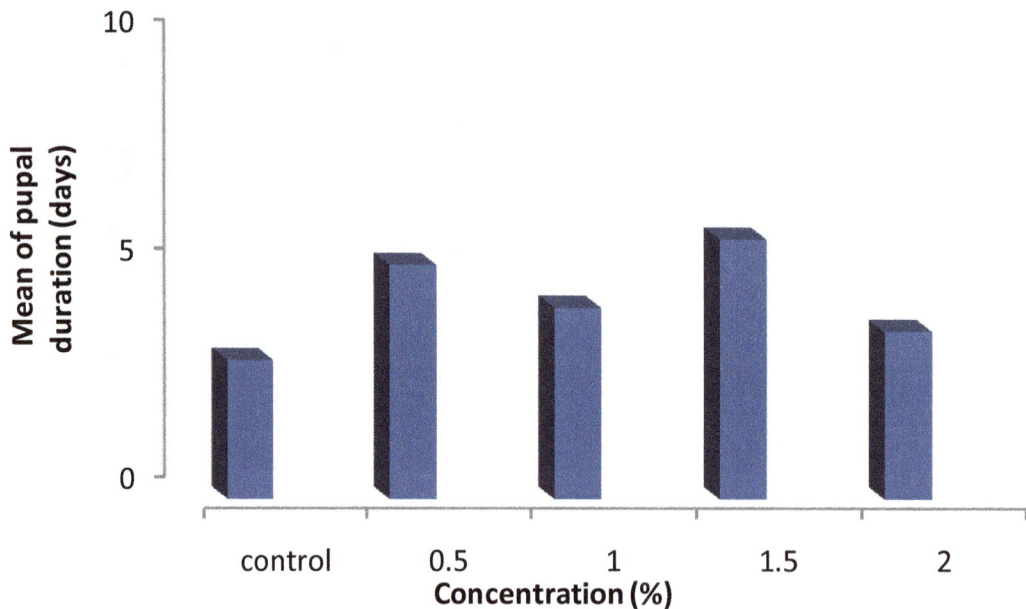

Figure 6. Effect of different concentrations of *Bacillus thuringiensis israelensis* on pupal duration (days) of *M. domestica* treated as 2nd instar larvae.

duration, pupation, pupal duration, pupal weight and pupal malformation were calculated. Also, adults emergence, longevity, fecundity, fertility and malformation were recorded.

Statistical analysis

Data were expressed as mean ± standard error; the statistical significance of differences between individual means was determined by student "t" test for paired observations. The level of significance of each experiment was stated to be non significant (P< 0.05) and highly significant (P< 0.05). The percentage reduction was calculated according to Khazanie (1979). The corrected mortality percentage was statistically computed according to Finney (1972), from which the corresponding concentration Probit lines (Ld –p line) of the 2nd instar larvae were estimated, in

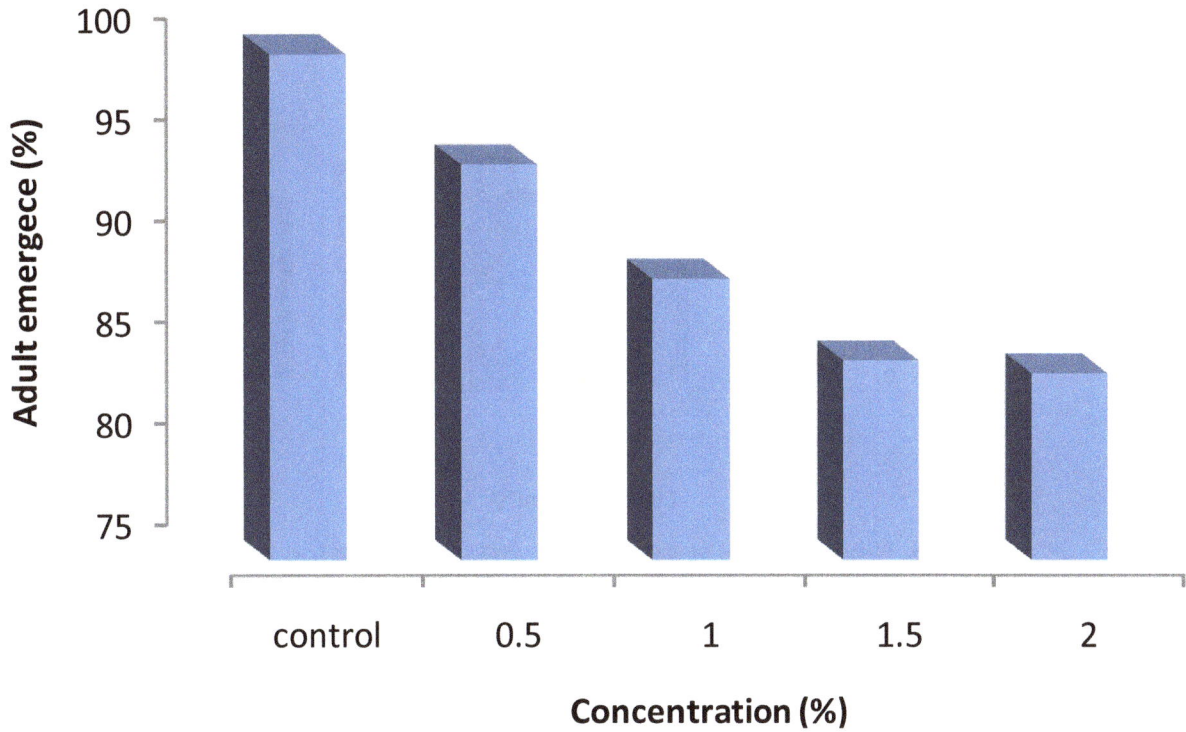

Figure 7. Effect of different concentrations of *Bacillus thuringiensis israelensis* on the percentage of adult emergence of *M. domestica* treated as 2nd instar larvae.

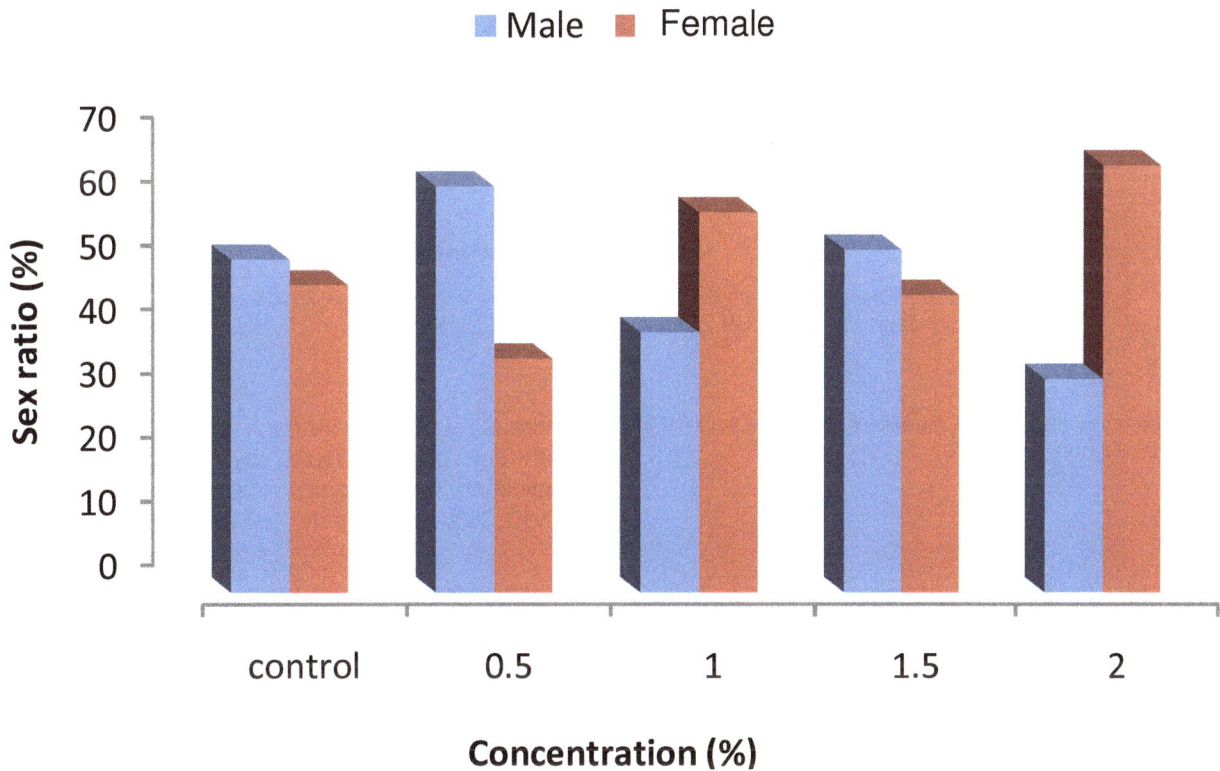

Figure 8. Effect of different concentrations of *Bacillus thuringiensis israelensis* on the percentage of sex ratio (male, female) of *M. domestica* treated as 2nd instar larvae.

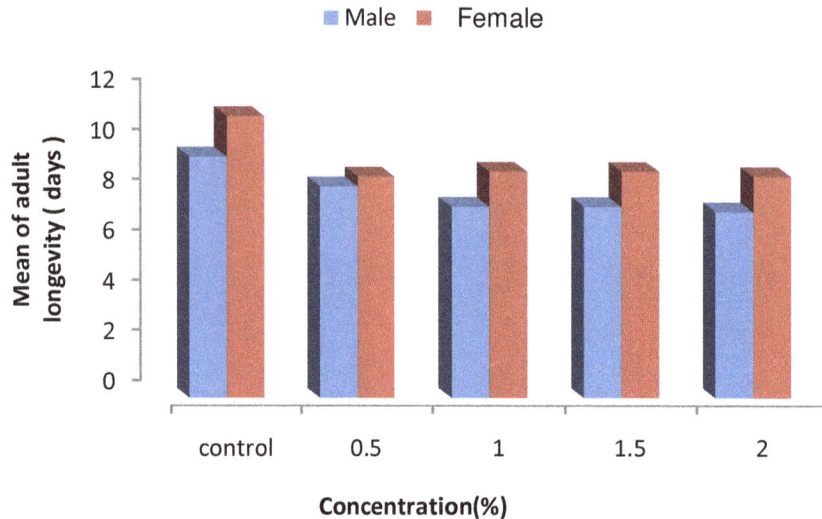

Figure 9. Effect of different concentrations of *Bacillus thuringiensis israelensis* on adult longevity (days) of *M. domestica* treated as 2nd instar larvae.

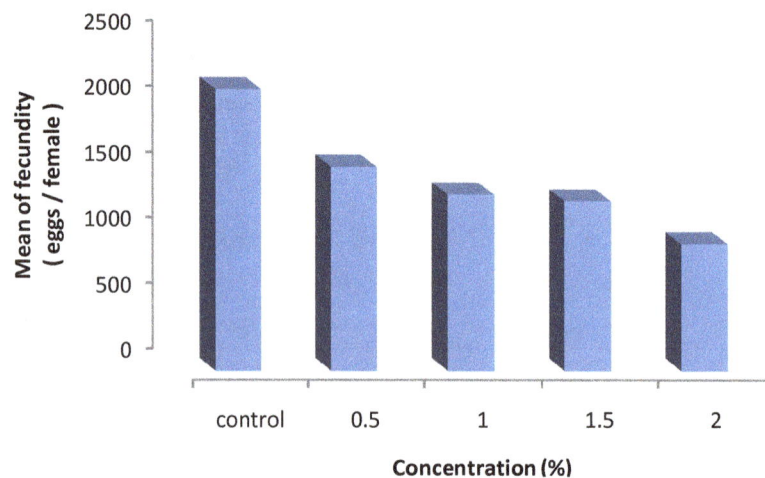

Figure 10. Effect of different concentrations of *Bacillus thuringiensis israelensis* on fecundity (eggs / female) of *M. domestica* treated as 2nd instar larvae.

addition to determine 30 and 50% mortalities and slope value of the tested material (Figures 11 and 12).

RESULTS

Susceptibility of 2nd larval instar of *M. domestica vicina* to different concentrations of *B. thuringiernsis israelensis* after 48 h of treatment

Data presented in Table 1 summarized the efficacy of B.t.i at different concentrations (0.5, 1.0, 1.5, and 2.0%) against the 2nd larval instar larvae *M. domestica*. It is clear that the pathogen affected the percentage of observed larval mortality, increasing gradually with the

increase in concentration. The response of larvae to different concentrations by straight regression lines indicating homogeneity. The LC_{30} and LC_{50} were 0.87 and 1.305, respectively. Figure 11 and 12

Impact of the tested bioinsecticide (B.t.i) at different concentration on some biological attributes of 2nd instar larvae of *M. domestica vicina*

The latent effects of the tested bioinsecticides on the 2nd instar larvae of *M. domestica vicina* are shown in Tables 2 and 3. Larval duration, pupal duration, pupal percentages, pupal weight, adults emergence percentage longevity of adults, number of eggs laid/ female

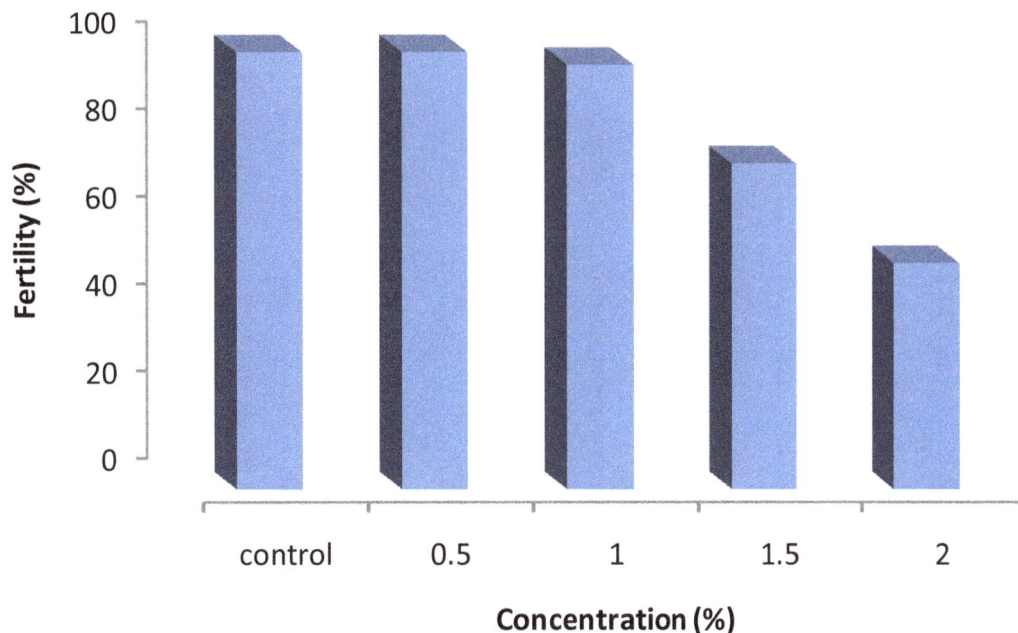

Figure 11. Effect of different concentrations of *Bacillus thuringiensis israelensis* on the percentage of fertility of *M. domestica* treated as 2nd instar larvae.

Table 1. Susceptibility of 2nd instar larvae of *M. domestica* vicina to different concentration of *B. thuringiensis israelensis* after 48 h of treatment.

Concentration (%)	Observed % of mortality	Expected % of mortality
Control	-	-
0.5	16	10.745
1.0	26	36.525
1.5	52	57.125
2.0	80	70.934

Fifty larvae were used for each concentration. P-value = 0.0017, slope of the regression line = 2.976, LC_{30} = 0.87, LC_{50} = 1.305.

(fecundity), hatchability percentage (fertility) and morphological malformations of pupae and emerged adults were investigated and recorded. Data in Table 2 showed that, all treatments caused significant prolongation in the total larval duration compared to control. The duration period ranged between 2.45 ±0.251 and 8.68 ± 0.158 days at concentrations of 0.5 and 2.0% respectively. Also, data clearly indicated that there was an inverse relationship between different concentrations and the pupation percentages, this criterion was 78.57 and 27.14% at concentrations of 0.5 and 2.0% respectively as compared with 100% in the check experiment. As well as all tested concentrations induced significant reduction in pupal weight. Also, the duration periods of pupal stage were increased as affected by treatment, this increase ranged between 3.12 and 5.68 days.

From our results presented in Table 3, it was clear that the percentage of adults emergence was insignificantly decreased with an increase in applied concentrations, it reach to 84.21 for 2.0% as compared with 100% in the check experiment. As for sex ratio, it was remarkable that the sex ratio was directed to the female side. Also, there was a significant reduction in the percentage of adult longevity of both sexes (8.8 ± 1.02) days for 2.0% as compared to 11.2 ± 0.37 days in the control group. The total number of eggs laid/ female were significantly reduced as compared to control (P < 0.01), these effects were increased with increasing the concentration. Concerning the effect of larval treatment at different concentrations on the fertility of eggs laid by the resulted females (Table 3) it is clear from the obtained results that, the hatchability percentage of *M. domestica* eggs laid by flies resulted from treated 2nd instar larvae ranged from

Table 2. Latent effects of different concentrations of *Bacillus thuringiensis israelensis* on some biological aspects of immature stages of *M. domestica*.

Concentration (%)	Larval duration (days) Means ± SE	+ or – (%)	Pupation (%)	+ or – (%)	Pupal weight (mg) Means ± SE	+ or – (%)	Pupal malformation (%)	Pupal duration (days) Means ± SE	+ or – (%)
Control	10.36 ± 0.127	-	100	-	73.5±8.85	-	0	3.04 ±0.115	-
0.5	2.45 ± 0.251**	20.17	78.57	-21.43	70.2 ±12.80	-1.21	5.46	5.12 ±0.119	0.81
1	4.92 ± 0.363 **	44.02	51.43	-48.57	66.4 ± 10.01*	-2.60	16.67	4.19 ±0.337**	23.72
1.5	8.58 ± 0.666**	79.34	47.14	-52.86	55.1 ±12.20**	-6.73	21.21	5.68 ±0.861**	73.40
2	8.68 ± 0.158 **	80.31	27.14	-72.86	54.7 ± 12.20*	-6.87	21.15	3.67 ±0.836**	40.12

Data expressed as Mean ± Standard Error (S.E). *Significant (P< 0.05); ** Highly significant (P< 0.01); + or – (%): $\dfrac{\text{Treated-Control}}{\text{Control}} \times 100$ = Percentage of increase or decrease as compared to the control

Table 3. Latent effects of different concentrations of *Bacillus thuringiensis israelensis* on some biological aspects of adult stage of *M. domestica* treaded as 2nd larval instar.

Conc. (%)	Adult emergence (%)	+ or – (%)	Sex ratio (%) Male	Female	Adult longevity (days) means ± SE Male	+ or – (%)	Female	+ or – (%)	Fecundity (eggs/ female) means ± SE	+ or – (%)	Fertility (%)	+ or – (%)	Adult malformation (%)	+ or – (%)
Control	100	-	52	48	9.6±0.51	-	11.2 ±0.37	-	2142.00 ±159.80	-	100	-	4	-
0.5	94.55	-5.46	63.46	36.54	8.4 ±0.40*	-12.50	8.8 ±0.49**	-21.43	1552.50 ±157.80**	-27.52	100	0	17.31	332.75
1	88.89	-11.11	40.63	59.38	7.6 ±0.68**	-20.83	9.0 ±1.18**	-19.64	1339.80 ±40.03**	-37.45	97.12	-2.88	21.88	447.00
1.5	84.85	-15.15	53.57	46.43	7.6 ±1.30**	-20.83	9.0±0.55**	-19.64	1289.75 ±40.03**	-39.79	74.49	-25.51	28.57	614.25
2	84.21	-15.79	33.33	66.67	7.4 ±1.21**	-22.92	8.8 ±1.02**	-21.43	964.50 ±111.06**	-54.97	51.70	-48.30	37.50	837.50

Data expressed as Mean ± Standard Error (S.E). *Significant (P< 0.05), ** Highly significant (P< 0.01). + or – (%): $\dfrac{\text{Treated-Control}}{\text{Control}} \times 100$ = Percentage of increase or decrease as compared to the control

51.70 to 97.12% at the concentrations of 2.0 and 1.0%, the hatchability percentage of deposited eggs was not affected by treatment (0.5%). Malformed individuals (pupae and adults) were observed, high concentrations induced dwarf larvae and prepupae with dark colour of the whole body positively directed with dose. It is clear that the abnormalities extended to the resulted pupae, they were characterized by the following; small size directed downward head capsule, undifferentiated body segments and the puparum was fully separated and all these malformations of pupae were concentration dependent. Flies of *M. domestica* resulted from treated 2nd instar larvae which showed also varying degrees of deformities as a side effect of bioinsecticide at different concentrations. The deformities include the attachment of pupal exuvium and the fly failed to emerge, enlargement in the abdominal region, severe shrinkage of appendages especially wings. Wings were folded and extremely reduced in size.

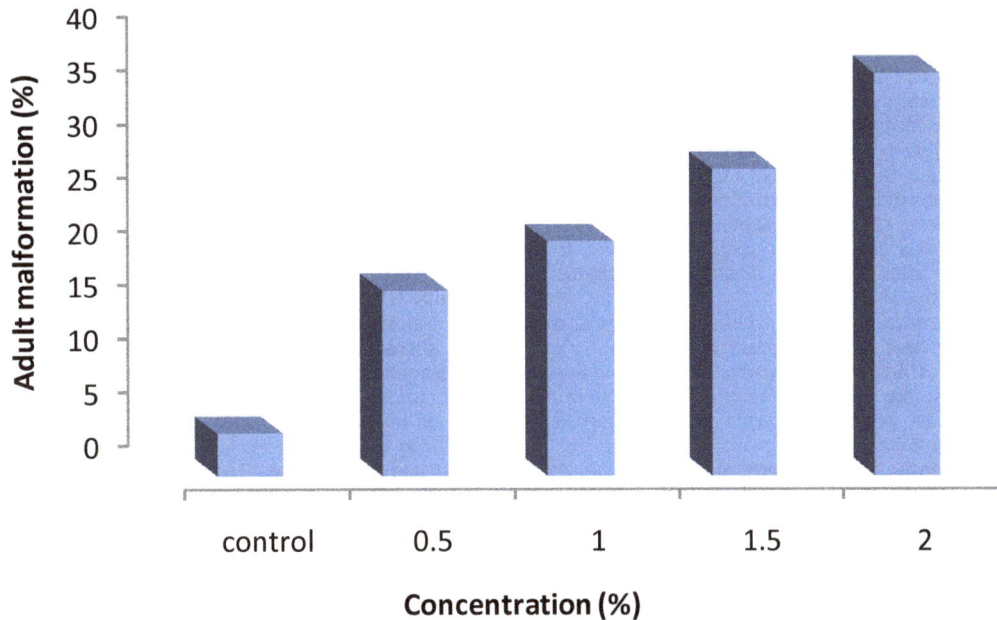

Figure 12. Effect of different concentrations of *Bacillus thuringiensis israelensis* on the percentage of adult malformation of *M. domestica* treated as 2nd instar larvae.

Finally, the resulted adults appeared unable to fly.

DISCUSSION

The biocides *(B. thuringiernsis)* used in the present study caused considerable toxic effects against 2nd instar larvae of *M. domestica vicina*. Our results clearly indicated that, the different applied concentration of the bioinsecticides clearly affected the percentage of larval mortality and the response of larvae is represented by straight regression lines indicating homogeneity. The present conclusion was in harmony with Gharib and Wyman (1991), El Zoghbey and Attalla (2003), Wang and Jall (2005).The obtained results showed that, there was an inverse relationship between the different concentrations under investigation and the pupation percentage, pupal weight (Ayanta et al., 1999; Attala et al., 2003; Dotton et al., 2003; Koja et al., 2006). Moreover, B.t.i showed an increase in the percentage of malformed pupae and adults. The reason of malformations may be due to the reduction in proteins, transaminase enzymes, carbohydrate hydrolyzing enzymes and lipids and these results and observations are in agreement with those of Attalla et al. (2003).

Adult mortality and longevity of both male and female was significantly decreased, these may be due to the latent toxic effects of the tested material, and this was agreed with that obtained by Mohamed et al. (2005), Koja et al. (2006) and Younes et al. (2008). The mean number of deposited eggs per female significantly decreased after the treated 2nd instar larvae of *M. domestica* with tested material at different concentrations due to the

inhibition of protein contents and its synthesis, which is necessary for the nutrition of eggs (El- Halim, 1993; Tawfik et al., 2002; El Bandary, 2004). Hatchability percentage (fertility of the deposited eggs decreased after treatments at different concentrations). These results are in harmony with those obtained by Sondos et al. (2000), El Gemiey (2002), Omar (2003) and Narayanan (2004). It may be possible in this instance to control flies by the use of this bacterium which incorporate spores and crystals of the appropriate strain of B.t.i. (Wang et al., 2009; Wirth et al., 2010; Fernandez et al., 2010).

REFERENCES

Ajanta C, Kaushik NC, Gupta GP, Chandra A (1999). Studies of *Bacillus thuringiernsis* on growth and development of *Helicoverpa armigere* Hiibner. Ann. Plant. Protection. Sci., 7(2): 154-158.

Attala FA, Shoeb MA, Ablas MST (2003). Evaluation of Agerin a commercial formulation of *Bacillus thuringiernsis* against certain insect pests of cabbage. Egyptian. J. Biol. Pest. Control., 13(112): 115-117.

Cohen D, Green M, Block C, Slepon R, Ambar R, Wasserman SS, Lavine MM (1991). Reduction of transmission of shigellosis by control of houseflies (*Musca domestica*). Lancent., 337: 993- 997.

Dutton A, Klein H, Romeis J, Bigler F (2003). Prey mediated effects of *Bacillus thuringiernsis* spray on predator *Chrysoperla earnea* in maize. Biol. Control., 26(2): 209-215.

El Bandary FE, Al-Khaalaf AA (2004). Laboratory studies of two commercial formulations of *Bacillus thuringiernsis* (Berliner) for activity *against 1st larval instar of Spodoptera littoralis* (Boisd.) Lepidoptera: Noctuidae). Ann. Agric. Sci. Moshtoher., 42(3): 1395-1404.

El-Gemeiy HM (2002). Impact of two formulations of *Bacillus thuringiernsis* var. *kurstaki* on the life table parameters of the spiny ballworm *Earias insulana* (Boisd.) Ann. Agric. Sci.Moshtohor., 40(3): 1817-1824.

El–Halim SMA (1993). Bioactivity of Dipel 2X, a commercial preparation of *Bacillus thuringiernsis* Berliner against the cotton leafworm *Spodoptera littoralis* (Boisd.) .Egypt. J. Agric. Res., 71(1): 175-183.

El-Zoghbey AA, Attalla FA, Mesbah AH (2003). Effect of two biocides in controlling *Cassida Vittata* (vill.) and *Spodoptera littoralis* (Biosd.) infesting plants. Ann. Agric. Sci, Moshtohor., 41(1): 339-346.

Fernandez-Luna MT, Tabashnik BE, Lanz-Mendoza H, Bravo A, Soberon M, Miranda-Rios J (2010). Single concentration tests show synergism among *Bacillus thuringiensis* subsp. *Israelensis* toxins against the malaria vector mosquito *Anopheles albimanus.* J. Invertebr. Pathol., 104(3): 231 -3 .

Finney DJ (1972). Probit analysis. A. statistical treatment of the sigmoid response curve 7th Ed., Cambridge Univ. press, England.

Gharib AH, Wyman JA (1991). Food consumption and survival of *Trichoplusia ni* (Lepidoptera: Noctuidae) larva following intoxication by *Bacillus thuringiernsis* var. *kurstaki* and var. *thuringiernsis.* J. Econ. Entomol., 89(2): 436-439.

Hodgman TC, Ziniu Y, Ming S, Sawyer T, Nicholls CM, Ellar DJ (1993). Characterization of a *Bacillus thuringiensis* strain which is toxic to the housefly *Musca domestica* . FEMS. Microbiol. Lett., 114: 17- 22.

Hogsette JA (1999). Management of ectoparasites with biological control organismis . Internat. J. Parasitol., 29: 147- 151.

Indrasith LS, Suzuki N, Ogiwara K, Asano S, Hori H (1992). Activated insecticidal crystal proteins from *Bacillus thuringiensis* seovars killed adult houseflies. Lett. Appl. Microbiol., 14: 174 – 177.

Hohnson C, Bishop AH, Turner CL (1998). Isolation and activity of strains of *Bacillus thuringiensis* toxic to larvae of the housefly (Diptera: Calliphoridae). J. Invertebr. Pathol., 71: 138- 144.

Khazanie R (1979): Eementary statistics (Good Year Publishing Co; California, U.S.A, 488P)

Koja SMT, Rezk GN, Madiha, Hanafy HEM (2006). Effect of *Bacillus thuringiernsis* (Berliner) commercial formolutions against the spiny ballworm *Earias insulana* (Boisd.) Ann. Agric. Sci. Cairo., 51(1): 261-269.

Labib IM, Rady M (2001). Application of *Bacillus thuringiensis* in poultry houses as a biological control agent the housefly, *Musca domestica sorbons.* J. Egypt. Soc. Parasitol., 31(2): 531 – 544.

Lelwallen LL (1954). Biological and toxicological studies of the little house fly. Econ. Entomol., 47: 1137 – 1141.

Lonc E, Lecadet M, Lachowicz TM, Panek E (1997). Description of *Bacillus thuringiensis wratislaviensis* (H-47), a new serotype origination from Wroclaw (Poland), and other *Bt* soil isolates from the same area. Lett. Appl. Microbiol., 24: 467 – 473.

Luga R, Alberto S, Ignazio F (2008). Immature housefly (*Musca domestica*) control in breeding sites with a new *Brevibacillus laterosporus* formulation. Environ. Entomol., 37(2): 505 – 509.

Mohammed EHM, Abd El- Haleem SM, El- Husseini MM (2005). Efficacy and residual effect of *Bacillus thuringiernsis* against larvae of the cotton leaf worm, *Spodoptera littoralis* (Biosd.) in Egyptian clover fields. Egyptian. J. Biol. Pest. Control. 15 (1/2): 81-83.

Moon RD (2002). Muscid flies, In. Mullen G, Durden L, (Eds). Medical and veterinary entomology. Academic, London, United Kingdom. pp. 279-301.

Narayanan K (2004). Insect defense: its impact on microbial control of insect pests. Curr. Sci., 86(6): 800 - 814.

Omar NAM (2003). Impact of *Bacillus thuringiernsis* on some biological, histological and physiological aspects of *Galleria mellonella L.* as a susceptible host. Ph. D. Thesis Fac. Agric. Cairo Univ.

Padmanabhan V, Prabakaran G, Paily KP, Balaraman K, (2005). Toxicity of a mosquitocidal metabolite of *Pseudomonas fluorescens* on larvae and pupae off the housefly, *Musca domestica.* Indian. J. Med. Res., 121: 116 – 119.

Sondos Mohamed A, Bader NA, El Hafez AA (2000). Efficacy of two formulations of pathogenic bacteria *Bacillus thuringiernsis* against the first instar larvae of *Spodoptera littoralis* (Boisd.) and *Agrotis ipsilon* (Hfn.) (Lepidoptera: Noctuidae). Egyptian. J. Agric. Res., 78(3): 1025-1040.

Tawfik SM, Farghali AA, Sokar A, Abd- El-Wahab IS, (2002). Biochemical studies of Dipel 2X and abamectin on the 4th instar larvae of cotton leafworm, *Spodoptera littoralis* (Boisd.). Egypt. J. Appl. Sci., 17(3): 371-386.

Wang LY, Jael Z (2005). Sublethal effects of *Bacillus thuringiernsis* H-14 on the survival rate, longevity, fecundity and f1 generation developmental period of *Aedes aegypti.* Denque- Bull., 29: 192-196.

Wang Y, Jin X, Zhu A, Chu F, Yang X, Ma Y (2009). Expression pattern of antibacterial genes in the *Musca domestica.* Sci. China. C. Life. Sci. 52 (9): 823 – 30 .

Wirth MC, Walton WE, Federici BA (2010). Evolution of resistance to the *Bacillus sphaericus* Bin toxin is phenotypically masked by combination with the mosquitocidal proteins of *Bacillus thuringiensis* subspecies *israelensis* . Environ. Microbiol., 12(5): 1154 – 60.

Younes MWF, El-Sayed YA, Hegazy MMA (2008). Effect of *Bacillus thuringiernsis* var. Kurstaki on some biochemical parameters of the cotton leafworm *Spodoptera littoralis* (Boisd.). 4th Int. Conf. Appl. Entomol. Fac. Sci. Cairo. Univ.

Zhong C, Ellar DJ, Bishop A, Johnson C, Lin S, Hart ER (2000). Characterization of a *Bacillus thuringiensis* delta-endotoxin which is toxic to insects in three orders. J. Invertebr. Pathol., 76: 131-139.

29

The influence of trichome characters of soybean (*Glycine max* Merrill) on oviposition preference of soybean pod borer *Etiella zinckenella* Treitschke (Lepidoptera: Pyralidae) in Indonesia

Agus D. Permana[1]*, Asni Johari[2,3], Ramadhani Eka Putra[1], Soelaksono Sastrodihardjo[1] and Intan Ahmad[1]

[1]School of Life Sciences and Technology Institut Teknologi Bandung, Ganesa 10 Bandung, Indonesia.
[2]Laboratorium PMIPA FKIP Universitas Jambi - Kampus Mendalo Darat Jambi, Indonesia.
[3]Politeknik Pertanian Negeri Kupang, Kupang, Indonesia.

Study of oviposition behavior of *Etiella zinckenella* Treitschke on several Indonesian local cultivars of soybean, *Glycine max* (L.) Merrill was conducted with multiple choice and no choice tests. This study confirmed that sequence of eggs oviposition of *E. zinckenella* on tropical region does not differ with other results on non-tropical region. Total number of egg deposited and oviposition frequency of *E. zinckenella* do not correlate with trichome density, trichome length and pod surface when they do not have a choice of host plants. On the contrary, on the condition of many host plants to choose, *E. zinckenella* prefers to deposit eggs on the cultivar with many short trichomes. The preference test based on soybean phenology showed that the *E. zinckenella* prefers R4 stage (fully filled pod) for oviposition.

Key words: Oviposition, *Etiella zinckenella*, *Glycine max,* phenology, trichome.

INTRODUCTION

The ability of female insect to correctly choose the perfect spot to oviposit its eggs is an important factor to ensure the survival of offspring. Host selection of insect consists of series of behavior which is essentially a response to external stimuli, resulting in the selection or rejection of a particular plant. The morphological characters of a plant could act as physical stimuli for insect feeding and its oviposition behavior (Hattori, 1988; Bailey and Smith 1991; Shanower et al., 1996). Variation in leaf size, shape, color, plant tissue rigidity, as well as the presence of hair (trichome) and papilla could determine preference of insect likely to a plant (Metcalf and Luckmann, 1982).

Soybean is one of the major agricultural commodities consumed in Indonesia. This plant species is mainly attacked by *Etiella zinckenella* and *Helicoverpa armigera*, but the former is considered as high priority pest of soybean in Indonesia as its larvae significantly destroy the seeds (Talekar and Lin, 1994). Report on *E. zinckenella* in Indonesia showed that this insect may attack 9% of total pods with total seed loss up to 12% per plant (van den Berg et al., 1998). In order to control the population of *E. zinckenella*, most farmers in Indonesia applied insecticide. However, it caused more problems as spraying does not affect the density of moth larvae or damage to pods and seeds. Rather, it lowers the population of parasitoid (van den Berg et al., 2000).

Referring to this condition, the resistant plant tactic has become the alternative methods to reduce the impact of *E. zinckenella* attack on soybean. This tactic is actually based on the characteristics of trichomes present on the stem, leaf, flower, and pod and they vary among different cultivars (Kogan and Herzog, 1980; PPPT, 1993). Trichome characteristics have been reported to correlate with damage level caused by pod borer (Talekar and Lin,

*Corresponding author. E-mail: agus@sith.itb.ac.id.

1994) and oviposition preference (Suharsono and Suntono, 2004; Suharsono, 2006).

In this study, we tested the effect of different trichome characteristics of various soybean cultivars on the oviposition behavior and preference of *E. zinckenella*. Unlike earlier research work carried out by Suharsono and Suntono (2004) and Suharsono (2006) where only no choice test was used, we applied both no choice and multiple test in the present study. The logical explanation for this is that local farmer use the combination of soybean cultivars in their fields. Further, we also observed the oviposition preference of *E. zinckenella* on different phenological stages of soybean in order to test the hypothesis whether *E. zinckenella* has preffered phonological stage.

MATERIALS AND METHODS

Etiella zinckenella rearing

Larvae of *E. zinckenella* were collected from Sanggar Penelitian Latihan dan Pengembangan Pertanian Universitas Padjajaran Unit Arjasari. The collected *E. zinckenella* larvae were kept in cylinder container (70 ml) and fed with natural food (soybean soaked in water for several hours) at Laboratory of Entomology, School of Life Science and Technology, ITB. The larvae conditions were checked daily, the food were added or renewed if necessary. Rearing containers were kept in a room with the temperature of 27°C, relative humidity of 70 to 80% and a photoperiod of 12:12.

When the larvae were about to complete the fifth-instar stage, they were transferred into separate container filled with tissue papers. When larvae pupated, the pupae were shifted into fresh container until imago emerged. The emerging imagoes were fed with 10% sucrose solution and placed in a cage (40 × 40 × 60 cm). In the following days, adults mated.

Host plant nurture

During this study, we used the common seeds of local soybean cultivar (Sindoro, Dieng, Wilis, Tidar, Galunggung, Leuser, dan Guntur cultivars). The seeds were obtained from Balai Penelitian Tanaman Kacang-Kacangan dan Umbi-Umbian (Balitkabi), Malang. The seeds were sown in a polybag, which were filled with soil that was enriched with NPK fertilizer (3 g/polybag), and kept at greenhouse of School of Life Science and Technology, Institut Teknologi Bandung. The soybean plants were watered after every 2 days and kept for 56 days. Before they were used in this study, the plants have started to produce flowers.

Oviposition behavior test

No-choice test

Seven soybean cultivars were placed in 7 different cages (each with size 40 × 40 × 60 cm). Then, 15 couples of experiment insects were released into the cage. During this study, 3 cages were used for each cultivar. Every day, numbers of oviposited eggs and oviposition behavior of the females were observed which included observing a series of behavior, starting from the host selection alighting of the insect on the plant and until the insect leaved the plant. The observations were taken 2 h prior to the dark period and throughout the dark period, with the aid of flashlight with low light

Table 1. Criteria of soybean stage stated by Indonesian center for food crops research and development (PPTP, 1993).

Stage	Explanation
V	Vegetative stage
R2	Fully flowered
R3	Start fruiting
R4	Full filled pod
R5	Start seeding

intensity and used indirectly to produce enough light to observe the behaviour.

Multiple choice test

Seven soybean cultivars were placed in a cage, each with the size of 85 × 75 × 100 cm. The cultivars were arranged in orderly manner, forming a hexagon with one cultivar in the middle. Then, 15 couples of experiment insects were released into the cage. This procedure were replicated four times. All oviposited eggs and frequency of oviposition were counted and recorded daily. In the same time we also switched each of the cultivar position randomly. The observation procedures for this test is same as procedures for no-choice test.

Effect of trichome characteristics to oviposition preference

The density of trichomes of each cultivar was measured under binocular microscope by placing small paper of 1.4 × 1.4 mm with grid under sample as guidance. Thus, the density of trichomes stated as number of trichomes per 1.96 mm^2. In addition, length of trichomes (mm) was also measured with the help of ocular guidance scale. Further, the surface area of the pods (mm^2) was measured using small scale paper. These procedures were carried out for 10 to 20 plants (depend on the numbers of available plants) of each cultivar. All the data were presented as average values.

Effect of phenological stage of plant to oviposition preference

No-choice test

The soybean cultivar that was preferred most by *E. zinckenella* as host plant (based on oviposition behavior test) was chosen to observe the effect of host plant phenology to its oviposition preference. We used 5 plants for each phenological stage (V, R2, R3, R4 and R5) based on criteria set by Pusat Penelitian dan Pengembangan Tanaman Pangan (PPTP) (Indonesian Center for Food Crops Research and Development, ICFORD) (1993) (Table 1). Each plant was placed in different cage having the dimensions of 40 × 40 × 60 cm. Then, 10 pairs of *E. zinckenella* were released for each repetition and numbers of oviposited eggs were counted daily.

Multiple choice test

Plants of different phenological stages, of the same cultivar that was used in no-choice test were placed inside 85 × 75 × 100 cm sized cage. These plants were arranged in orderly manner, forming a rectangle with 4 plants at each corner and 1 plant in the middle

Table 2. Differences in some of characters of soybean cultivar used in this study.

Cultivar	No. of trichomes	Length of trichomes (mm)	Pod's surface area (mm²)	Number of eggs oviposited	
				No-choice method	Multiple choice method
Sindoro	7.400 ± 0.21^a	1.60 ± 0.01^a	236.11 ± 7.72^a	63.50 ± 6.06^{ab}	107.00 ± 10.80^a
Dieng	7.373 ± 0.24^a	1.50 ± 0.01^b	232.84 ± 4.95^{ab}	78.75 ± 11.96^a	122.50 ± 8.91^a
Wilis	5.673 ± 0.11^b	2.00 ± 0.04^c	210.10 ± 8.30^b	72.75 ± 3.64^a	47.00 ± 1.87^{bcd}
Tidar	5.620 ± 0.20^{bc}	$1.60 \pm 0.01a$	221.82 ± 6.88^{ab}	45.25 ± 5.65^{bc}	66.25 ± 5.88^{cd}
Galunggung	3.147 ± 0.10^d	2.30 ± 0.04^f	239.67 ± 8.25^a	69.25 ± 4.64^a	42.50 ± 5.45^{bc}
Leuser	4.667 ± 0.17^e	$1.95 \pm 0.01c^d$	226.16 ± 10.04^{ab}	37.00 ± 4.88^{cd}	54.25 ± 3.17^{cdc}
Guntur	5.247 ± 0.15^b	$1.90 \pm 0.01d^c$	215.54 ± 10.10^{ab}	67.75 ± 10.23^a	72.50 ± 16.30^d

Average number of trichomes was measured for 1.96 mm² surface area. Average value followed by same letter are not significantly different (P < 0.05, n =16). Average number of eggs was measured from 4 females at both no-choice method and multiple choice method.

Then, 10 pairs of experimental insect were released into the cage. This test was replicated four times. The numbers of oviposited eggs were counted each day and the position of each plant was switched daily.

Data analysis

The data were analyzed by analysis of variance (ANOVA), continued by least significant difference test. We also carried out general linear model test to find the correlation of trichome density, trichome length, and pod surface area with number of eggs deposited. All the data were analyzed using Statistica 6.0 for Windows.

RESULTS AND DISCUSSION

Oviposition behavior

No-choice test method

Female insects began moving their antennae for about 15 min, just before the dark period began. After that, they began flying around the host plants randomly for about 30 to 60 min. Some female insects moved closer to host plant, flew around the plant, and mostly landed on the lower part of the stem, possibly aiming for the young pods. After alighting on pods, they touched for a short time the surface of the pods with their antennae, and they protruded their ovipositor to the lower surface of the pod, forming an angle of 45° relative to their body surface. Female insects moved forward dragging their ovipositor against plant surface while tapping it with their antennae.

The female stopped walking, lifted their antennae, and by raising the abdomen they laid an egg. They continued sensing the surface of the young pod with the end of their ovipositor and antennae sometimes only with the end of their ovipositor. Sensing was paused when they stopped walking, and by raising their abdomen, another egg was laid on new area. This series of behavior was performed repetitively, and an egg was laid in each oviposition attempt which lasted in 4 to 4.5 min. During the

oviposition process, some female took a brief rest or normal walk on the host plant or sometimes moved away. After completion of oviposition behavior, females often take a rest. The frequency of oviposition was higher at 2 h after dark and generally the female insects stayed till the end of the dark period.

Multiple choice test method

Female insects displayed the oviposition behavior same as shown in no choice test accept the number of eggs laid on each cultivar that was different (Table 2). They began by moving their antennae at the beginning of the dark period, and continued their flying around the host plant. This activity was shown for 30 to 60 min to find the perfect host to land and oviposit. Some females moved closer to the host plant, flew around them, and landed on it. During oviposition process, the females laid eggs individually on different pods in each visit.

The results obtained are in accordance with the previous work done on oviposition behaviour of *E. zinckenella* in Japan (Hattori, 1986). The finding indicates that oviposition behavior of *E. zinckenella* might be the result of evolutionary process and it is possibly not affected by differences in climate even though further researches are needed to confirm this. However, Hattori did not report the oviposition behaviour on different hosts. Our study showed that *E. zinckenella* has rigid oviposition behavior as it carries out the same behavioral pattern and spent relatively the same time on each available plant in the cage. It was also observed that even after more tapping on plants, the female insects did not always oviposit egg on each visit.

Effect of trichome characteristics to oviposition preference

Table 2 shows that Dieng and Sindoro cultivars had significantly higher number of trichomes (7.400 trichomes and 7.373 per 1.96 mm², respectively) than other

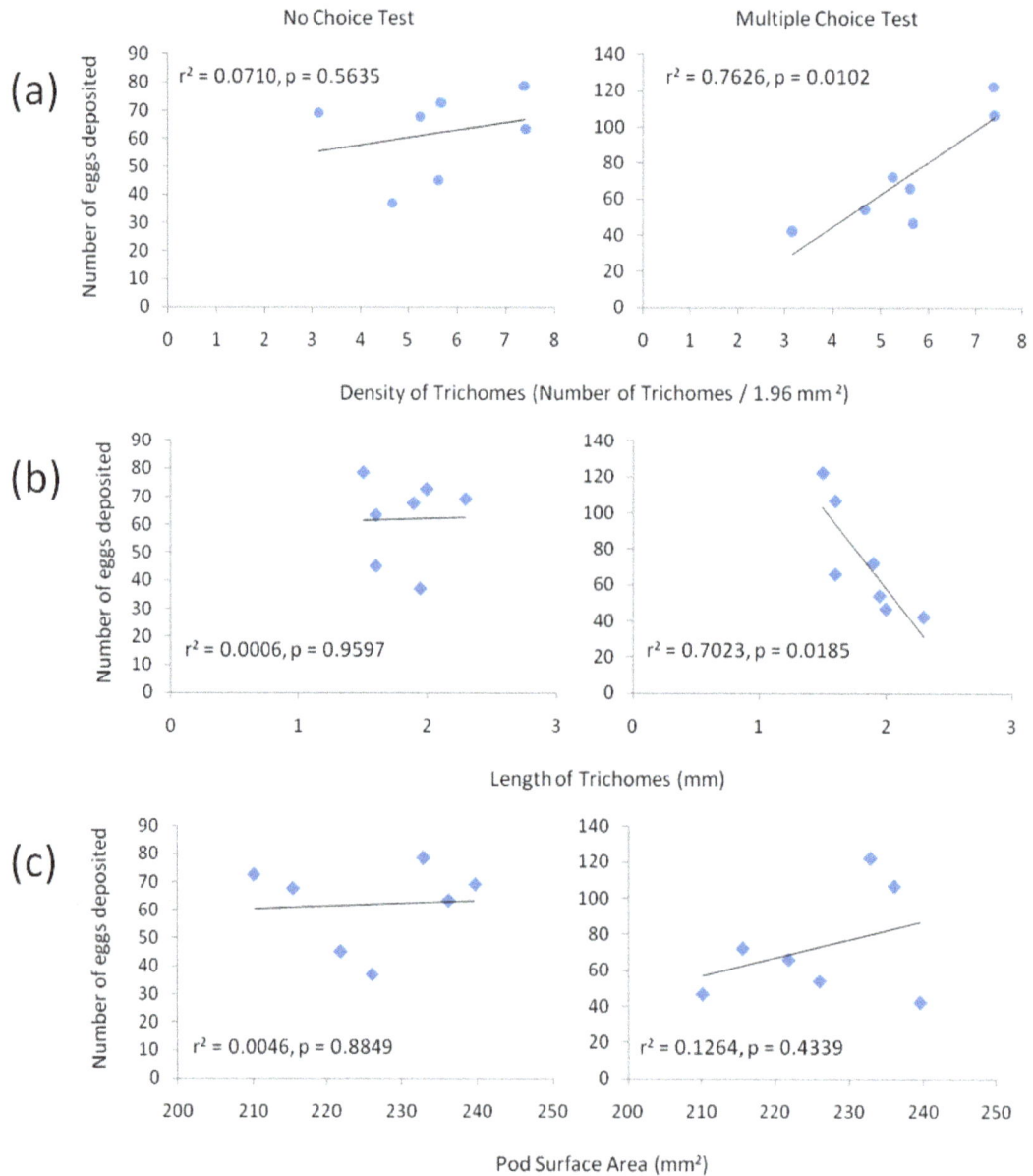

Figure 1. Correlation between trichomes characteristics and number of eggs deposited during no-choice and multiple choice tests. (a) Density of trichomes; (b) length of trichomes; (c) pod surface area.

cultivars. High variations were found in the average length of trichomes of each cultivar - Galunggung had the longest (2.3 mm) and Dieng had the shortest (1.5 mm). Further, Sindoro had significantly bigger pod (236.11 mm^2) and Wilis had significantly smaller pod (210.1 mm^2) (p <0.05) compare with other cultivars.

We did not find any significant difference in number of eggs deposited during no-choice test for all the soybean cultivars. On the other hand, significant preference of *E. zinckenella* was found to oviposit in Sindoro and Dieng cultivars (Table 2). This result indicated that local *E. zinckenella* population has preferred these cultivars over others. When these preferred soybean cultivars are absent, the insect species is also capable to use other cultivars for oviposition.

We found significant positive correlation between trichome density and number of eggs deposited (R^2 = 0.76, p <0.05) and negative correlation between the length of trichome and number of eggs deposited (R^2 = 0.70, p <0.05) only on multiple choice test. On the other hand, we did not find strong correlation between pod surface area and number of egg deposited (Figure 1a, b and c).

As shown in Figure 2, strong positive correlations of density of trichomes (R^2 = -0.72, p<0.05) and number of eggs deposited (R^2 = -0.9459, p<0.05) was observed with

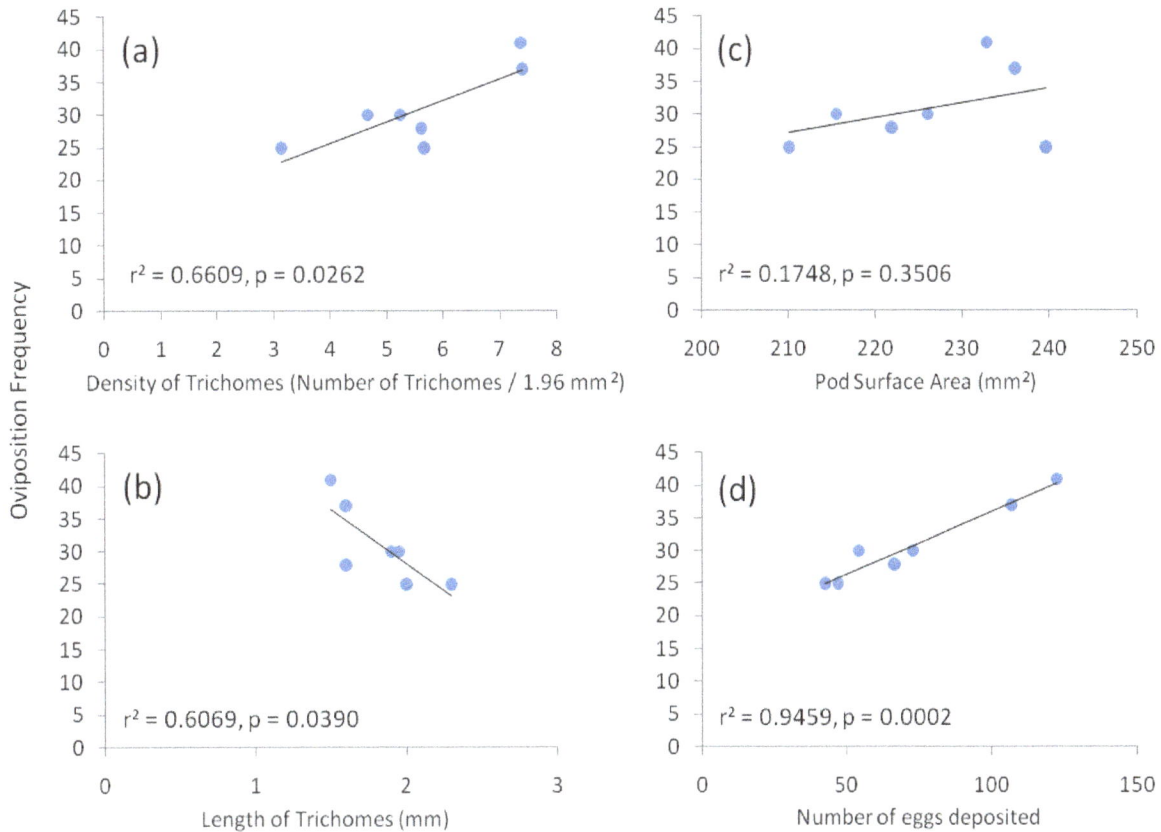

Figure 2. Correlation between (a), Density of trichomes; (b), length of trichomes; (c), pod surface area; (d), number of eggs deposited with oviposition frequency (data only gathered by multiple choice test).

oviposition frequency. On the other hand, length of trichomes showed negative correlation ($R^2 = -0.6069$, $p<0.05$) with oviposition frequency as pod surface area did not affect it ($R^2 = -0.1748$, $p>0.05$).

The present study showed that *E. zinckenella* prefers to oviposit on soybean pods having denser short trichomes which support the results of Talekar and Lin (1994) and Susanto and Adie (2008) who conducted their study, respectively, outside tropical region and in Indonesia. However, our finding does not support the result of another similar research work in Indonesia carried out by Suharsono and Suntono (2004). The difference in the method might be the reason for the deviation in results. Unlike us, Suharsono and Suntono (2004) performed only no-choice test method that could affect the oviposition behavior of female insects. Ohgushi (1992) suggested that the placement of only one host plant per cage could not give female insect the opportunity to choose the perfect place to oviposit. In this circumstance, insect could not fully utilize the host plant which it did not really prefer for its offspring's survival. Notably, Metcalf and Luckmann (1982) emphasized that in certain situations (also in the field) when no better choice was available, the insect species could eat and oviposit on a host plant which they did not really prefer.

The higher number of oviposited eggs by *E. zinckenella* on plant with denser short trichomes might correlate with the structure and function of soybean trichomes. The trichomes of soybean are glandular (Shanower et al., 1996) and secrete viscous liquid that harden as soon as it comes in the contacts of air (Metcalf and Luckmann, 1982) and acts like a sticky glue. Herbivores use this liquid to prevent their eggs from being attacked by natural enemies (Shanower et al., 1996). In addition, denser trichomes can negatively affect the walking speed and behavior of natural enemies (Kashyap et al., 1991).

Effect of plant phenology on oviposition preference of *E. zinckenella*

In this study, we found that *E. zinckenella* preferred to deposit eggs on the pod at developmental stage R4 (mean numbers of eggs for no-choice test and multiple choice test are 72.75 and 105, respectively). The number of eggs deposited increased constantly from stage V (having no eggs) to peak number in stage R4 followed by sharp decline in stage R5 (Table 3).

The present study confirmed other previous researches that suggest correlation between development stage of

Table 3. The average number of *E. Zinckenella* eggs on several stage of soybean development by no-choice and multiple choice tests.

S/N	Stage	Average number of the eggs	
		No-choice test	Multiple choice test
1	V	0.00 ± 0.00^a	0.00 ± 0.00^a
2	R2	23.75 ± 2.69^b	27.50 ± 4.17^{bc}
3	R3	41.75 ± 3.84^b	69.00 ± 6.82^c
4	R4	72.75 ± 12.89^c	105.00 ± 10.66^d
5	R5	38.00 ± 4.78^b	39.25 ± 7.49^e

Average value followed by same letter are not significantly different (P < 0.05, n = 16). Average number of eggs was measured from 4 females at both no-choice method and multiple choice method.

the pod and total number of eggs deposited (Hattori and Sato, 1983; Kobayashi and Oku, 1980; Kamandalu et al., 1997). High preference of *E. zinckenella* females to oviposit in the pods at stage R4 may correlate with the emergence time of larvae as female choose the best possible food for their offspring. Pods at stage R4 provide best physical texture and moisture also which act strongly on the oviposition response of *E. zinckenella* (Hattori and Sato, 1983). We found that the numbers of eggs deposited by *E. zinckenella* in the pods at stages R2, R3, and R5 during no-choice test and multiple choice test were more or less same. This finding indicates that *E. zinckenella* has similar oviposition behavior for less favored phenological stages of their host plant.

The difference in density and length of trichomes in seven soybean cultivars affect the oviposition frequency but not the oviposition behavior of *E. zinckenella*. Agarwala (1996) suggested that the way an insect species gets and uses its host is a historical product of evolutionary adaptation that is established by natural selection in the dimension of space and time. Insect behavior responses in selecting host plants depend on particular genotype pool and environmental condition. Our study showed that *E. zinckenella* can tolerate changes in wide range of environmental conditions, which is a key factor for its successful life as major pest in soybean plantation in Indonesia.

Most insect species choose plants based on the patterns of sensorial information complex combined with several modalities such as taste, smell, sight, and touch that will generate complex neural patterns in central nervous system. The selection of plants by insect is mainly affected by sensorial capture that, in turn, affects behavior of the insect (Rani, 1996). Hirai et al. (1980) reported that several chemical and mechanical stimuli by young pods of soybean could act as the factors attracting *E. zinckenella* for oviposition even though not contribute, at least, in the final step of oviposition behavior (Hattori and Sato, 1983). This fact could be explained by the observation, as mentioned earlier, that *E. zinckenella* protrude its ovipositor on the pod surface, and walked

with its ovipositor still contacting the pod surface before oviposition.

Our results are in agreement with Ramaswamy (1988) who proposed that generalist moths do not depend on volatiles for host finding. These results also support other research studies on *Helicoverpa virescens* (F.), *Epyphyas postvittana* (Walker) and *Spodoptera frugiperda* (Smith) (Ramaswamy et al., 1987; Foster and Howard, 1998; Rojas et al., 2003)

Conclusion

This study extends our understanding on the host finding behavior of *E. zinckenella*. First, it suggests that *E. zinckenella* females use the combination of trichome density, trichome length, and pod surface area condition to select particular soybean cultivar as host within the spatial limit of our study. Secondly, we found that host finding and eggs deposition process by *E. zinckenella* in tropical region is similar to that in non-tropical region. This finding raised other questions on the origin of host finding mechanism and egg deposition behavior of this insect species. Finally, *E. zinckenella* may have fixed behavioral pattern for oviposition during the period when the number of preferred hosts are scarce.

REFERENCES

Agarwala KB (1996). Host related behavioural responses in insect populations their genetic basis. In: Ananthakrishnan TN, editor, Biotechnological and Perspective in Chemical Ecology of Insects. Science Publishers, Inc., New Hampshire, USA. pp. 102-112.

van den Berg H, Shepard BM, Nasikin (1998). Damage incidence by *Etiella zinckenella* in soybean in East Java, Indonesia. Int. J. Pest. Manage., 44: 153-159.

van den Berg H, Aziz A, Machrus M (2000) On-farm evaluation of measures to monitor and control soybean pod-borer *Etiella zinckenella* in East Java, Indonesia. Int. J. Pest. Manage., 46(3): 219-224.

Bailey WJ, Smith JR (1991). Reproductive Behaviour in Insects-Individuals and Populations. Chapman and Hall., London (UK).

Foster SP, Howard J (1998). Influence of stimuli from *Camelia japonica* on oviposition behavior of generalist herbivore *Epiphyas postvittana* J. Chem. Ecol., 24: 1251-1275.

Hattori M (1986). Oviposition behavior of the Limabean Pod Borer, *Etiella zinckenella* TREITSCHKE (Lepidoptera: Pyralidae) on the Soybean. Appl. Entomol. Zool., 21: 33-38.

Hattori M (1988) Host plant factors responsible for oviposition behaviour in the limabean podborer, *Etiella zinckenella* Treitschke. J. Insect Physiol., 34: 191-195.

Hattori M, Sato A (1983). Substrate factors involved in oviposition response of the limabean podborer, *Etiella zinckenella* TREITSCHKE (Lepidoptera : Pyralidae). Appl. Entomol. Zool., 18: 50-56.

Kamandalu AANB, Samudra IM, Priyanto BH, Tengkano W (1997) [Identification of biophysic factors of food plant effected egg laying preference of *Etiella zinkenella* Tr and *Helicoverpa armigera* Hubner] Symposium of Entomology, Bandung, Indonesia. Bandung (Indonesia).

Kashyap RK, Kennedy GG, Farrar RR (1991). Behavioral response of *Trichogramma pretiosum* and *Telenomus sphingis* to trichome/methyl ketone mediated resistance in tomato. J. Chem. Ecol., 17: 543-556.

Kobayashi T, Oku T (1980). Sampling lepidopterous podborers on soybean. In: Kogan M, Herzog DC, editors, Sampling Methods in Soybean Entomology. Springer- Verlag., New York (USA) pp. 422-435.

Kogan M, Herzog DC (1980). Sampling Methods in Soybean Entomology. Springer-Verlag., New York (USA).

Metcalf RL, Luckmann WH (1982). Introduction to Insect Pest Management. John Wiley & Sons., New York (USA).

Ohgushi T (1992). Effect of Resource Distribution in Animal-Plant Interaction. New York (USA): Academic Press Chapter 8, Resource limitation on insect herbivore population, pp. 199-241.

Pusat Penelitian dan Pengembangan Tanaman Pangan (PPTP) (1993). (Indonesia). Soybeans. Bogor.

Ramaswamy SB (1988). Host finding by moths: sensory modalities and behavior. J. Insect. Physiol., 34: 235-249.

Ramaswamy SB, Ma WK, Baker GT (1987). Sensory cues and receptors for oviposition by *Heliothis virescens*. Entomol. Exp. Appl., 43: 15-18.

Rani UP (1996). Sensillar dynamics in insect-plant interactions. In: Ananthakrishnan TN, editor, Biotechnological Perspective Biotechnological and Perspective in Chemical Ecology of Insects. Science Publishers, Inc., New Hampshire, USA. pp. 149-157.

Rojas JC, Virgen A, Cruz-López L (2003). Chemical and Tactile Cues Influencing Oviposition of a Generalist Moth, *Spodoptera frugiperda* (Lepidoptera: Noctuidae). Environ. Ent., 32: 1386-1392.

Shanower GT, Romeis J, Peter AJ (1996). Pigeon pea plant trichomes: multiple trophic level interactions. In: Ananthakrishnan TN, editor, Biotechnological Perspective Biotechnological and Perspective in Chemical Ecology of Insects. Science Publishers, Inc., New Hampshire, USA., pp. 76-84.

Suharsono (2006). Morphological Antixenosis as one of the factors of soybean resistant against podborer pest. Bull. Pala., 12: 29-34.

Suharsono, Suntono (2004). Oviposition preference of limabean podborer to Soybean Genotypes. Pen Pert., 23(1): 38-48.

Susanto GWA, Adie MM (2008). Characteristic of morphological resistance on soybean genotype to podborer insect. Pen. Pert., 27: 95-100.

Talekar SN, Lin PC (1994). Characterization of resistance to limabean podborer (Lepidoptera: Pyralidae) in soybean. J. Econ. Entomol., 87: 821-824.

30

Stoneflies (Insecta: Plecoptera) in Malaysian tropical rivers: Diversity and seasonality

Suhaila Abdul Hamid* and Che Salmah Md. Rawi

School of Biological Sciences, Universiti Sains Malaysia. 11800 Minden, Penang, Malaysia.

Adult stoneflies (Plecoptera) were light-trapped monthl from January to December 2008 in Tupah River, Kedah, Malaysia. Two families of Plecoptera, Perlidae and Nemouridae were represented with nine species. More Plecoptera was collected in the wet seasons especially for the family Nemouridae. *Neoperla* asperata was a common species that occurred throughout the year. *Neoperla fallax* was equally common but this species was absent during early part of the year. The percentage of females was higher than male for all species of Plecoptera collected.

Key words: Upstream river, Plecoptera, season, sex ratio, peninsular Malaysia.

INTRODUCTION

Plecoptera are primitive group of insects also known as stoneflies or salmonflies. The diversity of Plecoptera declines rapidly from temperate Asian latitudes (nine families) to tropical latitudes (four or fewer families). The only diverse stonefly family in the Malaysian region is the Perlidae. Comparative to their temperate counterparts, tropical stoneflies are incompletely understood (Sheldon and Theischinger, 2009) although these regions have the highest diversity of stoneflies (Zwick, 2000). Asian stoneflies diversity is much greater than that of Europe or North America but the knowledge of the enduring Asiatic areas is extremely poor (Fochetti and Tierno de Figueroa, 2008). In Malaysia, no systematic work on Plecoptera has been undertaken. Sivec and Yang (2001) estimated there are approximately 350 Plecoptera species in countries forming the Oriental Region except for Southern China. One reason for the paucity of ecological studies of tropical river Plecoptera in the international literature is the fact that identification of tropical species is difficult for non-specialist. Many lower taxa (especially genera and species) have received limited study and relevant literature is scarce (Boyero, 2002).

Plecoptera provides a valuable food source for a wide variety of vertebrates. Nymph and adult stages are eaten by many species of fish and amphibians (Petersen et al., 1999). Actively dispersing adults' Plecoptera are a valued and plentiful food source for bats and birds that feed at dusk on flying insects (Fochetti and Tierno de Figueroa 2008). Moreover, according to Sweeney (1993), mammals such as shrews and raccoons also eat on Plecoptera nymphs as well as emerging adults.

Aquatic insect emergence is strongly influenced by season. Based on Gopal (2002), most of the Asian countries are affected by the monsoon and seasonal rainfall where monsoon behavior is nearly unpredictable. Most tropical rivers have an annual cycle, which like so many features of the river is dictated by the pattern of rainfall (Payne, 1986). Precipitation plays a major role in changing the benthic community (Robinson and Minshall, 1986) in the tropical rivers (Silveira et al., 2006). In this study, the diversity and seasonal distribution of stoneflies of a water catchment in a tropical river of northern peninsular Malaysia were described. Previously, studies by Che Salmah et al. (2001; 2007) emphasize functional roles of immature plecopterans. In the present study adults were identified to the species level and their species richness evaluated for conservation value in Tupah River. Factors that could influence the temporal distribution of adults were investigated.

MATERIALS AND METHODS

Sampling of adult Plectopera

The sampling was carried out in Tupah River located at N5°45 E100°26' in Kuala Muda district. It is situated within the Gunung

*Corresponding author. E-mail: sue_sj@yahoo.com.

Table 1. Succession of adult Plecoptera (stoneflies) at Tupah River.

Taxon	DS								WS			DS
	Jan	Feb	Mar	Apr	May	Jun	Jul	Aug	Sep	Oct	Nov	Dec
N. asperata	■	■	■	■	■	■	■	■	■	■	■	
N. fallax				■	■	■	■	■	■	■	■	
N. hamata				■	■							
K. trang		■	■	■				■	■	■		
K. jariyae	■			■								
K. curriei				■	■				■	■	■	
E. nigrogeniculatum						■	■		■	■	■	■
P. malayana				■	■							
Indonemoura spp.	■		■	■								■
Total	4	8	16	100	20	34	6	25	13	5	26	42

Key: DS – Dry season; WS – Wet season.

Jerai catchment area, a virgin forest reserve. The 5.6 km long, 0.32 ± 0.05 meter mean depth with mean width 4.14 ± 0.28 meter river enters Merbok River, which flows, into the Straits of Malacca. Tupah River has fast flow of 0.56 ± 0.16 m/s. The water pH in the Tupah River ranges from 5.03 to 6.66 whiles the yearly mean water temperature ranged from 22.8 to 25.7 ºC. Tupah River passes through a low land Dipterocarp Forest at 100-200 meters above sea level. Tree species, such as *Shorea leprosula*, *Shorea ovata*, *Dipterocarpus sp.*, *Dillenia sp.*, *Pometia pinnata*, *Pongamia pinnata*, *Dolichandrone spathacea* and *Sindora sp.* are dominant in the area. River substrates are predominantly cobble and gravel (65%), and the other 35% of river sediment is made up of boulder.

Light trap was used to collect adults Plecoptera in this study. Two white sheets (1.7 m long x 1.2 m wide) were hung adjacent to each other at right angle (90°). They were placed 5 m from the river edge. Mercury light bulb (250 watt) powered by portable generator (EM650Z) was provided between the sheets. White mercury light was used for the trap because it consisted of a combination of all the visible wavelengths (Ward, 1965). All adults landed on the sheets were collected either with forceps or by hands and placed them in an empty vial. All methods followed Bispo et al. (2002). On each sampling occasions, the light trap was deployed for 5 hours from 1830-2330 h. This duration was found to be an active flying period for the adults based on preliminary observations prior to the sampling. The collection was terminated at 0000 hr because no Plecoptera was found after that time. Number of trap, frequency of trapping and duration time of sampling was proposed by Bowden (1982). Samples were not separated on hourly basis because the total adults captured was small showing no clear pattern of Plecoptera composition. Adults were collected one day in a month for 12 months from January to December 2008. All collected specimens in one night were considered as one sample for that particular month. Sampling was done on a clear sky night, avoiding the full moon that might interrupt light trap catches.

In the laboratory, adult Plecoptera were placed in a universal bottle filled with 75% ethanol. They were identified using keys of Merritt and Cummins (1996), Wang and McCafferty (2004) and Triplehorn and Johnson (2005), counted and placed in the corresponding species. Species identifications were verified species by Prof. Ignec Sivec (Slovenian Museum of Natural History) and Prof. Bill P. Stark from Mississippi College, Mississippi, USA.

Data analysis

Independent samples t-test analysis was used to determine seasonal influence on Plecoptera abundance and diversity while the sex ratio was determined using the chi-square test for goodness of fit. The differences of adult Plecoptera abundance among months of sampling were determined using the Kruskal-Wallis test for non-normally distributed data. All analysis were tested using the SPSS software version 14®.

Monthly means of temperature and precipitation for Kuala Muda district were obtained from the Malaysian Meteorological Department headquarters in Kuala Lumpur. According to the department's database, annual precipitation in Kuala Muda district for the year 2008 was 2301.3 mm. Dry seasons was determined when mean monthly precipitation was less than 200 mm. The wet season was identified when mean monthly precipitation was more than 200 mm. Based on the amount of precipitation in the area, the dry season commenced from January to July and December, whereas the wet season started from August until November 2008.

RESULTS

Diversity and succession of adult Plecoptera (stoneflies) in Tupah River

In this study, the diversity of adult Plecoptera varied widely over the months. The result showed 299 individuals of Plecoptera belonging to two families, five genera and at least nine species were identified (Table 1). This assemblage represented large number of individuals considering the collection was made from a small area of low elevation and only single trap was operated. Out of four plecopteran families occurring in peninsular Malaysia (Yule and Yong 2004), two families (Perlidae and Nemouridae) were obtained from Tupah River. No representative of adults Peltoperlidae and Leuctridae was caught throughout the sampling period although Peltoperlidae nymphs were collected from the river.

Adult stoneflies were present at all times of the year in Tupah River. The largest abundance of perlid adults were collected in April (33.4 ± 1.10) while the lowest was recorded in January (1.33 ± 0.21) (Figure 1). Two main species of *Neoperla* were commonly collected in this

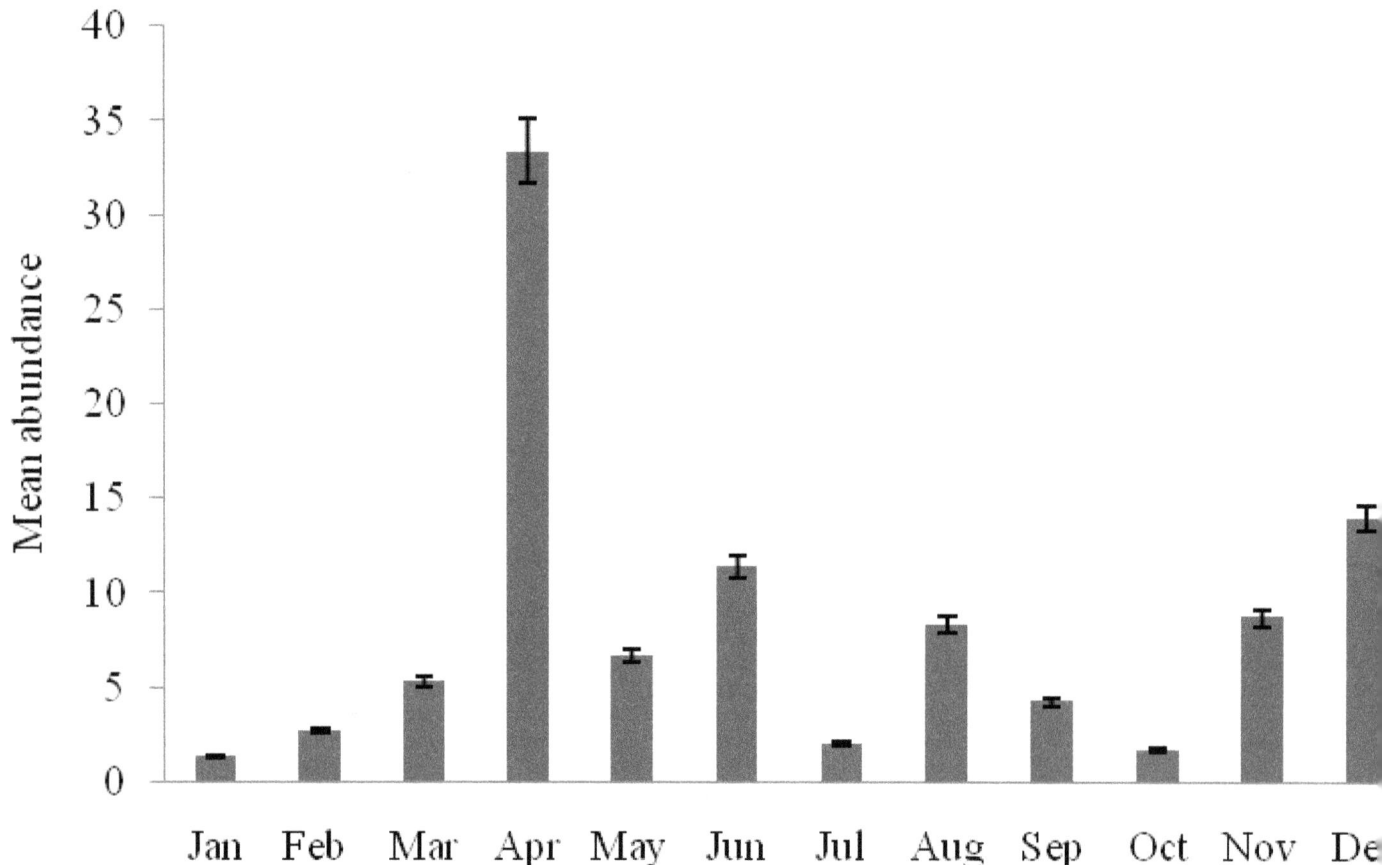

Figure 1. Mean abundance (±SE) of adult Plecoptera collected in Tupah River in the year 2008.

area. *Neoperla asperata* were found in all months of the year while *N. fallax* was found in most of the months especially in January, February and July 2008. *Phanoperla malayana* was spotted in April 2008 (dry season) in Tupah River. In November 2008, all three species of *Kamimuria* (*K. trang*, *K. jariyae* and *K. curriei*) were present. *Indonemuoura* spp. appeared in January, March and December 2008. The Kruskal-Wallis test indicated that there was no significant difference in number of individual Plecoptera among months of sampling ($X^2 = 3.255$, P > 0.05).

The highest diversity (seven species) of adult Plecoptera was recorded in April 2008 whiles the lowest number of species occurred in July 2008 (one species) (Figure 2). Relatively few species were present from May to August 2008 but species number began to increase in September until December 2008. There was no significant difference in diversity of adult Plecoptera among sampling months ($X^2 = 1.373$, P > 0.05).

Succession among plecopteran species in Tupah River is depicted in Table 1. Most species were absent at sometimes in a year. However, *N. asperata* was a common species that occurred throughout the year. *N eoperla fallax* was equally common although this species

was absent during early part of the year in February and June. Occurrence of other species were scattered in some months in the year. *Kamimuria trang* was more common than *K. jariyae* that was only found in January and November. Except for *N. asperata*, the distribution of adult plecopterans in general was rather restricted to certain time of months in either dry or wet seasons.

Perlid *P.malayana* and nemouridaes were present only in the dry season (January, March and December 2008). Other perlid species were present at least during part of both seasons. The lowest number of adults Plecoptera were collected in the months with mean rainfall less than 150 mm. The Mann-Whitney U test revealed that there was no significant difference in abundance of adult Plecoptera in both seasons (z=-0.444, P=0.657).

Distribution of male and female plecopteran adults varied during their flight period and more females were collected than males (Figure 3). The result of Chi-square test for goodness of fit revealed that there was a significant difference in the frequency of occurrence of males and females. Females were more abundant than males, X^2 (1, N = 309) = 79.77, P<0.05. However, the sex ratio remained 1:1. Throughout the sampling, only males *K. jariyae* and females *P.malayana* were collected but

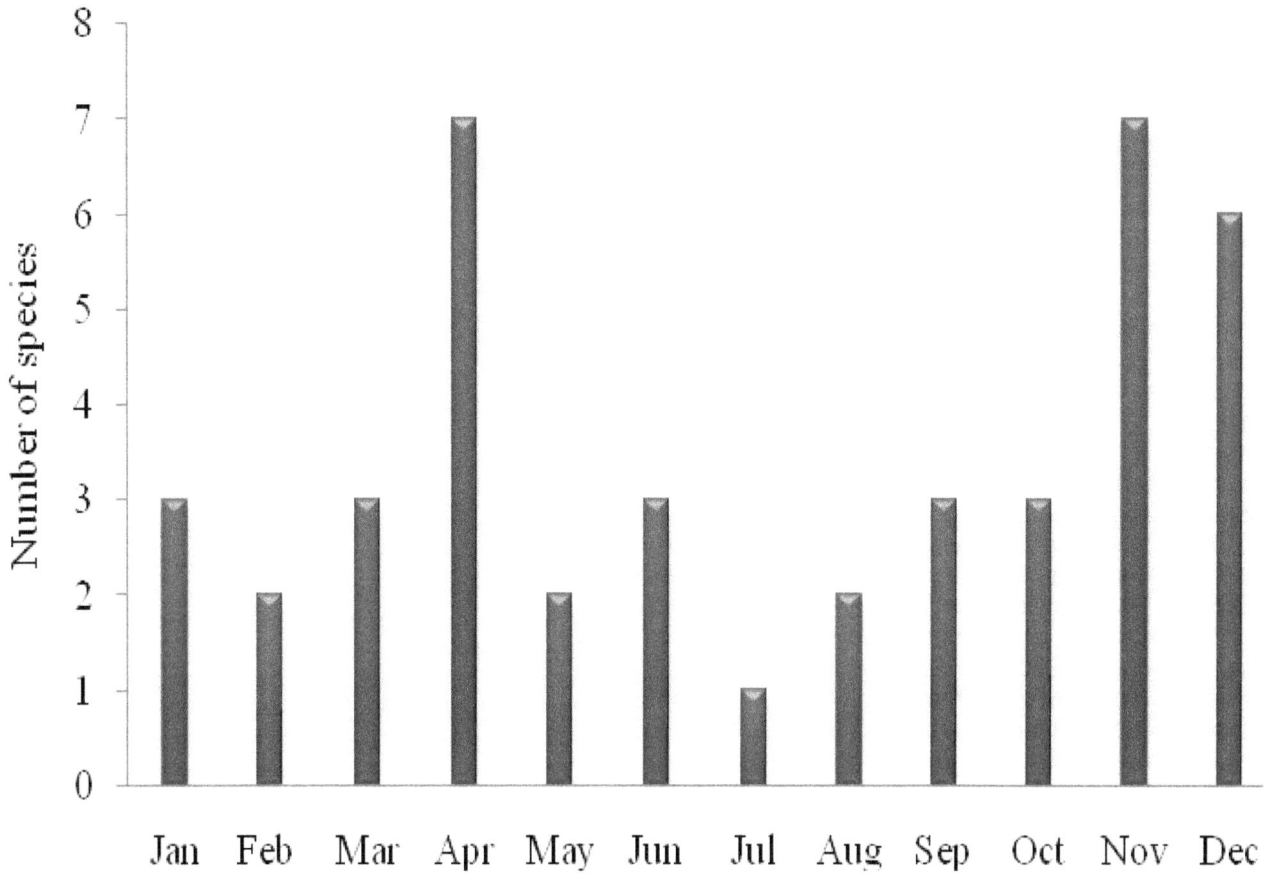

Figure 2. Monthly diversity of adult Plecoptera species collected Tupah River in 2008.

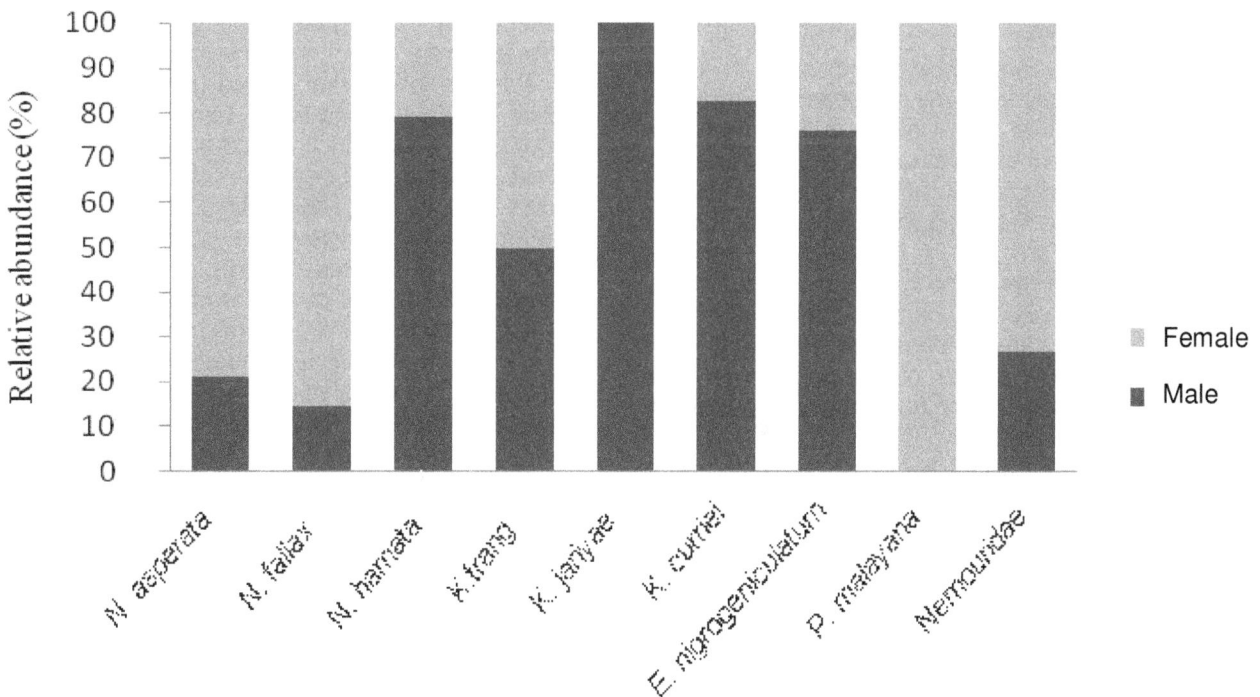

Figure 3. Relative abundance (percentage) of males and females of plecopterans species caught in light traps in Tupah River.

both sexes were represented in other species.

DISCUSSION

Naturally, the stoneflies are poorly represented in the tropics (Sivec and Yule, 2004). Different plecopteran species are active at different times of the day (Triplehorn and Johnson, 2005). Plecoptera caught in this study are nocturnal since the samplings were done at night. Study by Vaught and Stewart (1974) has confirmed the nocturnal emergence of Neoperla clymene. Meanwhile, Nemouridae is a diurnal stonefly as Hitchcock (1974) found at least two species of Nemoura emerged in the morning.

Petersen et al. (1999) suggest that differences in riparian vegetation may affect the distribution of adult Plecoptera. Sweeney (1993) supported Petersen's claim when he found many plecopteran species actively choose different streamside trees as preferred site to rest or mate. For instance, Nemouridae show high preference over deciduous trees and shrubs (Harper, 1973). In their studies, Sweeney (1993) and Baas and Mennen (1996) reported that loss of riparian trees can increase predation pressure. In Tupah River, the unshaded water surface led to higher water temperature thus decreased the abundance of plecopteran nymphs because stoneflies reach their maximum richness in cool, riffle area streams (Hynes, 2000). Apart from having naturally low population in tropical streams, the habitats of Tupah River may have been less suitable for this order of insects.

In Tupah River the substrate was dominated with boulders and cobbles that was not suitable for families such as Leuctridae (Pescador et al., 2000; Stewart and Stark, 2002). Moreover, fewer gravel substrates (5%) which were optimally embedded in the sand in Tupah River were less suitable for many plecopteran nymphs.

According to Duffield and Nelson (1990), the flight period of aquatic insect adults depends on the duration of the adult life. Furthermore, adult Nemouridae can live for a few days while the life span of Perlidae last for four weeks (Lees and Ward, 1987). Hynes (1976) reported that Nemouridae may be found lesser in the tropics but numerous in Western Europe than in North America. The reason behind this varied abundance is based on a speculation that Atlantic climate of Europe is milder, thus providing suitable temperature for flight activity (Hynes, 1976).

Many authors reported that the emergence pattern of Plecoptera corresponds to weather conditions. Most emergences occur during sunny days and cease during cloudy days (Flannagan and Lawler, 1972; Masteller and Buzby, 1993; Collier et al., 1997). In New Zealand, the peak emergence that generally occurs in summer and their flight activity is related to air temperature (Collier et al. 1997). Seasonal precipitation has been suggested as a factor influencing composition and temporal abundance of aquatic insect in Puerto Rico (Masteller and Buzby,

1993). In addition, DeWalt and Donald (1998) confirms that lower richness of adult Plecoptera is related to lower air and water temperature.

In this study, Plecoptera population was naturally low in both seasons especially in the wet season. Predation and behavior of the stoneflies could also contribute to the paucity of the population. Bird and Hynes (1980) found that most of plecopterans especially Capniidae move away from the water towards the woods. Collier and Scarsbrook (2000) reported that many Plecoptera emerge on land and because of this, they are prone to carabid (Coleoptera) predation. In their study in Isar River, Germany, Hering and Plachter (1997) found large proportion of recognizable Plecoptera preys in the guts of Nebria picicornis (Carabidae), More Plecoptera species can be seen during the dry season especially in April (seven species). The diversity began to decrease in July at the onset of the wet season. Bottorff and Bottorff (2007) found that colder temperatures have delayed stonefly emergence in Sagehan Creek but warmer temperatures hastened stonefly emergence in Irish Gulch Creek, California.

In this study, two most dominant species, N. asperata and N. fallax showed non-seasonal emergence. Similarly, in southeastern Brazil (subtropical areas) by Froehlich and Oliveira (1997) have found Macrogynoplax sp. (Perlidae) flew throughout the year. Anacroneuria sp. (Perlidae) in the tropics of Pedregulho, Sao Paulo State, Brazil fly all year round in high abundance (Bispo et al., 2002).

Both males and females were found in the plecopteran population in the Tupah River. The percentage of females was higher than males for N. hamata, K. jariyae, K. curriei and E. nigrogeniculatum. Only females P. malayana were caught in this study probably because the females live longer than males (Hynes 1976; Smith and Collier, 2005). Long-lived female plecopterans are important for reproduction and species dispersal. Those females may be important in colonization of new areas (Bunn and Hughes, 1997). In contrast to P. malayana, only males K. jariyae were caught in Tupah River. According to Hynes (1976), normally the males emergence is ahead of the females although they could overlap. Consequently, more males were caught at certain period of time. In contrast, Petersen et al. (1999) collected more females than males Plecoptera because many females may disperse inland to mate and rest. They later move back to the stream to oviposit whereas the males do not. However, this study was very preliminary. Conclusion on the plecopteran sex ratio must remain tentative. More sampling variations could be explored, for example variation in time of the day and collection using various sampling gears.

Conclusion

Plecoptera were only represented by two families,

Perlidae and Nemouridae in the study area. The low number of plecopteran population may have caused by the terrestrial taxon-specific predation either by riparian arthropods or riparian birds. Furthermore, few plecopteran individuals belonging to two families which reported in the Tupah River is confirming the statement of their rarity in the tropical streams.

Abundance of Plecoptera was more affected in the wet seasons especially for the family Nemouridae. *Neoperla aspearata* (Perlidae) was the most common species of Plecoptera found in Tupah River and based on sample collections; the abundance of this species may not have seasonal variation. This study also revealed only males *Kamimuria jariyae* and females *Phanoperla malayana* were represented in light traps. The main reason for the low stonefly numbers and richness reported in this study could be related to the collection technique.

ACKNOWLEDGEMENTS

We thank the School of Biological Sciences for providing various facilities to carry out this study. We appreciate many individuals from the Laboratory of Aquatic Entomology (USM) who tirelessly helped us both in the field and in the laboratory. We are grateful to Kedah Forest Department for granting us permission to work in Gunung Jerai Forest Reserve. This study was supported by the Fundamental Research Grant Scheme (203/PBIOL/671060), Ministry of Science, Technology and Inovation (MOSTI), Malaysia.

REFERENCES

Baas SFJ, Mennen JBT (1996). Aspects of microclimate in contrasting land use and the light climate along stream continua in New Zealand. NIWA Internal Report.

Bird GA, Hynes HBN (1980). Movements of immature aquatic insects in a lotic habitat. Hydrobiologia, 77: 103-112.

Bispo PC, Froehlich CG, Oliveira LG (2002). Stonefly (Plecoptera) fauna of streams in a mountainous area of Central Brazil: Abiotic factors and nymph density. Rev. Bras. Zool., 19(1): 325-334.

Bottorf RL, Bottorff LD (2007). Phenology and diversity of adult stoneflies (Plecoptera) of a small coastal stream, California. Illiesia, 3(1): 1-9.

Bowden J (1982). An analysis of factors affecting catches of insects in light-traps. Bull. Entomol. Res., 72: 535-556.

Boyero L (2002). Insect biodiversity in freshwater ecosystems: Is there any latitudinal gradient? Mar. Freshwater Res., 53: 753-755.

Bunn SE, Hughes JM (1997). Dispersal and recruitment in streams: evidence from genetic studies. J. N. Am. Benthol. Soc., 16: 338-346.

Che Salmah MR, Amelia ZS, Abu Hassan A (2001). Preliminary Distribution of Ephemeroptera, Plecoptera and Trichoptera (EPT) in Kerian River Basin, Perak, Malaysia. Pertanika J.Trop. Agric. Sci., 24: 101-107.

Che Salmah MR, Abu Hassan A, Jongkar G (2007). Diversity of Ephemeroptera, Plecoptera and Trichoptera in various tributaries of Temenggor Catchment, Perak, Malaysia. Wetland Sci., 5(1): 20-31.

Collier KJ, Brian JS, Brenda RB (1997). Summer light-trap catches of adult Trichoptera in hill-country catchments of contrasting land use, Waikato, New Zealand. New Zealand J. Mar. Freshwater Res., 31: 623-634.

Collier KJ, Scarsbrook MR (2000). Use of riparian and hyporheic habitats. In Collier KJ, Winterbourn MJ (eds.). New Zealand stream invertebrates: Ecology and implications for management. New Zealand Limnological Society, Christchurch, New Zealand.

DeWalt RE, Donald WW (1998). Summer Ephemeroptera, Plecoptera and Trichoptera (EPT) taxa richness and Hilsenhoff Biotic Index at eight stream segments in the lower Illinois river basin. Technical Report. Illinois Natural History Survey. Center for Biodiversity.

Duffield RM, Nelson CH (1990). Seasonal emergence patterns and diversity of Plecoptera on Big Hunting Creek, Maryland with a checklist of the stoneflies of Maryland. Proc. Entomol. Soc. Wash., 92(1): 120-126.

Flannagan JF, Lawler GH (1972). Emergence of caddisflies (Trichoptera) and mayflies (Ephemeroptera) from Heming Lake, Manitoba. Can. Ent., 104: 173-183.

Fochetti R, Tierno de Figueroa JM (2008). Global diversity of stoneflies (Plecoptera; Insecta) in freshwater. Hydrobiologia, 595: 365-377.

Froehlich CG, Oliveira LG (1997). Ephemeroptera and Plecoptera nymphs from riffles in low-order streams in Southeastern Brazil. In Landolt P, Sartori M (eds.). Ephemeroptera and Plecoptera: Biology-Ecology-Systematics. MTL. Fribourg.

Gopal B (2002). Conserving biodiversity in Asian wetlands: issues and approaches. In Ali, A., Che Salmah, M.R., Mashor, M., Reiko, N., Sundari, R. and Taej, M. (eds.). The Asian Wetlands: Bringing Partnerships into Good Wetland Practices. Penerbit Universiti Sains Malaysia. Pulau Pinang.

Harper PP (1973). Life histories of Nemouridae and Leuctridae in Southern Ontario (Plecoptera). Hydrobiologia, 41: 309-356.

Hering D, Plachter H (1997). Riparian ground beetles (Coleoptera: Carabidae) preying on aquatic invertebrates: A feeding strategy in alpine floodplains. Oecologia (Berlin), 111: 261-270.

Hitchcook SW (1974). The Plecoptera or stoneflies of Connecticut. Conn.State Geol. Nat. Hist. Surv. Bull., 107: 1-262.

Hynes HBN (1976). Biology of Plecoptera. Ann. Rev. Ent., 21: 135-153.

Hynes HBN (2000). Introduction. In Stark BP, Armitage BJ (eds.) Stoneflies (Plecoptera) of Eastern North America. Pteronarcyidae, Peltoperlidae and Taeniopterygidae. Ohio Biological Survey, Columbus, Ohio, Volume 1.

Lees JH, Ward RD (1987). Genetic variation and biochemical systematics of British nemouridae. Biochem. Syst. Ecol., 15(1): 117-125.

Masteller EC, Buzby KM (1993). Composition and temporal abundance of aquatic insect emergence from a tropical rainforest stream, Quebrada Prieta, at El Verde, Puerto Rico. J. Kansas Entomol. Soc., 66(2): 133-139.

Merritt RW, Cummins KW (1996). An introduction to the aquatic insects of North America. 3rd edition. Kendall/Hunt Publishing Company. Dubuque.

Payne AI (1986). The Ecology of Tropical Lakes and Rivers. John Wiley and Sons, Great Britain.

Pescador ML, Rasumussen AK, Richard BA (2000). A guide to the stoneflies (Plecoptera) of Florida. Florida Department of Environmental Protection Division of Water Resource Management, Tallahassee, Florida.

Petersen I, Winterbottom JH, Orton S, Friberg N, Hildrew AG, Spiers DC, Gurney WSC (1999). Emergence and lateral dispersal of adult Plecoptera and Trichoptera from Broadstone Stream, U.K. Freshwater Biol., 42: 401-416.

Robinson CT, Minshall GW (1986). Effects of disturbance frequency on stream benthic community structure in relation to canopy cover and season. J. N. Am. Benthol., Soc., 5: 237-248.

Sheldon AL, Theischinger G (2009). Stoneflies (Plecoptera) in a tropical Australian stream: Diversity, distribution and seasonality. Illiesia, 5(6): 40-50.

Silveira MP, Buss DF, Nessimian JL, Baptista DF (2006). Spatial and temporal distribution of benthic macroinvertebrates in a southeastern Brazilian River. Braz. J. Biol., 66 (2B): 623-632.

Sivec I, Yang PS (2001). Stoneflies of Taiwan within the oriental stonefly fauna diversity. In Domı́nguez E (ed.) Trends in Research in Ephemeroptera and Plecoptera. Kluwer Academic/ Plenum Publishers, New York.

Sivec I, Yule CM (2004). Insecta: Plecoptera Freashwater invertebrates of the Malaysian region. In Yule CM, Yong HS. Freshwater invertebrates of the Malaysian Region. Academy of Science Malaysia. Kuala Lumpur.

Smith BJ, Collier KJ (2005). Tolerances to Diurnally Varying Temperature for three species of adult aquatic insects from New Zealand. Environ. Entomol., 34(4): 748-754.

Stewart KW, Stark BP (2002). Nymphs of North American Stonefly Genera (Plecoptera). Second edition. The Caddis Press. Columbus, Ohio.

Sweeney BW (1993). Effects of streamside vegetation on macroinvertebrate communities of white clay creek in eastern North America. Proc. Acad. Natl. Sci. Philad.,. 144: 291-340.

Triplehorn CA, Johnson NF (2005). Study of Insects (7th ed.). Thomson Brooks.

Vaught G, Stewart KW (1974). The life history and ecology of the stonefly *Neoperla clymene* (Newman). Ann. Entomol. Soc. Am., 67: 167-178.

Wang TQ, McCafferty WP (2004). Heptageniidae (Ephemeroptera) of the World. Part 1: Phylogenetic Higher Classification. Trans. Am. Entomol. Soc., 130(1): 11-45.

Ward PH (1965). Some Ephemeroptera, Neuroptera and Trichoptera collected by mercury vapour light trap in a Hertfordshire Garden. Entomol. Gazette, 16: 169-174.

Yule CM, Yong HS (2004). Freshwater invertebrates of the Malaysian Region. Academy of Science Malaysia. Kuala Lumpur.

Zwick P (2000). Phylogenetic system and zoogeography of the Plecoptera. Annu. Rev. Entomol., 45: 709-746.

Permissions

List of Contributors

Hassan Ghahari
Department of Agriculture, Islamic Azad University, Shahre Rey Branch, Tehran, Iran

Shaaban Abd-Rabou
Plant Protection Research Institute, Ministry of Agriculture, Dokki-Giza, Egypt

Jiri Zahradnik
Podebradova 498, 512 51, Lomnice nad Popelkou, Czech Republic

Hadi Ostovan
Department of Entomology, Islamic Azad University, Fars Science and Research Branch, Iran

Amel Ben Hamouda
Life Sciences Laboratory, Faculty of Sciences of Sfax, Route de la Soukra km 3.5 - B.P. n° 1171 – 3000, Sfax, Tunisia
Plant Protection laboratory, High Institute of Agronomy of Chott-Mériem, 4042 Chott Mériem, Sousse, Tunisia

Seiji Tanaka
Locust Research Laboratory, National Institute of Agrobiological Sciences at Ohwashi (NIASO), Tsukuba, ibaraki 305-8634, Japan

Mohamed Habib Ben Hamouda
Plant Protection laboratory, High Institute of Agronomy of Chott-Mériem, 4042 Chott Mériem, Sousse, Tunisia

Abderrahmen Bouain
Life Sciences Laboratory, Faculty of Sciences of Sfax, Route de la Soukra km 3.5 - B.P. n° 1171 – 3000, Sfax, Tunisia
Ibaraki 305-8634, Japan

R. Sapkota
Institute of Agriculture and Animal Science Chitwan, Nepal

K. C. Dahal
Institute of Agriculture and Animal Science Chitwan, Nepal

R. B. Thapa
Institute of Agriculture and Animal Science Chitwan, Nepal

Abdul Nasser Trissi
Aleppo University, Faculty of Agriculture, Aleppo, Syria

Mustapha El Bouhssini
International Center for Agricultural Research in the Dry Areas, P. O. Box 5466, Aleppo, Syria

Mohammad Naif Al Salti
Aleppo University, Faculty of Agriculture, Aleppo, Syria

Mohammad Abdulhai
General Commission for Scientific Agricultural Research, Aleppo Center, Aleppo, Syria

Margaret Skinner
Entomology Research Laboratory, University of Vermont, Burlington, VT, USA. 05405-0105, USA

Bruce L. Parker
Entomology Research Laboratory, University of Vermont, Burlington, VT, USA. 05405-0105, USA

S. L. Deore
Government College of Pharmacy, Kathora Naka, Amravati- 444604, (M.S), India

S. S. Khadabadi
Government College of Pharmacy, Kathora Naka, Amravati- 444604, (M.S), India

R. K. Arora
Division of Entomology, Faculty of Agriculture, Shere-Kashmir University of Agricultural Sciences and Technology-J, Jammu-180 002, India

R. K. Gupta
Division of Entomology, Faculty of Agriculture, Shere-Kashmir University of Agricultural Sciences and Technology-J, Jammu-180 002, India

K. Bali
Division of Entomology, Faculty of Agriculture, Shere-Kashmir University of Agricultural Sciences and Technology-J, Jammu-180 002, India

Sainudeen Pattazhy
Department of Zoology, S.N. College, Punalur, Kerala, India

Dokurugu Maayiem
University for Development Studies, Faculty of Renewable Natural Resources, Department of Forestry and Forest Resources Management, P.O. Box TI 1882, Tamale

Baatuuwie Nuoleyeng Bernard
University for Development Studies, Faculty of Renewable Natural Resources, Department of Forestry and Forest Resources Management, P.O. Box Tl 1882, Tamale

Aalangdong Oscar Irunuoh
University for Development Studies, Faculty of Renewable Natural Resources, Department of Range and Wildlife Management, P.O. Box Tl 1882, Tamale

B. K. Shivanna
Department of Agricultural Entomology, University of Agricultural Sciences (UAS), AINRP(T), ZARS, Shimoga -577204, Karnataka, India

B. Gangadhara Naik
Department of Plant Pathology, University of Agricultural Sciences (UAS), College of Agriculture, Shimoga -577204, Karnatak, India

R. Nagaraja
Department of Agricultural Microbiology, University of Agricultural Sciences (UAS), KVK, Shimoga -577204, Karnataka, India

S. Gayathridevi
Department of Agricultural Entomology, College of Agriculture, University of Agricultural Sciences (UAS), Shimoga -577204 Karnataka, India

R. Krishna Naik
Department of Computer Science, College of Agriculture, University of Agricultural Sciences (UAS), Shimoga -577204 Karnataka, India

H. Shruthi
Department of Agricultural Microbiology, College of Agriculture, University of Agricultural Sciences (UAS), Shimoga -577204 Karnataka, India

Zohair I. F. Rahemo
Department of Biology, College of Science, University of Mosul, Mosul, Iraq

Sagida S. Hussain
Department of Biology, College of Science, University of Mosul, Mosul, Iraq

M. D. Shoko
Department of Agronomy, Stellenbosch University, Box X1, Matieland, RSA

M. Zhou
School of Plant, Environmental and Soil Sciences, Louisiana State University, Baton Rouge, LA 70803, USA

B. R. Guruprasad
Kannada Bharthi College, Kushalnagar, Madikari. Atreya Ayruvedic Medical College, Bangalore Department of Zoology and Genetics, University of Mysore, Mysore, India

Pankaj Patak
Kannada Bharthi College, Kushalnagar, Madikari. Atreya Ayruvedic Medical College, Bangalore Department of Zoology and Genetics, University of Mysore, Mysore, India

S. N. Hegde
Kannada Bharthi College, Kushalnagar, Madikari. Atreya Ayruvedic Medical College, Bangalore Department of Zoology and Genetics, University of Mysore, Mysore, India

H. S. Blei
Laboratory of Biocatalysis and Bioprocessing, University of Abobo-Adjame, Abidjan, Côte d'Ivoire

S. Dabonné
Laboratory of Biocatalysis and Bioprocessing, University of Abobo-Adjame, Abidjan, Côte d'Ivoire

Y. R. Soro
Laboratory of Biotechnology, University of Cocody, Abidjan, Cote d'Ivoire

L. P. Kouamé
Laboratory of Biocatalysis and Bioprocessing, University of Abobo-Adjame, Abidjan, Côte d'Ivoire

B. N. Iloba
Department of Animal and Environmental Biology, Faculty of Life Sciences, University of Benin, Benin City, Nigeria

T. Ekrakene
Department of Basic Sciences, Faculty of Basic and Applied Sciences, Benson Idahosa University, P. M. B. 1100 Benin City, Nigeria

K. Osei
CSIR- Crops Research Institute, P. O. Box 3785, Kumasi, Ghana

R. Moss
West Africa Fair Fruit, P. M. B. KD11, Kanda, Accra, Ghana

A. Nafeo
West Africa Fair Fruit, P. M. B. KD11, Kanda, Accra, Ghana

R. Addico

A. Agyemang
CSIR- Crops Research Institute, P. O. Box 3785, Kumasi, Ghana

J. S. Asante
CSIR- Crops Research Institute, P. O. Box 3785, Kumasi, Ghana

A. D. Banjo
Department of Plant science and applied zoology, Olabisi Onabanjo University, P. M. B. 2002, Ago-Iwoye, Ogun State, Nigeria

M. S. Doddaswamy
Department of Sericulture Science, University of Mysore, Manasagangotri, Mysore, Karnataka, India

G. Subramanya
Department of Sericulture Science, University of Mysore, Manasagangotri, Mysore, Karnataka, India

E. Talebi
Department of Sericulture Science, University of Mysore, Manasagangotri, Mysore, Karnataka, India
Department of Animal Science, Islamic Azad University, Darab, Fars, Iran

J. W. Zheng
Southern Research and Outreach Center, University of Minnesota, Waseca, MN 56093; USA
Institute of Biotechnology, College of Agriculture and Biotechnology, Zhejiang University, Hangzhou 310029, P. R. China

S. Y. Chen
Southern Research and Outreach Center, University of Minnesota, Waseca, MN 56093; USA

Shabir Ahmad Bhat
Temperate Sericulture Research Institute Mirgund, Sher-e-Kashmir University of Agricultural Sciences and Technology, Kashmir, Srinagar-190001, India

B. Nataraju
Central Sericultural Research and Training Institute, Mysore-570 008, India

Ifat Bashir
Sericulture Development Department Jammu and Kashmir, Srinagar, India

Nighat Sultana
Pharmaceutical Research Center, Pakistan Council of Scientific and Industrial Research (PCSIR) Laboratories Complex, Karachi-75280, Pakistan

Musarrat Akhter
Food and Marine Resources Research Centre, Pakistan Council of Scientific and Industrial Research (PCSIR) Laboratories Complex, Karachi-75280, Pakistan

Muhammad Saleem
Department of Chemistry, the Islamia University of Bahawalpur, Pakistan

Yousaf Ali
Pharmaceutical Research Center, Pakistan Council of Scientific and Industrial Research (PCSIR) Laboratories Complex, Karachi-75280, Pakistan

M. B. Mochiah
Entomology Section, CSIR-Crops Research Institute, P. O. Box 3785, Kumasi, Ghana

B. Banful
Department of Horticulture, College of Agriculture and Natural Resources, Kwame Nkrumah University of Science and Technology, Kumasi, Ghana

K. N. Fening
Entomology Section, CSIR-Crops Research Institute, P. O. Box 3785, Kumasi, Ghana

B. W. Amoabeng
Entomology Section, CSIR-Crops Research Institute, P. O. Box 3785, Kumasi, Ghana

K. Offei Bonsu
Entomology Section, CSIR-Crops Research Institute, P. O. Box 3785, Kumasi, Ghana

S. Ekyem
Entomology Section, CSIR-Crops Research Institute, P. O. Box 3785, Kumasi, Ghana

H. Braimah
Entomology Section, CSIR-Crops Research Institute, P. O. Box 3785, Kumasi, Ghana

M. Owusu-Akyaw
Entomology Section, CSIR-Crops Research Institute, P. O. Box 3785, Kumasi, Ghana

Anne le Mellec
Landscape Ecology Section, University of Göttingen, Goldschmidtstr 5, D-37077 Göttingen, Germany

Jerzy Karg
Research Centre for Agricultural and Forest Environment, Polish Academy of Sciences, Field Station Turew, Szkolna 4, Pl-4-000 Ko_cian, Poland

Jolanta Slowik
Centre for Nature Conservation (CNC), University of Göttingen, von Sieboldstrasse 2, D-7075 Göttingen, Germany

Ignaczy Korczynski
Department of Forest Entomology, Poznan University of Life Sciences, Ul Wojska Polskiego 71c, PL-60-637 Poznan, Poland

Andrzej Mazur
Department of Forest Entomology, Poznan University of Life Sciences, Ul Wojska Polskiego 71c, PL-60-637 Poznan, Poland

Timo Krummel
Landscape Ecology Section, University of Göttingen, Goldschmidtstr 5, D-37077 Göttingen, Germany

Zdzislaw Bernacki
3Centre for Nature Conservation (CNC), University of Göttingen, von Sieboldstrasse 2, D-7075 Göttingen, Germany

Holger Vogt-Altena
Landscape Ecology Section, University of Göttingen, Goldschmidtstr 5, D-37077 Göttingen, Germany

Gerhard Gerold
Landscape Ecology Section, University of Göttingen, Goldschmidtstr 5, D-37077 Göttingen, Germany

Annett Reinhardt
Landscape Ecology Section, University of Göttingen, Goldschmidtstr 5, D-37077 Göttingen, Germany

Abir Sulaiman Al-Nasser
Faculty of Applied Sciences for Girls, Umm Al-Qura University, Makkah, Kingdom of Saudi Arabia

Gehad T. El-Sherbini
Department of Parasitology, Faculty of Pharmacy, October 6 University Cairo, Egypt

Eman T. El-Sherbini
Department of Zoology, El Nahda University, Beni Sweif, Egypt

B. K. Shivanna
Department of Agricultural Entomology, UAS, AINRP (T), ZARS, Shimoga -577204, Karnataka, India

S. Gayathridevi
Department of Agricultural Entomology, UAS, College of Agriculture, Shimoga -577204 Karnataka, India

R. Krishna Naik
Department of Computer Science, UAS, College of Agriculture, Shimoga -577204 Karnataka. India

B. Gangadhara Naik
Department of Plant Pathology, UAS, College of Agriculture, Shimoga -577204, Karnatak, India

H. Shruthi
Department of Agricultural Microbiology, UAS, College of Agriculture, Shimoga -577204

R. Nagaraja
Department of Agricultural Microbiology, UAS, KVK, Shimoga -577204, Karnataka, India

Pooja Chadha
Department of Zoology, Guru Nanak Dev University, Amritsar, Punjab India

Anupam Mehta
Department of Zoology, Guru Nanak Dev University, Amritsar, Punjab India

Gantigmaa Chuluunbaatar
Institute of Biology, Mongolian Academy of Science,Ulan Bataar, Mongolia

Kamini Kusum Barua
Centre for Nature Conservation, Georg-August University, Gottingen, Germany

Michael Muehlenberg
Centre for Nature Conservation, Georg-August University, Gottingen, Germany

Y. Abozinadah Najlaa
Department of Biology, Faculty of Science, King Abdul Aziz University, Jeddah, Saudi Arabia

Faten, F. Abuldahb, Nawal
Department of Biology, Faculty of Science, King Abdul Aziz University, Jeddah, Saudi Arabia

S. Al-Haiqi
Department of Biology, Faculty of Science, King Abdul Aziz University, Jeddah, Saudi Arabia

Agus D. Permana
School of Life Sciences and Technology Institut Teknologi Bandung, Ganesa 10 Bandung, Indonesia

Asni Johari
Laboratorium PMIPA FKIP Universitas Jambi - Kampus Mendalo Darat Jambi, Indonesia
Politeknik Pertanian Negeri Kupang, Kupang, Indonesia

Ramadhani Eka Putra
School of Life Sciences and Technology Institut Teknologi Bandung, Ganesa 10 Bandung, Indonesia

Soelaksono Sastrodihardjo
School of Life Sciences and Technology Institut Teknologi
Bandung, Ganesa 10 Bandung, Indonesia

Intan Ahmad
School of Life Sciences and Technology Institut Teknologi
Bandung, Ganesa 10 Bandung, Indonesia

Suhaila Abdul Hamid
School of Biological Sciences, Universiti Sains Malaysia.
11800 Minden, Penang, Malaysia

Che Salmah Md. Rawi
School of Biological Sciences, Universiti Sains Malaysia.
11800 Minden, Penang, Malaysia